Principles of Light Microscopy: From Basic to Advanced

Volodymyr Nechyporuk-Zloy
Editor

Principles of Light Microscopy: From Basic to Advanced

 Springer

Editor
Volodymyr Nechyporuk-Zloy
Nikon Europe B.V.
Leiden, The Netherlands

ISBN 978-3-031-04476-2   ISBN 978-3-031-04477-9   (eBook)
https://doi.org/10.1007/978-3-031-04477-9

This Springer imprint is published by the registered company Springer Nature Switzerland AG
The registered company address is: Gewerbestrasse 11, 6330 Cham, Switzerland

Preface

Initially, tools for visual perception of small objects, invisible to the naked human eye, nowadays microscopes are complex instruments controlled by sophisticated software and integrated in data management, analysis, and inference systems. Adequate support and the usage of them require substantial amounts of knowledge and expertise. Now, the best way to achieve this is to combine instruments in imaging core facilities where users can obtain advice, support, and guidance in sample preparation and imaging.

Having been an imaging core facility manager for a long time, I have observed that the main difficulty of the user education and induction is not the lack of knowledge or teaching materials but rather a proper logical organization of them, viz. building consistent systems of knowledge starting from easy to understand basic terms. The latter allows users to quickly operate microscopes in an independent manner, change imaging systems from different vendors, and apply different imaging technologies. Keeping in mind this idea, we deliver the current book which will be useful in everyday life of imaging microscopy facilities and labs where microscopy is the main research tool. Minimum amount of mathematical concepts is used, and main epistemological instruments are geometry and illustrations; useful protocols are added to the theoretical parts. The book has 13 chapters covering basic and advanced optical microscopy topics, and it is written by 24 authors who are active microscopists working as biomedical researchers, imaging and digital technologies developers, and imaging core facility managers, with substantial teaching experience.

I would like to thank all authors who during pandemic time delivered their time and attention to write the book and the publishers for their full support and patience.

Leiden, The Netherlands   Volodymyr Nechyporuk-Zloy

Contents

The Physical Principles of Light Propagation and Light–Matter Interactions

1

James Norman Monks

Contents

What You Will Learn in This Chapter

This chapter outlines a basic and introductory understanding of the light, its behavior, and its interaction with matter. A definition of light is noted, as well as a description of the particle and wave properties, and the polarization, refraction, reflection, diffraction, and absorption of light. The diffraction limit and optical light filtering have been discussed with respect to microscopy. This chapter provides an introduction education into the underlining fundamental physics to which all microscopy is built upon.

1.1   Definitions of Light

Light exhibits properties of both waves and of particles. This is known as wave-particle duality and states the concepts that every quantum entity can be described either as a particle or as a wave [1].

J. N. Monks (✉)
School of Computer Science and Electronic Engineering, Bangor University, Bangor, UK

When dealing with light as a wave, wavelengths are assigned to describe the single wave cycle. The entire electromagnetic spectrum, from radio waves to gamma rays, is defined as "light." Yet often when discussed, people generalize light to identify just the visible spectrum, i.e., the light accepted by human vision. Visible light has wavelengths in the range of 380–740 nm (1 nm equals 10^{-9} m). Surrounding the visible spectrum is ultraviolet radiation, which has shorter waves, and infrared radiation, which has longer waves. For microscopy purposes, the ultraviolet, visible, and infrared spectrums provide a vital role. Naturally, light is never exclusively a single wavelength, but rather a collection of waves with different wavelengths. In the visible spectrum, the collection of waves manifests as white light, with each wavelength providing a unique color from violet to red.

When characterizing light by wavelength, it is often intended within a vacuum state, as vacuum has a unity refractive index. When traveling through a vacuum, light remains at a fixed and exact velocity of 299,792,458 m/s, regardless of the wavelength and movement from the source of radiation, relative to the observer. The velocity is usually labeled c, which is descriptive for the Latin word "celeritas," meaning "speed." Wavelengths are designated with the Greek letter lambda, λ.

Light, as previously mentioned, is the electromagnetic spectrum consisting of electromagnetic waves with variation in the one wave cycle distance, and as the name implies are synchronized with oscillating electric and magnetic fields. Another property of light that is often referred to is the frequency. This describes the number of periods per time unit the electric field waves oscillate from one maximum to another maximum. Frequency is conventionally designated f, and within vacuum follows the relation of $c = \lambda f$. As light passes through matter (*anything including air but excluding a vacuum*), the velocity and wavelength decrease proportionally. Occasionally the wavenumber, $1/\lambda$, can be found to characterize light and is symbolized by $\bar{v}$ with a common unit of cm^{-1}.

When referring to light as a particle, namely a photon, it is assigned an amount of energy, E. The energy is intrinsically connected to the wave properties of light by the momentum of the photon and its wavelength, or inversely, its frequency; $E = hv = hc/\lambda$. Where h is Planck's constant with the value $6.626,070,150 \times 10^{-34}$ Joules per second.

1.2   Particle and Wave Properties of Light

Previously, I mentioned the duality nature of light and described them independently of one another. This seems a strange concept that light can exhibit properties of both waves and of particles. The notion of duality is entrenched in a debate over the nature of light and matter dating back to the 1600s, where Huygens and Newton proposed conflicting theories of light. The work by Schrödinger, Einstein, de Broglie, and many others has now recognized that light and matter have both wave and particle nature and is currently an appropriate interpretation of quantum mechanics [2].

One such experiment that demonstrates this strange duality behavior is the Young's double slit experiment [3]. The American physicist and Nobel prize winner

Richard Feynman said that this experiment is the central mystery of quantum mechanics. To explain really how bizarre this concept is, I will try to describe the experiment.

Imagine that there is a source of light shining against a screen with two slits. Importantly, this source of light must be monochromatic light (*light of a particular wavelength or color*). The light exits the source in waves like ripples in a rain puddle, that is the nature of wave-like behavior, and as the light hits the screen and exits through the two slits, it allows each individual slit to act almost like a new source of light. As such, the light extends out through the process of diffraction, and as the waves of these two "new" light sources overlap they interfere with one another. This creates crests and troughs within the diffraction pattern, such that when a crest hits a trough, the light cancels, and where a crest hits a crest, they will amplify. This results in an interference pattern displaying on the back screen a series of light and dark fringes when the light waves have either canceled or amplified in phase. This in itself is a rather simple process to understand and this wave-like property has been known since the early nineteenth century.

Imagine conducting the same experiment again, but rather than using waves, instead let us consider particles. Envisage pouring grains of sand through two slits, rather than waves propagating out and constricting through the individual slits. Each particle of sand would either go through one slit or the other, and the output imaginable would be two mounds of sand underneath each slit. Therefore, two distinct peaks are reminiscent of particle-like behavior, whereas a multiple peaks pattern is a wave-like behavior. However, what really happens when the sand particles are replaced with photons, particles of light? Let us first consider blocking of a single slit and project a stream of photons through the single open slit. Nothing particularly strange happens here, and the profile displayed on the back screen appears as a single mound, particle-like. The first mystery of quantum mechanics arrives when we open that second slit. The results now produce something very similar to the interference pattern obtained when considering wave-like behaviors (Fig. 1.1). Rather than having two bands where the photons have gone through the two slits (Fig. 1.2), *as described by the grain of sand experiment*, the photons have gone through the slits behaving like waves. If we did not know anything about the photons, we could assume that there is some force between them that allows them to coordinate their actions which could give rise to the interference pattern. However, we could now adjust the experiment to try and force this understanding. Therefore, instead of sending the photons through all at once, they are sent one at a time, leaving enough of a time interval between each photon to allow those that can pass through the slits and reach the back screen. If we run the experiment slowly, gradually we can visualize the photons passing through the slits and hitting the back screen one at a time. At first, they appear to be randomly arriving at the screen, but as time goes on, that same interference wave-like pattern appears. Consequently, each photon by some means is influencing a small part to the complete wave-like behavior.

On the surface of things, this is an extremely odd function. As the photons pass through a single slit, they produce a particle-like pattern, but if they pass through two slits, even individually, they produce a wave-like pattern. Let us again adjust the

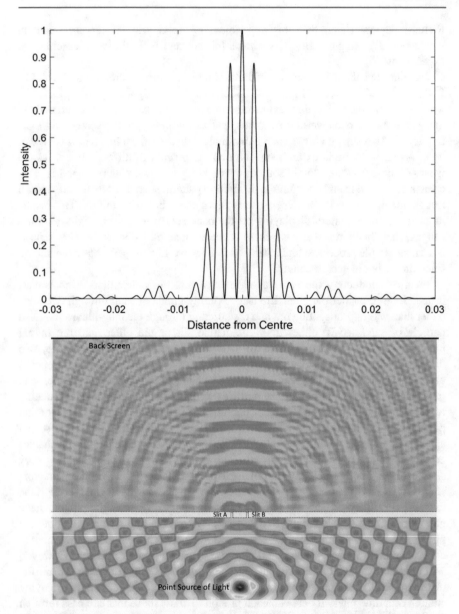

Fig. 1.1 (Top) A computational output of the interference pattern generated in Young's double slit experiment. The x-axis is dimensionless in this case to demonstrate the produced pattern. (Bottom) The computer simulation illustrating light propagating from the point source and impinging on the screen with two slits open. The two splits act like independent point source that interfere with one another. This image is intended to simplify the notion of interference phenomenon and only produces a relative distribution of light

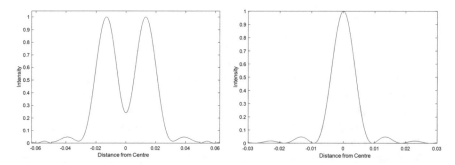

Fig. 1.2 (Left) Two-mound example of sand granules traversing through two slits suggestive of particle-like behavior. (Right) Single-mound pattern profile produced from a photon traveling through a single slit. The x-axes are relative to the open slits

experiment to observe the photon to see the path it follows by placing a detector at one of the slits. If the photon goes through the slit with the detector there will be a record; if it does not then the assumption is that the photon must have gone through the other slit. Like the previous setup, we send the photons through one at a time; 50% of the time the detector will record the photon, meaning the other 50% of photons must have gone through the other slit. Now the only adjustment made to the experiment is the addition of a detector and as such it has detected the photons that have either gone through one slit or the other. Interestingly though, the results are now different and present in a particle-like pattern (Fig. 1.2), two distinct mounds. Finally, if we adjust the experimental again, but this time switch the detector off but leave it in place. When we run the experiment again, we obtain a wave-like pattern, almost as if the photons are aware that observation of their paths is no longer happening.

This is the complexity and puzzling properties of wave-particle duality. The act of observation can influence the result. Young's experiment demonstrates that each photon *must be aware* of both slits, and the presence of the detector, and must travel through both slits at the same time when unobserved and travel through only one or the other when being observed. Clearly, our knowledge of light is incomplete, and the wave and particle description are both just models of an incomplete picture.

For microscopy purposes, the wave-particle duality is subjugated in electron microscopy [4], *transmission and scanning electron microscopy,* and *scanning tunneling microscopy.* The electrons are used to generate small wavelengths that can be used to observe and distinguish much smaller features than what visible microscopy can achieve. However, the penetration depth of electron microscopes is unable to resolve sub-surface characteristics.

1.3   Polarization

Light waves propagate in a transverse manner, with the electric and magnetic fields oscillating perpendicular to the propagation direction of light. Additionally, the electric and magnetic fields are perpendicular to one another. When all components of the electric fields oscillate in a parallel style, the wave is said to be *linearly* polarized. The plane of polarization can be grouped by the *transverse electric* (*TE* or *s*-) and the *transverse magnetic* (*TM* or *p*-), depending on which field is absent from the direction of propagation (Fig. 1.3).

The electric field directions can also spiral along the propagation line and are known as *circular* polarization which can exist as left-handed (anticlockwise rotation) or right-handed (clockwise rotation) polarizations and maintain constant magnitude in all directions of rotation. *Elliptical* polarization can also exist but does not manage to maintain a constant magnitude due to a phase quadrature component.

Natural light (sunlight) and man-made light (light bulbs) produce unpolarized light. The unpolarized light provides a random mixture of all possible polarizations, known as *incoherent* light, which do not contain phonons with the same frequency or phase. Opposing this, *coherent* light (laser) is a beam that contains the same, or a narrow band or similar frequencies, phase, and polarization. Incoherent light can become partially polarized upon reflection.

1.4   Refraction

Snell's law formulates the relationship between the angle of incidence and refraction when light meets an interface between two mediums with different refractive indices, n_1 and n_2. Figure 1.4 provides an example of how light rays refract at the interface, assuming $n_1 < n_2$. Since the velocity is lower in the second medium, the angle of refraction β is less than the angle of incidence α. Therefore, the relative velocity of light in the medium can be regarded as the inverse refractive index.

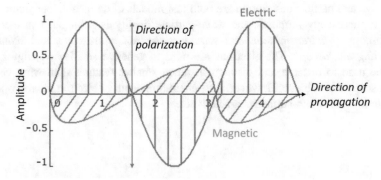

Fig. 1.3 Representation of linearly polarized light

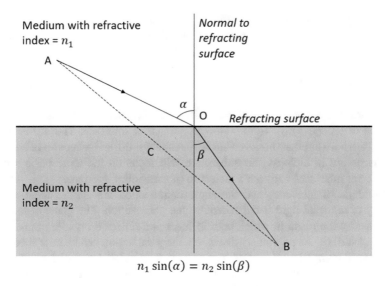

$$n_1 \sin(\alpha) = n_2 \sin(\beta)$$

Fig. 1.4 The refraction path of light entering a transparent material where $n_1 < n_2$

Snell's law permits that the path of light takes the fastest route between any two points. In comparison to the dashed line in Fig. 1.4, the solid line represents the path that the light ray has take between points A and B. As such, it can be observed that the light trails a further distance in the upper medium (n_1), due to a lower index equaling a higher velocity, than the lower medium (n_2) that has a higher refractive index resulting in a lower velocity. Thus, $AO > AC$ and $OB < CB$.

The refractive index of a given material is governed by the permittivity and permeability. These material parameters describe the density of the electric and magnetic dipole moments that have been induced by an external field. However, at optical frequencies, the magnetic component is neglected and described by unity. As such, the refractive index becomes a product of the permittivity. Thus, nonmetallic and semi-conductive materials are given the name "dielectrics." The refractive index for dielectric materials often decreases with wavelength and is called *dispersion*. However, there are spectral absorption bands where the index steeply increases with wavelength, and this is referred to as *anomalous dispersion*. Generally, the refractive index for dielectric materials is given by a real number. Yet, at the anomalous regions, they do display a complex, two-dimensional number.

In some special case materials, more notably crystals and some biological material [5] (*chitin for example*), the refractive index becomes anisotropic and varies conditionally on the direction and polarization of light. This term is known as *birefringence* and is an important consideration when studying cellular plant biology and other lengthened molecules. Isotropic materials (*with refractive index properties the same in all directions*) may convert to birefringent properties with external treatments. As light propagates and transmits through a birefringent material, a phase difference between the different polarizations develops, resulting in the transmitted light becoming circular or elliptically polarized.

1.5   Reflection

Reflections can either be specular or diffusive. The specular type is the standard example of reflected rays from a smooth surface or interface. The diffusive type arises from a textured and uneven surface. Although diffuse reflection is an important discussion in biology, this subsection will focus on the more basic specular reflection at interfaces between a dielectric (nonmetallic) interface.

The angle of incidence and the angle of reflection are always identical. The quantity of reflected light is dependent on the polarization. The plane of incidence is defined by the plane in which both incident and reflected rays lie normal to the reflecting surface. As previously discussed, the polarization can be described by the electric field component. Light with an electric field parallel to the plane is denoted by $\parallel$, and with an electric field perpendicular to the place of incidence by $\perp$. Fresnel's equations, where α is the angle of incidence and equal to the angle of reflection, and β is the angle of transmission, provide the polarization reflection fractions ($R_{\parallel}$ and $R_{\perp}$).

$$R_{\parallel} = \left(\frac{\tan(\alpha - \beta)}{\tan(\alpha + \beta)}\right)^2$$

$$R_{\perp} = \left(\frac{\sin(\alpha - \beta)}{\sin(\alpha + \beta)}\right)^2$$

The mean of the two ratios is the reflected fraction of unpolarized light. At normal incidence, $\alpha = \beta = 0$; thus, the above equations result in a division by zero. Therefore, another equation must be employed for normal incidence where there is no differentiation between the parallel and perpendicular cases.

$$R = \left(\frac{n_1 - n_2}{n_1 + n_2}\right)^2$$

Let us look at a glass microscope slide as an example for the latest equation. Microscope slides are often soda-lime glass which has a refractive index of $n_2 = 1.52$ and let us consider the glass slide in air, $n_1 = 1$. When the light hits the glass slide at normal incidence, $R = [(1 - 1.52)/(1 + 1.52)]^2 = 0.0426 = 4.26\%$. As the light passes through the glass and reaches the second interface (*glass to air*) the reflection percentage matches the 4.26% from the first interface (*air to glass*). Therefore, because the light passes through two interfaces of the same materials, the total transmitted light ($1 = R + T$) for soda-lime glass would be 91.6615%. In more practical terms, it can be estimated that clean glass slides, dishes, and plates have a reflection loss at normal incidence of approximately 8%. However, this value may

Fig. 1.5 Amplitude of light reflected for different incidence angles for air ($n_1 = 1$) and glass ($n_2 = 1.52$). (Top) Reflection values for when light goes from less dense to denser medium. (Bottom) Reflection values for when light goes from denser to less dense medium. Both inserts for the ray paths for the respective examples

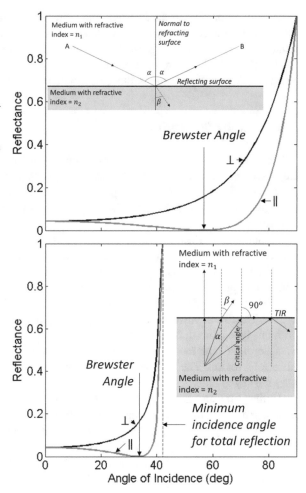

not hold true for plastics and glasses with filters or antireflection coatings. Additionally, if the glass substrates are not clean, then this 8% value would also not true (Fig. 1.5).

When examining the case of $\alpha < 90°$, it can be found that ($R_{\parallel}/R_{\perp}$) $= [\cos(\alpha - \beta)/\cos(\alpha + \beta)]^2$ provides a ratio of >1, meaning $R_{\parallel} > R_{\perp}$. Consequently, light with an electric field perpendicular to the plane of incidence and parallel to the interface will be reflected more easily than its counterpart factor. This allows interfaces to be used as a polarizer. The reflected light can become entirely polarized with the condition of $\tan\alpha = n_2/n_1$. This special condition is called the *Brewster angle* and is the angle at which the $\parallel$ polarized light becomes perfectly transmitted. Taking the soda-lime glass microscope example, the Brewster angle would be at $\tan^{-1}(1.52/1) = 56.7°$.

According to Snell's law, if a light ray hits a flat interface from a high index medium to a low index medium at an increasing oblique angle, the reflection will eventually be parallel to the interface and there will be total reflection. The *critical angle* is the minimum angle of incidence at which entire reflection occurs. For example, if light was exiting a soda-lime glass microscope slide at an angle greater than critical angle, then light would be guiding and tunnel between the two interfaces. The critical angle in this case would be at $\sin^{-1}(1/1.52) = 41.1°$.

1.6   Light Scattering

While reflection and refraction are strictly both types of scattering, the term scattering is used in a broader sense to illustrate phenomena that tend to change the order of light propagation to a random state [6]. There are three types of scattering that can be identified. The three types are *Mie scattering, Rayleigh scattering*, and *Raman scattering*.

Particles of the same size as the wavelength of the light while displaying a refractive index different from its surrounding generate Mie scattering. Due to the barriers between cells and the different portions of the cell make-up, almost all animal and plant tissue produce a strong Mie scattering effect.

When the scattering particle is smaller than the wavelength, Rayleigh scattering occurs. The fourth power of the wavelength of light has an inverse relationship, $1/\lambda^4$, with this type of scattering. All matter has a natural oscillating frequency that results in absorption and most substances having a strong absorption band in the far ultraviolet region. The most common natural example of Rayleigh scatter explains why the sky is blue.

Unlike Mie and Rayleigh scattering where the scattered wavelength remains unchanged, Raman scattering operates slightly different. Under Raman scattering one of two things happens, either some additional energy is taken up by the particle or the scattering particle gives off part of the photon energy. The energy disparities between vibrational states in the particle correspond to the quantity of energy taken up or given out. Raman scattering applications for a biological setting can be found in fluorescence analysis.

1.7   Diffraction and Its Limits

It is often thought that when there is nothing in its path, light travels in a straight line. However, as previously mentioned, Young's double slit experiment demonstrated that this is not always the case. Diffraction is a term that describes a variety of phenomena that occurs when a wave collides with an obstruction or an opening. It is described as the bending of waves around the corners of a barrier or through an aperture into the region of the obstacle's geometrical shadow. This facilitates the diffracting item or aperture effectively becoming the propagating wave's secondary source. Understanding some biological phenomena, such as biological crystallography [7], necessitates taking diffraction into account.

Optical resolution imaging can be restricted by such issues like imperfections in the lenses or lens misalignments. Nonetheless, due to diffraction, there is an elementary limit to the resolution of any optical systems [8]. The resolution of an optical system is comparative to the size of the objective, and inversely comparative to the wavelength of the observed light.

The resolution limit to far-field objects was initially determined by Ernst Abbe in 1873. Abbe identified that light waves cannot be detained to a minimum resolvable distance in order to image anything less than one half of the input wavelength. This led to the formulation of the optical diffraction limit.

$$d_{\min} = \frac{\lambda}{2 \ N.A.},$$

where d is the minimum resolvable distance, λ is the input wavelength of light, and $N.A$ is the numerical aperture of the optical system. The numerical aperture is usually stated on the objective of a microscope, making it an easy calculation to estimate the resolution of the optical system.

The theory surrounding the diffraction limit is founded on the Heisenberg uncertainty principle. The Heisenberg uncertainty principle engages the position and moment of a photon.

Classic physics described the free propagation waves and how they intimately relate to the diffraction limit in the far-field regions. With relation to near-field optics, there are two primary aspects that establish the resolution limit, the first being diffraction and the second being the loss of evanescent waves in the far-field regions. Evanescent waves develop at the border of two dissimilar mediums that have different refractive indices. These waves decompose significantly within a distance of a small number of wavelengths. This means that they are only detectable in the areas close to the interface of the mediums.

The evanescent waves transmit the sub-diffraction limited information, and the amplitudes of these evanescent waves weaken hastily in no less than one direction. As such, the diffraction limit could diminish if the evanescent waves become significant; this is known as super-resolution imaging [9].

For optical microscopy much of the "tiny" world is hidden. The current resolution limit for an optical microscope, unassisted from super-resolution technology, is about 200 nm. This makes live observations for ribosome, cytoskeleton, cell wall thickness, virus', proteins, lipids, atoms, and much more impossible to view with a conventional optical microscope setup.

The numerical aperture of a system can be improved with immersion oil placed between the objective and the imaging object [10]. With a high-quality immersion oil, an N.A of 1.51 can be achieved. Taking into account the point spread functions coefficient of 1.22, which describes the response of an imaging system to a point object, the resolution of an optical microscope at 550 nm would be $d = 1.22 \times (550 \ nm/(2 \times 1.51)) = 222$ nm. In most biological specimens, using a longer wavelength like infrared has the advantage of better penetration and less scattering but lower resolution [11].

1.8   Photon Absorption, Fluorescence, and Stimulated Emission

Photon absorption is the process by which matter absorbs a photon's energy and converts it into the absorber's internal energy which is often dissipated through heat or a release of a photon. Absorption is achieved when bound electrons in an atom become excited as the photon's frequency matches the natural oscillation, or resonant, frequency of a particular material. This results in the light intensity attenuating as the wave propagates through the medium. Wave absorption is normally independent of the intensity under linear absorbing behaviors (Fig. 1.6). However, under some circumstances, particularly in the realm of optics, a material's transparency can fluctuate as a function of wave intensity resulting in saturable absorption. This process is a nonlinear effect.

The simultaneous absorption of two photons of the same or difference frequencies to excite a molecule or atom from one state, typically the ground state, to a higher energy is known as *two-photon absorption* (TPA) [12]. This is a nonlinear absorption effect. The change in energy from the lower and upper state is equal to or less than the sum of the photon energies of the two absorbed photons. This phenomenon is achieved by creating a virtual state between the ground and excited states where excitation is achieved by two separate one-photon transitions (Fig. 1.7). An example of applications involving TPA can be found in

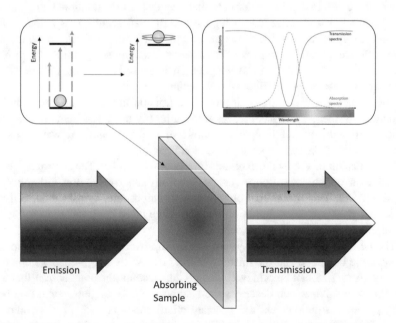

Fig. 1.6 A schematic diagram showing an overview of electromagnetic radiation absorption. As a white light source illuminates the sample, the photons that match the correct energy gap are absorbed and excited to a higher energy state. The other photons remain unaltered and transmit through the material

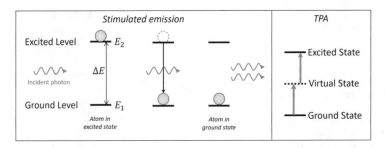

Fig. 1.7 (Left) The energy transition diagram for stimulated emission. Starting from a high energy state, as the photon with the correct energy interacts with the material, it stimulated the electron to decay and emit an identical photon. (Right) The energy levels for a two-photon absorbing system demonstrating the virtual state

fluorescent molecules where the excited state electrons decay to a lower energy state via spontaneous photon emission.

Spontaneous emission is the process where an excited energy state transitions to a lower energy state and releases the energy in the form of a photon. Emission can also be a stimulated process where a delivered photon interacts with a pre-excited electron to force or stimulate the electron to drop to a lower energy level. The stimulated process obtains its original photon and creates a new photon identical to the input photon, in wavelength, polarization, and direction of travel. The emission of photons from stimulated emission is a consequence of the conservation of energy and the wavelengths are well defined by quantum theory. As an electron transitions from a higher energy level to a lower energy level, the difference in energy between the two transition energy levels is released to conserve the energy of the system. The energy difference between the two levels regulates the emitted wavelength. This is the fundamental process of how lasers operate.

In terms of fluorescence, the electron absorbs a high-energy photon that is excited from the ground state. The electron then relaxes vibrationally and drops state levels through non-radiative transitions. The electron is still in a high state than the pre-absorption state, but a lower energy state compared to post-absorption. As such, when the electron falls back to the ground state, the material fluoresces at a longer wavelength (Fig. 1.8). Fluorescence has become an important function for the field of microscopy such as the use in fluorescence-lifetime imaging microscopy [13]

1.9   Filtering Light

It is often the case with optical instruments that in many circumstances the user may not want to use light directly from the light source. It may be beneficial to delete particular sections of the spectrum or pick only a restricted spectral band, or select

Fig. 1.8 A more detailed Jablonski energy diagram for fluorescent behavior. As the photon is absorbed, the electron is forced to a higher energy band. The material can then be excited electronically and vibrationally to relax and transition the electron to a lower state, allowing it eventually to fluoresce at a longer wavelength

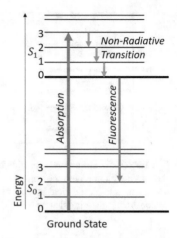

light with a specific polarization, or modulate the light in time, such as swiftly changing from darkness to light or obtaining a sequence of light pulses. All this can be achieved through optical filters, which provide use for a number of applications including microscopy, spectroscopy, chemical analysis, machine vision, sensor protection, and more. A variety of filter types are available and used according to the application requirements.

Similar to other optical components, filters possess many of the same specifications. However, there are several specifications unique to filters and require an understanding in order to determine what filter is suitable for a particular application. These include *central wavelength, bandwidth, full width-half maximum, blocking range, slope, optical density,* and *cut-off and cut-on wavelengths.* There are two main filter types to consider, *absorption* and *interference filters.*

The light-absorbing properties of some substances can be utilized to exploit blocking unwanted light. Light that is prevented through absorption is retained within the filter rather than reflecting off it. For example, certain types of glass can be excellent absorbers of UV light. Another common absorbent is cellulose acetate, although can be unstable and does bleach with time. Water can be another useful absorbing filter, particularly for infrared radiation. Potassium dichromate solutions are superb in absorbing light with wavelengths shorter than 500 nm. Although this is a useful solution, particularly for studying fluorescence, the liquid is also carcinogenic, and caution should be upheld.

Interference filters, often referred to dichroic filters, work by reflecting undesired wavelengths while transmitting the desired portion of the spectrum. The interference characteristic of light waves is exploited by adding numerous layers of material with variable indices of refraction. Only light of a specified angle and wavelength will constructively interfere with the incoming beam and pass through the material in interference filters, while all other light will destructively interfere and reflect off the material. These types of filters have much great flexibility in design, but the theory

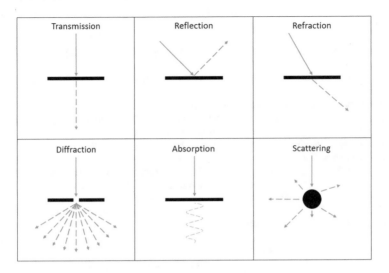

Fig. 1.9 Basic diagram outlining some of the fundamentals discussed in this chapter

can be complicated. Thankfully, there are many experts and companies that offer a range of dichroic filters.

Some solutions provide far superior filtering but may come at a cost. For desired optical qualities, thick layers may be required, and liquid filters may become large. Additionally, some of the most valuable-colored compounds are cancerous or hazardous in other ways. Therefore, great precaution is needed when selecting a desirable filter, particularly with absorbing ones. The more costly and expensive interference filters are often chosen due to these factors. However, the dichroic filters are extremely angle sensitive, unlike absorbing filters.

Take-Home Message
- Light is a beam of energy that travels at the universal speed limit.
- Light is not only the visible light we see but extends the entire electromagnetic spectrum from radio waves down to gamma rays.
- Light is a wave of altenating electric and magnetic fields.
- Light can be described as both a wave and a particle.
- Different wavelengths of light interact differently with matter (Fig. 1.9). For example, how plants turn sunlight into energy through the process of photosynthesis.
- Light can be manipulated for our advantage through reflection, refraction, polarization control, and scattering.
- Optical systems such as microscopes have an optical resolution limit that restricts the resolvable features. For example, molecules are undetectable by white light micrscopy.
- We can take advantage of optical filters and fluorescence to advance our imaging quality and contrast.

References

1. Roychoudhuri C, Creath K, Kracklauer A. The nature of light. 1st ed. Boca Raton, FL: CRC Press; 2008.
2. Greenberger D, Hentschel K, Weinert F. Compendium of quantum physics. 1st ed. Berlin: Springer; 2009.
3. Ananthaswamy A. Through two doors at once. 1st ed. Richmond: Duckworth; 2020.
4. Egerton R. Physical principles of electron microscopy. 2nd ed. Berlin: Springer; 2018.
5. Murphy D, Spring K, Fellers T, Davidson W. Principles of birefringence, Nikon's MicroscopyU. [Online]. https://www.microscopyu.com/techniques/polarized-light/principles-of-birefringence (2022).
6. Kokhanovsky A. Light scattering reviews. 1st ed. Berlin: Springer Praxis Books; 2006.
7. Baker E. New perspectives in biological crystallography. IUCrJ. 2014;1(2):82–3.
8. Silfies J, Schwartz S, Davidson M. The diffraction barrier in optical microscopy. Nikon's MicroscopyU. [Online]. https://www.microscopyu.com/techniques/super-resolution/the-diffraction-barrier-in-optical-microscopy (2022)
9. Jing Y, Zhang C, Yu B, Lin D, Qu J. Super-resolution microscopy: shedding new light on in vivo imaging. Front Chem. 2021;9:795767.
10. Chen X, Zheng B, Liu H. Optical and digital microscopic imaging techniques and applications in pathology. Anal Cell Pathol. 2011;34(1–2):5–18.
11. Zhang H, Salo D, Kim D, Komarov S, Tai Y, Berezin M. Penetration depth of photons in biological tissues from hyperspectral imaging in shortwave infrared in transmission and reflection geometries. J Biomed Opt. 2016;21(12):126006.
12. Rubart M. Two-photon microscopy of cells and tissue. Circ Res. 2004;95(12):1154–66.
13. Lichtman J, Conchello J. Fluorescence microscopy. Nat Methods. 2005;2(12):910–9.

Further Reading

Giusfredi G. Physical optics. 1st ed. Cham, Switzerland: Springer; 2019.
Rossing T, Chiaverina C. Light science. 2nd ed. Cham, Switzerland: Springer; 2019.
Shaw M. Your microscope hobby: how to make multi-colored filters. 1st ed. Scotts Valley: CreateSpace Independent Publishing Platform; 2015.
Weiner J, Ho P. Light matter interactions. 1st ed. Hoboken, NJ: Wiley; 2007.

Design, Alignment, and Usage of Infinity-Corrected Microscope

2

Sanjukta Sarkar

Contents

What You Will Learn in This Chapter

This chapter discusses how infinity-corrected microscopes work, as well as the principles of optics that are applied to their development. Proper designing and good alignment of an optical microscope are essential for accurate studies of cells, observation of cellular growth, identification, and counting of cells. To design an optical microscope the two important aspects are, namely, a better understanding of the function of each component and how their control influences the resulting images. The design of the infinity-corrected optics is routinely incorporated into multiple lenses, filters, polarizers, beam-splitters, sensors, and illumination sources. This chapter discusses the development of microscope with infinity optics and the design of infinity-corrected optics with optical ray diagrams. The microscope design parameters and aberrations are discussed to understand the necessity of multiple lens

S. Sarkar (✉)
Queen Mary University of London, London, UK

© The Author(s), under exclusive license to Springer Nature Switzerland AG 2022   17
V. Nechyporuk-Zloy (ed.), *Principles of Light Microscopy: From Basic
to Advanced*, https://doi.org/10.1007/978-3-031-04477-9_2

objective system. To get the best resolution and contrast, the condition for Köhler illumination should be maintained within a microscope. The unstained sample is unable to image in bright field microscopy. The chapter also discusses label-free techniques with infinity-corrected optics, such as dark field microscopy, Zernike phase contrast microscopy, differential interference contrast (DIC), and digital holographic microscopy and their applications to study the various type of specimens without dye or label.

2.1   Development of Compound Microscope with Infinity Optics

A microscope is an instrument that produces a magnified two-dimensional image of an object which is not normally resolvable by the human eye. The human eye cannot resolve the object smaller than 150 μm [1]. A decent optical microscope can resolve the object which is near 0.2 μm, so it has multiple times better resolution over independent eye [2, 3].

Antonie van Leeuwenhoek (1632–1723) was a Dutch scientist who was first to document microscopic observation of bacteria using a single-lens microscope which consisted of a small, single converging lens mounted on a brass plate, with a screw mechanism to hold the sample to be examined. In the single-lens system, the object is magnified with a convex lens that bends light rays by refraction. The rays from the object are converged behind the lens to form a focused image. The distance from the object to the lens divided by the distance of the image from the lens determines the magnification of this system.

The disadvantage of Leeuwenhoek single-lens microscope is that it has to be placed very near to the eye to get a magnified image. The *compound microscope*, built by Robert Hooke (1635–1702), overcame this problem by using two thin lenses: the *objective* and the *eyepiece*. The highest magnification achieved by Leeuwenhoek microscope was limited to only 300×, whereas in Hooke's compound microscope, magnification up to 1500× was possible. As the application of microscope in studying biological samples grew by the beginning of the twentieth century, there was an increasing need to add multiple optics between objectives and eyepiece to enhance the contrast. However, the introduction of extra optical elements in the Hooke's microscope deteriorated the quality of images. In the early 1930s, the German microscope manufacturer Reichert introduced an infinity-corrected optical configuration for microscope to overcome this drawback [3]. The infinity-corrected microscope is a three-lens system where *objective, tube lens*, and *eyepieces* are used to get an image. Most of modern optical microscopes today use the infinity-corrected optical configuration due to the improved functionality of the microscope.

In microscope a specimen under observation is uniformly illuminated through a lens by an incoherent light source. The compound microscope (Fig. 2.1) is a two-step imaging system where the first lens (objective lens) is placed close to the object and creates a real, inverted, and magnified image of the object at the focal plane of the second lens (eyepiece).This image is also known as an intermediate

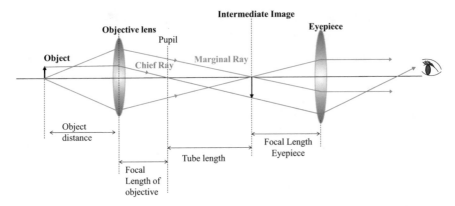

Fig. 2.1 Ray diagram for image formation in general compound microscope

Fig. 2.2 Different tube lengths in compound microscope

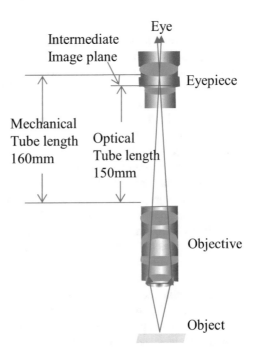

image and the plane is the intermediate image plane. The eyepiece produces a virtual image that projects an image to infinity, and the human eye creates the final magnified image. In most of the compound microscope systems, the final image is located at the minimum focusing distance of the human eye, which is generally 250 mm from the eyepiece [3, 4].

The distance between the objective and eyepiece is known as mechanical tube length. As illustrated in Fig. 2.2, the distance from the rear focal plane of objective to the intermediate image plane is named as optical tube length. The value of finite

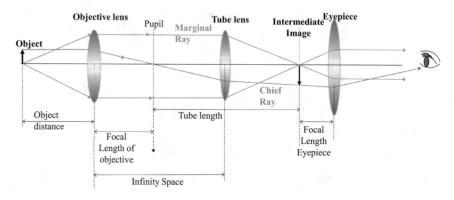

Fig. 2.3 Ray diagram for image formation in infinity-corrected compound microscope configuration

mechanical tube length is standardized to 160 mm. It is observed that adding auxiliary optical components such as beam-splitters, polarizers filters into the light path of a fixed tube length microscope increases the effective tube length and deteriorates the quality of images. But in the contrast-enhancing methods (DIC, fluorescence microscopy) these auxiliary optics are essential for a microscope. The introduction of infinity optics enables the use of auxiliary optics. Figure 2.3 show the schematic diagram of infinity optics where the objective collects the light transmitted through or reflected from the specimen and produces a parallel bundle of rays. The parallel light rays are then focused at the intermediate image plane by the tube lens, and subsequently magnified intermediate image is seen by the eyepiece. The area between the objective rear plane and the tube lens is called *infinity space*, where auxiliary components can be introduced without producing focus artefacts or optical aberrations. The term *infinity space* refers to the production of a bundle of parallel light rays between the objective and tube lens of a microscope.

The three main advantages of the infinity optical system are:

- The magnification and location of the intermediate image remains constant even when the distance between the objective lens and tube lens is altered.
- There is no image aberration even when prisms, filters are interposed between the objective lens and the tube lens.
- It allows users to switch between objectives with different magnification without needing to refocus the specimen (Parfocality), even when extra elements are added.

2.2   Design Parameters: Resolution Limit, Numerical Aperture, Magnification, Depth of Field, and Field of View

Resolution The resolution is one of the key parameters of the microscope. Resolution is defined by the minimum spacing between two specimen points which will be distinguished as separate entities. The resolution of an optical system is limited by diffraction and aberration [4]. The high resolution of an optical instrument means the ability to see a structure at a high level of detail.

Light propagates as a wave, when light passes through the objective it is diffracted and spread out in the focal region, forming a spot of light which is about a minimum 200 nm wide and minimum 500 nm along the optical axis [4, 5]. The diffraction pattern consists of a central maximum (zeroth order) surrounded by concentric first, second, third, etc., maxima with decreasing brightness, known as Airy disc or point spread function (Fig. 2.4). The size of the central maxima of the Airy disc is dependent on the numerical aperture of lens and wavelength of light. The limit of resolution of a microscope objective refers to its ability to distinguish between two closely spaced Airy discs in the image plane. The two principal ways of recognizing and calculating resolving power and the limit of resolution in the microscope are Rayleigh's resolution criterion and Abbe's resolution criterion. In Rayleigh's resolution criterion we consider how closely two finely detailed points can lie to one another and remain distinguishable as discrete entities, whereas Abbe's resolution criterion considers how many diffracted orders of light can be accepted by the objective to form an image.

Rayleigh Criterion Two-point sources are regarded as just resolved when the central zeroth order maximum of one image coincides with the first minimum of the other (Fig. 2.5). The minimum distance between the two points is

$$R = \frac{0.61\lambda}{n \cdot \sin\theta} = \frac{0.61\lambda}{\text{NA}}, \qquad (2.1)$$

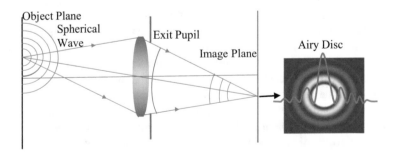

Fig. 2.4 Airy disc in a lens system as broadening the image of a single object point

Fig. 2.5 Rayleigh Resolution
criterion in diffraction-limited
system

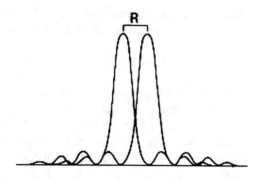

where λ is the average wavelength of illumination. The numerical aperture is defined
by the refractive index of the immersion medium (the medium between objective
and specimen, n; usually air, water, glycerine, or oil) multiplied by the sine of the
half aperture angle ($\sin(\theta)$).

$$NA = n \cdot \sin(\theta). \tag{2.2}$$

The numerical aperture is used to define the resolution and light gathering ability
of the lens. The values of NA range from 0.1 for very low magnification objectives to
1.4 for high-performance objectives utilizing immersion oils in imaging medium.

Abbe's Resolution Criterion The Abbe diffraction limit depends on the number of
diffraction order from object are accepted by objective and the NA of the lens
involved for image formation. It assumes that if two adjacent diffraction orders of
two points incident on the aperture of objective, these two points are resolved.
Therefore, the resolution depends on both imaging (objective lens) and illumination
apertures (condenser lens) and is

$$\text{Abbe Resolution}_{x,y} = \frac{\lambda}{NA_{condenser} + NA_{objective}}. \tag{2.3}$$

The above two equations indicate that the resolution of an optical system
improves with an increase in NA and decreases with increasing wavelength λ.

Due to the large difference between the refractive indices of air and glass; the air
scatters the light rays before they can be focused by the lens. To minimize this
difference a drop of oil can be used to fill the space between the specimen and an
immersive objective. The oil has a refractive index very similar to that of glass, it
increases the maximum angle at which light leaving the specimen can strike the lens.
This increases the light collected and, thus, the resolution of the image as shown
Fig. 2.6. For example, the image spot size produced by a $100\times$ magnification dry
objective of NA 0.95 in green light (550 nm) is approximately 0.34 µm, whereas the
spot size for $100\times$ oil immersion objective of NA 1.4 is approximately 200 nm or
0.2 µm.

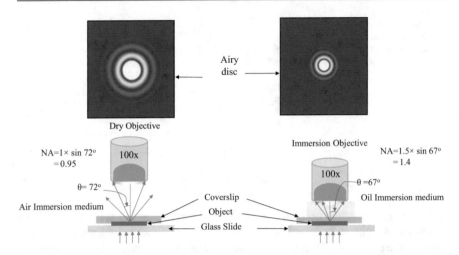

Fig. 2.6 Numerical aperture and resolution of dry objectives of NA 0.95 and oil immersion objective of NA 1.4

Magnification In compound microscopy, the magnification depends on the objective magnification (M_o) and the eyepiece magnification (M_e), where M_o is the ratio of tube length and object distance and M_e is the ratio of the virtual image distance and the focal length of eyepiece. So total magnification of the microscope (M_T) is

$$M_T = \frac{\text{Finite tube length}}{\text{Focal length}_{\text{objective}}} \times \frac{250 \text{ mm}}{\text{Focal length}_{\text{Eyepiece}}} = M_o \times M_e. \qquad (2.4)$$

So, if the objective magnification is 4× and the eyepiece magnification is 10×, $M_T = 40×$

In a microscope with infinity-corrected optics, the magnification of intermediate image is defined by the ratio of focal length of tube lens and objective lens. The focal length of the tube lens varies from 160 to 250 millimetres depending upon the manufacturer and model.

$$\text{Magnification}_{\text{infinity-corrected objective}} = \frac{\text{Focal length}_{\text{tubelens}}}{\text{Focal length}_{\text{objective}}}. \qquad (2.5)$$

Depth of Field and Depth of Focus Depth of field is the axial depth of the *object plane* within which the object plane is possible to shift without loss of sharpness in the image plane while the image plane position is fixed. Depth of focus is that the axial depth of the *image plane* within which the position of the image plane is allowed to move and the image appears acceptably sharp while the positions of the object plane and objective are maintained (Fig. 2.7).

Various authors [6, 7] have proposed different formulas for calculation of the depth of field. Suggested by Rudolf Oldenbourg and Michael Shribak [6] the total depth of field is

Fig. 2.7 Depth of field and
depth of focus of an
objective lens

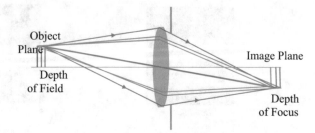

$$d_z = \frac{\lambda \cdot n}{\mathrm{NA}^2} + \frac{n \cdot p_{\mathrm{sens}}}{M \cdot \mathrm{NA}}, \tag{2.6}$$

where d_z is the depth of field (or axial resolution), λ is the wavelength of the illuminating light, n is the refractive index of the medium between the specimen and the objective lens, NA is the objective lens numerical aperture, M is the objective lens magnification, and p_{sens} is the pixel size of the image sensor placed in the intermediate image plane.

Two important aspects of depth of field and depth of focus are:

1. High magnification objectives with large aperture have extremely limited depth of field at object or specimen's plane and relatively large depth of focus at eyepiece or camera plane. This is why at high magnification focusing of specimen is very sensitive and accurate.
2. Low magnification objectives with small aperture have a relatively large depth of field at object plane and extremely shallow depth of focus at the eyepiece or camera plane. That is why eyepiece setting is critical to being properly adjusted.

Field of View The field of view on a microscope determines the size of the imaged area. The size of the field of view depends on the objective magnification; the greater the magnification smaller the field of view. In an eyepiece-objective system, the field of view from the objective is magnified by the eyepiece for viewing. In a camera-objective system, that field of view is relayed onto a camera sensor. The maximum field of view of the microscope is affected by the objective lens, the tube-diameter of the microscope's internal optical system, the eyepieces, and the scientific camera sensor size. Equations (2.7) and (2.8) can be used to calculate the field of view in the aforementioned systems.

$$\text{Field of View}_{\text{Camera}-\text{Objective}} = \frac{\text{Camera Sensor size}}{\text{Magnification}_{\text{objective}}}. \tag{2.7}$$

$$\text{Field of View}_{\text{eyepiece}-\text{Objective}} = \frac{\text{field stop diameter}_{\text{eyepiece}}}{\text{Magnification}_{\text{objective}}} \tag{2.8}$$

2.3   Optical Aberrations and Their Corrections

Aberrations in optical systems are often defined as the failure of getting a faithful image of an object. True diffraction-limited imaging is usually not achieved due to lens aberrations. Aberrations fall into two classes: monochromatic and chromatic. Monochromatic aberrations are caused by the geometry of the lens and the refraction of light through the lens. The monochromatic aberrations are: spherical aberration, coma, astigmatism, curvature of field, and distortion. Chromatic aberrations are caused by lens dispersion, the variation of a lens' refractive index with wavelength.

Spherical Aberration occurs when light waves passing through the periphery of a lens are not brought into identical focus with those passing near the centre. Light rays passing near the centre of the lens are refracted slightly, whereas rays passing close to the periphery are refracted to a greater degree resulting in different focal points along the optical axis (Fig. 2.8). The resolution of the lens system is degraded by this aberration because it affects the coincident imaging points along the optical axis, which will seriously affect specimen sharpness and clarity.

A simple way of reducing the spherical aberration is to place an aperture or lens stop over the entrance pupil of the lens to block out some of the peripheral rays. In biological imaging, the most common approach to correct spherical aberration is adjusting the objective correction collar which axially translates a movable lens group within the objective. The effect of spherical aberration in biological imaging has been corrected in several ways including altering the tube lens [8, 9], adjusting the rear pupil aperture of the objective lens [10], optimizing the immersion medium [11], and optical refocusing, which involves the use of multiple objectives lenses [12].

Chromatic Aberration is a result of the fact that white light consists of various wavelengths. When white light passes through a lens, the rays are refracted according to their wavelengths. Blue light rays are refracted with the greater angle followed by green and red light, this phenomenon commonly referred to as

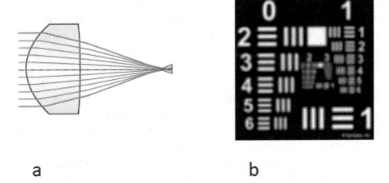

a   b

Fig. 2.8 (**a**) Representation of spherical aberration. (**b**) Simulation of spherical aberration by Zemax optical design software

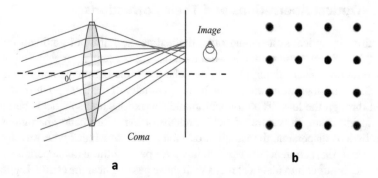

Fig. 2.9 (a) Representation of coma. (b) Simulation of coma by Zemax optical design software

dispersion. The inability of the lens system to bring all of the colours into a common focus results in the increase of image size and colour fringes surrounded the image.

By combining a special type of glass, crown glass and flint glass (each type has a different dispersion of refractive index) in the lens system it is possible to bring the blue rays (486 nm) and the red rays (656 nm) to a common focus, near but not identical with the green rays (550 nm). This combination is known as achromat doublet where each lens has a different refractive index and dispersive properties. This is the most widely used lens system in microscopes. The chromatic aberration is reduced in the doublet system by properly adjusting the lens thickness, curvature, and refractive index of glass.

Coma is an "off-axis aberration" that causes point objects to look like comets with a tail extending towards the periphery of the image plane (Fig. 2.9). Usually, coma affects the points located near the periphery of the image, resulting in a sharp image in the centre of the field and blurred towards the edges. When a bundle of oblique rays is incident on a lens, the rays passing through the edge of the lens may be focused at a different height than those passing through the centre. Coma is greater for lenses with wider apertures. Correction of this aberration is done by accommodating the object field diameter for a given objective.

Curvature of Field is another serious off-axis aberration. Field curvature indicates that the shape of the image plane is a concave spherical surface as seen from the objective (Fig. 2.10). In the infinity-corrected system, the field curvature is corrected by accommodating different lens systems (doublet, triplet) in the objective and tube lens.

Distortion is a monochromatic aberration produced mainly by the eyepiece in the microscope. Distortion changes the shape of the image while the sharpness maintained. If the image of an off-axis point is formed farther from the axis or closer to the axis than the actual image height given by the paraxial expressions, then the image is said to be distorted. The distortion provides a nonlinear magnification in the image from the centre to the edge of the field. Depending on whether the gradient in magnification is increasing or decreasing, the aberration is termed as pincushion or barrel distortion (Fig. 2.11). Corrections are made as described for field curvature.

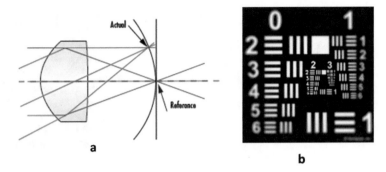

Fig. 2.10 (**a**) Representation of field of curvature. (**b**) Simulation of field of curvature

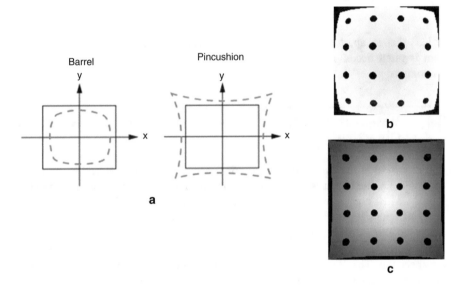

Fig. 2.11 (**a**) Representation of distortion. (**b**) Simulation of barrel distortion aberration. (**c**) Simulation of pincushion distortion aberration

Astigmatism Light rays lying in the tangential (T1) (planes contain chief ray and optic axis) and sagittal plane (S1) (planes contain only chief ray) are refracted differently. Therefore, both sets of rays intersect the chief ray at different image points, resulting in different focal lengths for each plane. This discrepancy in focal length is a measure of the astigmatism and will depend on the inclination angle of the light rays and the lens. The off-axis rays enter the optical system at increasingly oblique angles, resulting in larger focal length differences (Fig. 2.12).

These rays fail to produce a focused image point, but rather produce a series of elongated images ranging from linear to elliptical, depending upon the position within the optical train. Astigmatism errors are usually corrected in objectives through the precise spacing of individual lens elements with the appropriate choice

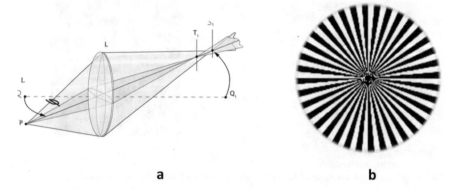

a   b

Fig. 2.12 (a) Representation of astigmatism. (b) Simulation of astigmatism

of lens shapes, aperture sizes, and indices of refraction. The correction of astigmatism is often accomplished in conjunction with the correction of field curvature aberrations.

2.4 Design Specifications of the Infinity-Corrected Microscope Objective

The objective is the eye of the microscope. Modern objective lenses are infinity-corrected, i.e. the object is placed in the front focal plane and its image is formed at infinity. Most compound microscopes have four or five objectives usually of $4\times$, $10\times$, $40\times$, and $100\times$ (or $5\times$, $10\times$, $20\times$, $50\times$, $100\times$) which revolve on a nosepiece (turret) to allow different magnifying powers. The $4\times$, $10\times$, and $40\times$ are generally "dry" objectives which operate with air imaging medium between the objective and the specimen. The $100\times$ may be a "wet" objective which operates with immersion oil between the lens and the specimen. The three design specifications of the objective set the ultimate resolution limit of the microscope. These include the wavelength of light, the numerical aperture, and the refractive index of the imaging medium. Higher magnifications yield higher numerical apertures, but shorter working distances with smaller fields of view. Lower magnifications yield lower numerical apertures, but longer working distances with larger fields of view. In modern microscopes, both objectives and eyepieces are formed by many different groups of lenses; by assembling lenses in the right way, very high magnification values may be obtained [13]. Nikon CF (Chrome Free) and Zeiss Jena objectives are fully corrected so as not to require additional chromatic correction by tube lens or eyepiece.

The most common types of objectives are: plan achromats, plan apochromats, and plan fluorite. "Plan" designates that these objectives produce a flat image plan across the field of view. "Achromat" refers to the correction for chromatic aberration featured in the objective design. The achromats are colour corrected for two wavelengths red (656 nm) and blue (486 nm) and are corrected for spherical aberration in the green wavelength (546 nm). Plan achromats are particularly used for monochromatic applications.

Plan apochromats are corrected for three or four wavelengths (red, green, blue, and violet), and the chromatic aberration is comparatively well corrected for other wavelengths. These objectives are corrected for spherical aberration for three or four wavelengths and have a high degree of flat field correction. They contain more lens elements than achromats. It is also possible to get very large NAs (up to 1.49) with this objectives design for high-resolution and low light applications. With the most effective colour correction and highest numerical apertures, plan apochromat objectives deliver brilliant images in bright field, DIC, and fluorescence techniques.

Fluorite or semi-apochromat objectives are made by glass materials, i.e. fluorite or fluorspar (CaF2) or synthetic lanthanum fluorite, giving high transmission and low colour dispersion. These objectives are corrected for chromatic aberrations at two wavelengths (red and blue) and spherical aberrations at two to three wavelengths.

The objective illustrated in Fig. 2.13a is 250× long working distance (LWD) infinity-corrected plan-apochromat objective, which contains 14 optical elements that are cemented together into three groups of lens doublets, a lens triplet group, and three individual internal single-element lenses. The objective has a hemispherical front lens and a meniscus second lens which allows to capture the light rays at high numerical aperture with minimum spherical aberration. The internal lens elements are carefully designed and properly oriented into a tubular brass housing that is encapsulated by the objective barrel. For infinity corrected objectives, an infinity symbol will be written on the body. The other parameters such as NA, magnification, optical tube length, refractive index, coverslip thickness, etc., are engraved on the external portion of the barrel (Fig. 2.13b).

The working distance is the distance between the surface of the front lens element of the objective and the top surface of the coverslip nearest to the objective. Working distance also depends on focal length and NA of the objective. Long working distance (LWD) objectives allow focusing over an extended range of up to several mm. For example, 40×/1.3 NA oil immersion objectives have a short working distance of just 0.2 mm, whereas 40×/1.0 LWD water immersion objective has a working distance of over 2 mm.

It is necessary to use a coverslip to protect the sample like bacteria, cell cultures, blood, etc., and microscope components from contamination. The light path from sample to objective depends on coverslip thickness and immersion medium. A cover slip, or glass microscope slide, affects the refraction angle of the rays from the sample. As a result, the objective needs proper optical corrections for coverslip thickness to provide the best quality image. Objective denotes a range of cover slip thicknesses for which they are optimized and it is imprinted after the infinity symbol (infinity-corrected design) on the objective barrel. The coverslip thickness ranges from zero (no coverslip correction) to 0.17 mm. The thickness of coverslip is different for upright and inverted microscope configuration. The upright microscope objective images the specimen between the coverslip and slide glass. The standardized value of the cover slip thickness for upright configuration is 0.17 mm

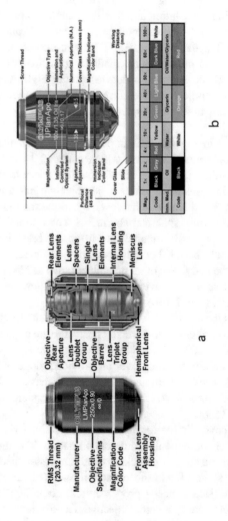

Fig. 2.13 (a) LWD infinity-corrected apochromat objective and its internal configuration (Source: Olympus Microscopy Resource Center). (b) NA, magnification, optical tube length, refractive index, and coverslip thickness are inscribed on the barrel of objective. The colour-coded ring, farthest from the thread, denotes the type of immersion medium (Source: D. B. Murphy and M. W. Davidson, Fundamentals of Light Microscopy and Electronic Imaging, Wiley-Blackwell 2012)

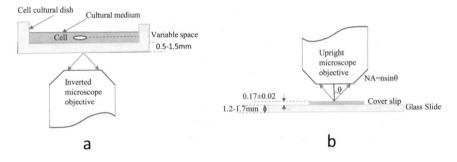

Fig. 2.14 (**a**) Inverted microscope observation. (**b**) Upright microscope observation

in biological applications (Fig. 2.14). The thickness of slide glass usually varies between 1.2 and 1.7 mm for different applications. The cover glass thickness typically has a tolerance of ±0.02 mm because the performance of the objective with NA of 0.95 is reduced by 71% for 0.02 mm thickness error [14]. Therefore, the objectives are designed in such a way so that it can also compensate for the induced coverslip aberration in the high NA system. If the objective has "–" (dash) inscribed on the barrel instead of 0.17 which indicates that the coverslip does not require to examine the sample. For example, the coverslip is not used for water glass dipping objectives.

The inverted microscope objective images the cell floating in the culture medium through the bottom of the cell culture dish. Therefore, the distance between the cell and the bottom surface of the dish is not fixed. The bottom thickness of the dish is varying between 0.5 and 1.5 mm. Furthermore, the inverted objectives often work with slide glass or without substrate. Consequently, the conventional inverted microscope objective with NA > 0.4 must be designed with correction collar for large range (0–2 mm) of cover glass (CG) correction. So, the objective must be flexible for a large scale of working distance for the large correction range. Thus, the conventional inverted objectives were mostly designed with relatively longer working distance. The LWD objectives always bring with more difficult correction of chromatic aberration, spherical aberration, and coma.

Immersion objectives generally have higher NA greater than 1.0 and less coverslip aberration. Oil immersion objectives require the use of a drop of immersion oil between and in contact with the front lens of the objective and the cover glass of the sample. These are very common on both upright and inverted configuration. They need to be treated with care, in order that immersion oil does not drip down into the objective. Sometimes small plastic protective covers will be placed around them to catch excess oil. Water immersion objectives are designed to work best with a drop of water between the objective and specimen, while water-dipping objectives are designed to interface directly with the specimen and it has long working distance. These objectives are not suitable for inverted microscopes; they are usually used in upright microscopes, where they can dip directly into the culture dish. Immersion media should never be used on dry objectives. This means that if immersion oil is on a sample, we cannot use a dry objective of more than 10× magnification on that sample without cleaning it, or the objective lens will touch the oil. Dry objectives have no protection shields against oil penetration and are easily destroyed. Note that if an immersion or dipping objective is used without the immersion medium, the image resolution becomes poor.

Another characteristic of the objective is Brightness. The ratio of NA to lateral magnification (M) determines the light gathering power (F) of an objective. The light gathering power determines the image *brightness* (B). F is defined as

$$F = 10^4 \times \left(\frac{NA}{M}\right)^2 \text{ (Transmission mode)}$$

$$\text{and, } F = 10^4 \times \frac{NA^4}{M^2} \text{ (Epi-illumination or reflection mode).}$$

(2.9)

The 60×/1.4 NA apochromatic objective gives the brightest images because its image is well chromatic corrected across the entire visual spectrum and substantially free from other aberrations (flat field and spherical aberration), it is popular in fluorescence microscopy. The 40×/1.3 NA fluorite objective is significantly brighter, but is less well corrected. The brightness of the image does not only depend on the geometry of the lens, but also on the number of reflecting surfaces and the material of the optical glasses used to construct the lens elements in objective.

The last most significant parameter that influences the objective structure is the parfocal length, which is the distance between the object and the objective shoulder. The parfocal length basically determines the amount of space for integrating different lens elements. By designing microscope objectives with identical parfocal length, the focus position is fixed when changing the objectives with different magnifications. A system with smaller parfocal length typically has a smaller number of elements but more critical sensitivity. Utilizing longer parfocal length, although more elements are used, better tolerance and reduced cost could be achieved. 60×/1.48 45 mm parfocal objective from Olympus used two cemented triplets in the middle group, whereas the 60×/1.45 60 mm parfocal objective utilized four doublets. These two objectives have similar functionality in spherical and chromatic aberration correction. But the triplet setup could relatively save space, resulting in the overall length of microscope is reduced.

2.5 Critical and Köhler Illuminations

The first stage of the light microscope is the illumination unit. Illumination is a critical determinant of optical performance in the microscopy. Two different types of illumination systems are commonly employed in a standard microscope for the illumination: (1) *Critical* illumination, (2) *Köhler* illumination. With critical illumination, an incoherent source such as a filament lamp is used to illuminate the object through a condenser lens. As illustrated in Fig. 2.15 the image point S′ of the light

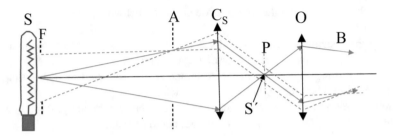

Fig. 2.15 Principal of critical illumination

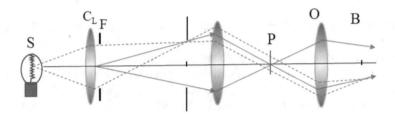

Fig. 2.16 Principal of Köhler illumination

source S is formed in the object plane P by the condenser lens (Cs). A field diaphragm F, placed close to the light source, controls the area of the illumination field, so it acts as a field stop. Simultaneously the iris diaphragm A in the front focal plane of the condenser controls the lights illuminating the object and entering the objective, i.e. it acts as an aperture stop. So, an image of the filament lamp is focused directly onto the sample in critical illumination system. A disadvantage of the critical illumination is that the source filament lamp is imaged onto the object plane and this type of source generates significantly highly non-uniform illumination.

In *Köhler illumination system* (Fig. 2.16) the light source and collector lens (C$_L$), and condenser lens (Cs) are responsible for establishing the primary illumination conditions for the microscope. In Köhler illumination, light from a source S is focused by a collector lens C$_L$ onto the aperture diaphragm (A) that lies in the front focal plane of the condenser lens Cs. The light from this diaphragm passes through the object plane P as parallel rays inclined to the optic axis. These rays enter the microscope objective O and are brought to focus in its back focal plane B. Simultaneously the condenser forms an image of the field diaphragm F, which lies at the back focal plane of collector lens, in the object plane P. The field diaphragm controls the diameter of the light beam emitted by the illumination system before it enters the condenser aperture. This system allows to optimize light quality and resolution in the image plane by aligning and adjusting each component of this optical system. It minimizes internal stray light, and allows for control of contrast and depth of an Image.

The advantages of Köhler illumination are listed below:

- Only the specimen area viewed by a given objective/eyepiece combination is illuminated; no stray light or "noise" is generated inside the microscope.
- Even, uniform illumination of the specimen area is achieved by distributing the energy of each source point over the full field.
- Full control of the illumination aperture (condenser field diaphragm) provides for best resolution, best contrast, and optimal depth of field.

2.6   Components of Infinity-Corrected Microscope System

The infinity-corrected microscope typically consists of an illuminator (including the light source and collector lens), a substage condenser, specimen, objective, tube lens, eyepiece, and detector, which is either some form of camera or the observer's eye (Fig. 2.17). There are two configurations of compound infinity-corrected system, based on the positions of the light source and the objective. With an inverted microscope, the source for transmitted light and the condenser are placed on the top of the sample stage, pointing down towards the stage. The objectives are located

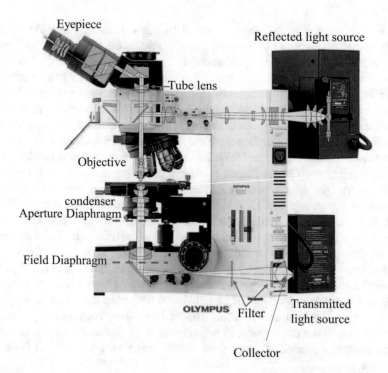

Fig. 2.17 Olympus upright microscope (Source: D. B. Murphy and M. W. Davidson, Fundamentals of Light Microscopy and Electronic Imaging, Wiley-Blackwell 2012)

below the sample stage pointing up. The cells are observed through the bottom of the cell culture vessel. With an upright microscope, the source of transmitted light and the condenser are located below the sample stage, pointing up and the objectives are placed on top of the stage, pointing down. The optical microscope design depends on two diaphragms to control the passage of light through the microscope. A diaphragm or *stop* is an opaque gate with a circular opening (often adjustable) that controls light flow through the microscope. Mainly, two diaphragms are utilized in the microscope: the *aperture diaphragm*, which adjusts the aperture angles within the microscope, and the *field diaphragm* that controls the dimension of the field imaged by the microscope. The primary role of diaphragms is to prevent light rays with aberration and stray light from reaching the image planes, and to balance the resolution against the contrast and depth of field of the image.

Light Source Most common light sources for optical microscopy are incandescent lamps, such as tungsten-argon and tungsten (e.g. quartz halogen) lamps. A tungsten-argon lamp is popular for bright field, phase contrast, and some polarization imaging. Halogen lamps are less costly and a convenient choice for a variety of applications that require a continuous and bright spectrum.

Xenon (XBO) and mercury (HBO) arc lamps are usually brighter than incandescent lamps but these are difficult to align and more expensive. Arc lamps are appropriate for high-quality monochromatic illumination when it is combined with the appropriate filter. Their spectral range starts in the UV range and continuously extends through visible to the infrared. Another popular light source is the gas-arc discharge lamp, which includes mercury, xenon, and halide lamps. About 50% of the spectral range of mercury arc lamp is located in the UV range. For imaging of biological samples using mercury arc lamp, the proper selection of filters is important to protect living cell samples and micro-organisms from the UV rays. (e.g., UV-blocking filters/cold mirrors). The xenon arc lamp can provide an output power greater than 100 W, and is often used for fluorescence imaging. However, over 50% of its power falls into the IR; therefore, IR-blocking filters (hot mirrors) are necessary to prevent the overheating of samples. Metal halide lamps were recently introduced for high-power sources (over 150 W). Light-emitting diodes (LEDs) are a new, alternative light source for microscopy applications. The characteristic features of LEDs include a long lifetime, a compact design, high efficiency, and easy to align.

Filter Microscopy filter is mainly used to increase the contrast of the image by allowing or blocking selective wavelengths of light. Two common types of filters are absorption filter and interference filter. Absorption filters normally consist of coloured glass which selectively absorbs wavelengths of light and transfers the energy into heat. The interference filters selectively transmit the wavelengths based on the interference effect. The other filters are neutral density (ND) filters that reduce the light intensity without changing wavelength and heat filters that absorb the infrared radiation to prevent the specimen from heating.

Collector Lens This lens is used to create an image of the filament onto the front focal plane of the condenser lens (Kohler illumination) or onto the specimen itself (critical or confocal illumination). The diameter of the field illuminated by the light source is controlled by the field diaphragm which is placed just behind the collector lens.

Condenser Imaging performance by a microscope depends not only on the objective lens but also on the light delivery system, which includes the illuminator, collector lens, and condenser lens. There are three types of condenser lens: (1) Abbe condenser, (2) Aplanatic condenser, and (3) Achromatic aplanatic condenser. The Abbe condensers have two lenses, and they are usually uncorrected for spherical and chromatic aberrations. The three-lens aplanatic condenser is superior to Abbe condenser. This type of condenser is corrected for spherical aberration and field curvature but still exhibits chromatic aberration. The highly corrected achromatic aplanatic condenser has five lenses including two achromatic doublet lenses, provides NAs up to 1.4, and is essential for imaging fine details using immersion-type objectives. These condensers are corrected for chromatic aberration at red and blue wavelengths, spherical aberration at green wavelength, and field curvature. Achromatic aplanatic condenser is suitable for both types of objectives (dry and oil immersion). Note, however, that for maximal resolution, the NA of the condenser must be equal to the NA of the objective, which requires that both the condenser and the objective should be oiled.

Tube Lens To create an image with an infinity-corrected objective, a tube lens must be used to focus the image. A typical infinity-corrected microscope employs a doublet pair as a tube lens (Fig. 2.18). The first doublet is Plano convex. It provides three features: optical power, correction of spherical aberration, and correction of axial colour. The second doublet is a meniscus lens with little optical power. It provides correction of coma and lateral colour. The distance between the objective and the tube lens (L) can be varied, but this will affect the image field diameter. In infinity-corrected system the tube lengths between 200 and 250 mm are considered optimal, because longer focal lengths will produce a smaller off-axis angle for

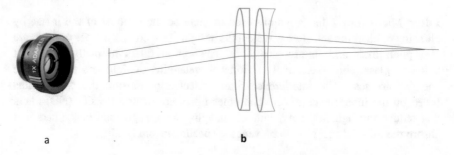

a   **b**

Fig. 2.18 (a) Standard 1× tube lens. (b) Doublet-pair tube lens layout

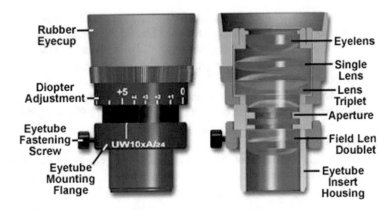

Fig. 2.19 Ultra-wide 10× eyepiece with its internal lens system

diagonal light rays, reducing system artefacts. Longer tube lengths also increase the flexibility of the system with regard to the design of accessory components.

Eyepieces The eyepieces are the multi-lens system at the top of the microscope that the viewer looks through; they are usually 10× or 15×. The magnification of an eyepiece is defined as 25 cm divided by the focal length of eyepiece. The eyepiece illustrated in Fig. 2.19 is marked with UW, which suggests it is an ultra-wide view field. Often eyepieces will have an H designation, which indicates a high eyepoint focal plane that permits microscopists to wear glasses to look at samples. Other inscriptions often found on eyepieces include WF for Wide-Field; UWF for Ultra-Wide-Field; SW and SWF for Super Wide-Field; HE for High Eye point; and CF for Chrome Free. The eyepiece magnification in Fig. 2.19 is 10× and also the inscription A/24 indicates the field number is 24, which refers to the diameter (in millimetres) of the fixed diaphragm in the eyepiece. Good eyepiece should also have the widest possible field of view. This is often helpful in estimating the actual size of objects. The field-of-view numbers vary from 6.3 to 26.5.

Digital Camera Nowadays a charged coupled device (CCD) camera is used to capture and store the images in microscopy. The CCD consists of a large matrix of photosensitive elements (referred to as "pixels") that capture an image over the entire detector surface. The incident light-intensity information on each pixel is stored as an electronic charge and is converted to an analogue voltage by a readout amplifier within CCD. This analogue voltage is subsequently converted to a numerical value by a digitizer in CCD chip resulting in the visualization of the digital image in the computer. The spatial and brightness resolution of digital image gives the information of fine details that were present in the original image. The spatial resolution depends on the number of pixels in the digital image. By increasing the number of pixels within the same physical dimensions, the spatial resolution becomes higher. The digital spatial resolution should be equal or higher than the optical resolution, i.e. the resolving power of the microscope. To capture the smallest degree of detail, two pixels or three pixels are collected for each feature. This criterion is called as Nyquist criterion, is expressed by this equation: $R * M = 2 *$ pixel size (ref), where R is the optical resolution of the objective; M is the resulting magnification at the camera sensor and it is calculated by the objective magnification multiplied by the magnification of the camera adapter. Consider a $10\times$ Plan Apochromat objective having NA 0.4 and the wavelength of the illuminating light is 550 nm, so the optical resolution of the objective is $R = 0.61 * \lambda/\text{NA} = 0.839$ µm. Assuming further that the camera adapter magnification is $1\times$, so the resulting camera magnification $M = 10\times$. Now, the resolution of the objective has to be multiplied by a factor of 10 to calculate the resolution at the camera, i.e. $R \times M = 0.839$ µm $* 10 = 8.39$ µm. Thus, in this setup, we have a minimum distance of 8.39 µm at which the line pairs can still be resolved, this is equivalent to $1/8.39 = 119$-line pairs per millimetre. The pixel size is calculated by the size of the CCD chip or CMOS sensor divided by the number of pixels. If 0.5-inch chip has a dimension of 6.4 mm $* 4.8$ mm, the total number of pixels for this chip needs to meet the Nyquist criterion with 2 pixels per feature is $(1/(R \times M)) \times$ chip size $\times 2 = 119$ line pairs/mm $\times 6.4$ mm $\times 2 = 1526$ pixels in horizontal direction and 1145 pixels in vertical direction. If we take 3 pixels per line pair, the result is 2289 pixels in horizontal direction. The system with a higher magnification objective has a small field of view so the number of pixels is reduced.

2.7   Alignment for Designing the Infinity-Corrected Bright Field Microscopy

Two basic types of microscopic optical illumination are possible: those using reflected light (episcopic or epi-illumination) and those using transmitted light (diascopic). In transmitted illumination system light allows to pass through the specimen, whereas in reflection illumination method light reflects from the specimens. Reflected light microscopy is used for imaging the opaque specimens, e.g. metals, minerals, silicon wafers, wood, polymers, and so on and transmitted light microscopy is for transparent samples such as bacteria, cell, etc. Today, many microscope manufacturers offer advanced models that permit the user to alternate or simultaneously conduct investigations using both reflected and transmitted illumination.

Figure 2.20 shows the two optical ray paths of imaging and illuminating for bright field transmissive infinity-corrected configuration. The design of a microscope must ensure that the light rays are precisely guided through the microscope. The knowledge of optical ray paths under Köhler illumination is important for proper designing and aligning the optical microscope [15]. Generally, the microscope contains two groups of optical planes which belong together. Within a group, the planes are always imaged one on the other, so they are known as conjugate planes. The first group of conjugate planes in the path of illuminating light rays includes the lamp filament, aperture diaphragm, the rear focal plane of the objective, and pupil of

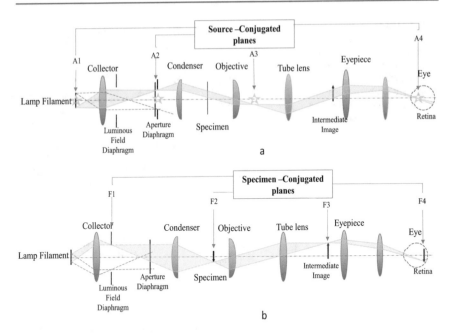

Fig. 2.20 (a) Illuminating light path consists of four conjugate planes A1, A2, A3, A4, known as source-conjugated planes. (b) Image-forming light path has four conjugate planes F1, F2, F3, F4, known as specimen-cojugated planes

observer's eye. As illustrated in Fig. 2.20, in the illuminating light ray path, from the light source to eyepoint, there are four images of the light source. These are known as source-conjugated planes, i.e. A1, A2, A3, A4. The final source image in the exit pupil of the eyepiece is located in the same plane as the entrance pupil of the observer's eye. The second group of conjugate planes in the image-forming light path includes the luminous field diaphragm, the specimen, the intermediate image plane, and the retina of the eye. Thus, from field stop to final image, there are again four specimen-conjugated planes, i.e. F1, F2, F3, F4, respectively.

Generally, these two conjugate groups occur simultaneously in microscopy. Suggested by Hammond [16], the complete symmetrical ray diagram, combing two conjugate groups, is shown in Fig. 2.21. Under the Köhler illumination condition the lamp collector and auxiliary lenses focused the illuminating rays to form an image of the filament (A1) in front of the condenser. The focused illuminating rays, from the conjugate plane A2, incident on the object as a series of parallel bundles of light and converged at the back focal plane of the objective (conjugate plane (A3)). The final image of filament is produced at the exit pupil of the eyepiece (conjugate plane (A4)). In the specimen conjugate optical path, the light rays from the filament (as shown in fig the green, red, and blue rays) is focused at different points in the field diaphragm plane (F1) which is front focal plane of the auxiliary lens. The parallel bundle of rays generated from the auxiliary lens is focused on the specimen plane (F2) in front of the objective. The infinity-corrected optics, the objective and

Fig. 2.21 Optical ray diagram of transmissive configuration in bright field mode (Source: C. Hammond, "A symmetrical representation of the geometrical optics of the light microscope," Journal of Microscopy)

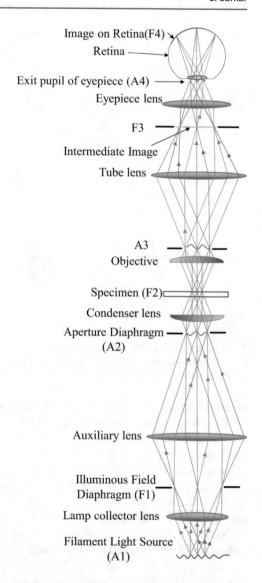

the tube lens, generate the intermediate image (F3) at front focal plane of the eyepiece and the final image of specimen is formed on observer's retina (F4). Under this alignment of microscopy if we decrease the size of the field diaphragm, a narrow bundle of rays will illuminate a smaller region of the specimen. If we decrease the size of the aperture diaphragm, then a smaller area of the filament contributes to the illumination at the object and the angles of aperture of the condenser will be smaller which causes a decrease in resolution and increase in contrast. Figure 2.22 shows the ray diagram in epi-illumination or reflection mode of microscopy. In reflection configuration, the positions of the field and aperture

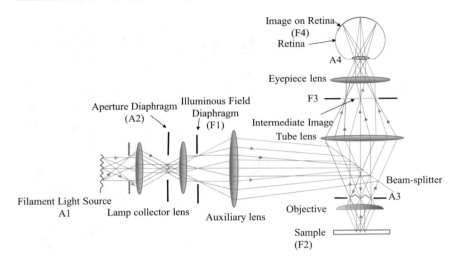

Fig. 2.22 Optical ray diagram for reflection configuration in bright field mode (Source: C. Hammond, "A symmetrical representation of the geometrical optics of the light microscope," Journal of Microscopy)

diaphragms are reversed and the objective performs the dual operations which are focusing the illuminating lights onto the specimen and collecting the imaging lights from the specimen. Fluorescence microscopy is usually performed using reflected light, even on microscopes where the bright field examination is done using transmitted light.

The fundamental step to get a good image in the microscope is to align the illuminating system correctly. The alignment of the illumination system depends on three factors: (1) Proper adjustment of field diaphragm, (2) Focusing and centre of condenser, and (3) Adjusting aperture diaphragm. The Steps for alignment in the illumination system in bright field (BF) observation mode are:

1. Open the aperture diaphragm and field diaphragm.
2. Place the 5 to 10× objective and focus the specimen.
3. Set intensity to a comfortable level (varying lamp intensity setting and/or neutral density filters).
4. Adjust the interpupillary distance of the eyepiece.
5. Close down the field diaphragm until its image is just visible in the field of view.
6. Rotate the condenser height adjustment knob to bring the field iris diaphragm image into focus, along with the specimen.
7. Centre the condenser to align the image of the field diaphragm concentric with the circular field of view.
8. Open the field diaphragm lever until its image inscribes the field of view. If using a camera, the field diaphragm should be adjusted to disappear just beyond the field of view of the camera image.

Fig. 2.23 Bright field
microscopy image of stained
cheek cell (Source: M.K. Kim,
Digital Holographic
Microscopy: Principles,
Techniques, and Applications,
Springer Series in Optical
Sciences)

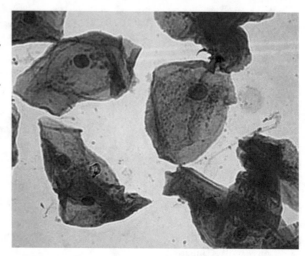

9. For optimal imaging the adjustment of aperture diaphragm is the last step of
   alignment. The aperture diaphragm is not able to be imaged directly through the
   microscope eyepieces or to a camera as it is in an intermediate plane in the optical
   path. By using phase telescope or Bertrand lens or removing the eyepiece the
   aperture diaphragm is possible to image. Normally, if the aperture diaphragm is
   closed to 70–80% of the numerical aperture of the objective, then a clear image
   with reasonable contrast will be obtained. The scale on the condenser aperture
   diaphragm ring shows numerical aperture, so adjust the condenser aperture
   diaphragm ring in accordance with the scale. Once we are able to see the aperture
   stop adjust it to the proper size and return the eyepiece or remove the Bertrand
   lens for normal imaging (Fig. 2.23).

2.8   Label-Free and Quantitative Phase Microscopy

Unstained biological samples, such as bacteria or cells, are phase objects/samples.
This type of object does not absorb incident light, it only alters the phase of light.
Conventional bright field microscopy gives only the information about the intensity
changes or amplitude changes not the phase changes introduced by the object. The
human eye also relies on changes in amplitude of a light wave, cells can be difficult
to visualize using a light microscope without dyes/labels which enhances cell
contrast. Therefore, phase sample or unstained (without dyes/labels) biological
samples are problematic for analysing in conventional bright field microscopy.
Such samples may be either transmissive or reflective in nature [17]. Rather than
using contrast-enhancing dyes/labels, label-free solutions rely on components of the
optical setup that use cells' inherent contrast characteristics (thickness and refractive
index (RI)) to create image contrast.

Here some of the popular label-free imaging techniques such as dark field illumination, phase contrast, differential interference contrast, digital holographic microscope are discussed.

2.8.1 Dark Field Microscopy

Dark field illumination requires blocking out the central zone of light rays and allowing only oblique rays to illuminate the specimen. This is a simple and popular method for imaging unstained specimens, which appear as brightly illuminated objects on a dark background.

Dark field conditions are created when bright field light from the source is blocked by an opaque dark field stop (annular stop) in the condenser. This stop must be of sufficient diameter to block the direct light (zeroth order illumination) passing through the condenser, but it must also be open around the edges, letting light pass by the outside of the stop. So, a hollow cone of light from the condenser lens illuminates the specimen. With the direct light blocked from entering the objective, the central zone of light fails to reach at image plane causing the background field of view becomes black instead of white. Only the interference of scattered light from the specimen contributes to image formation. When, objects, e.g., small particles of bacteria, are in the object plane, light is laterally diffracted away. Provided that this diffracted lights within the aperture cone of the objective, it is gathered by the objective and forms an image. The object becomes brightly visible in front of a dark background. If there is no sample, the image seen in the eyepieces remains completely dark. For dark field microscopy, it is necessary for the objective aperture to be smaller than the inner aperture of the condenser.

To design a dark field microscope, we need standard light source (halogen lamp or LED), condenser with dark field stop, infinity-corrected objective, tube lens, eyepiece, and CCD camera. The turret condenser is the best condenser option because the dark field stop is placed in exactly the same location as the condenser aperture as illustrated in Fig. 2.24.

While specimens may look washed out and lack detail in bright field, protists, metazoans, cell suspensions, algae, and other microscopic organisms are clearly

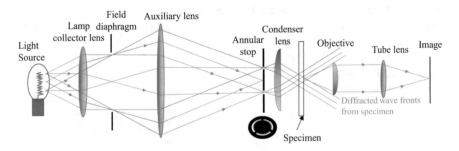

Fig. 2.24 Principal of dark field microcopy

Fig. 2.25 Dark field
microscopy image of cheek
cells (Source: M.K. Kim,
Digital Holographic
Microscopy: Principles,
Techniques, and Applications,
Springer Series in Optical
Science)

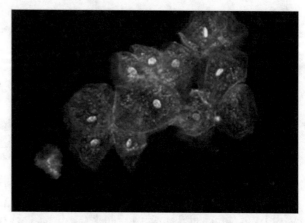

distinguished in dark field mode. Using 100× objective in darkfield mode we can see
bacteria and some structures (rods, curved rods, spirals, or cocci) and their
movement.

Alignment steps in transmitted dark field condenser:

1. Engage the 10× objective and bring the specimen into focus.
2. While looking through the eyepieces and using the condenser height adjustment
   knob, carefully adjust the height of the condenser until a dark circular spot
   becomes visible.
3. Turn the condenser centring screws to move the dark spot to the centre of field of
   view. This completes the centration.
4. Engage the desired objective. Using the condenser height adjustment knob, adjust
   until the dark field spot is eliminated and a good dark field image is obtained
   (Fig. 2.25).

2.8.2 Zernike Phase Contrast Microscopy

According to Ernst Abbe, a microscope objective can form the image of an object, by
superposing all the diffracted object beams in the image plane. Basically, the
resultant image is an interference pattern generated by the diffracted beams. Frits
Zernike (Nobel prize in Physics, 1953) invented phase contrast microscopy using
Abbe's image formation theory (Fig. 2.26).

Fig. 2.26 Abbe principle for
image formation in Coherent
illumination (m=diffraction
order)

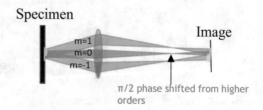

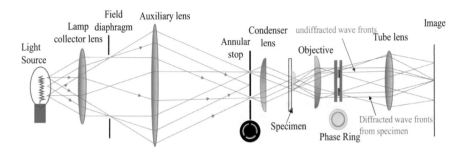

Fig. 2.27 Schematic diagram of Zernike phase contrast microscopy

For phase contrast microscopy two elements are needed. One is an annular aperture insert for the condenser, the other is special objectives that contain a phase plate. Light incident on a specimen emerges as two components: an un-diffracted wave and a diffracted wave that pass through the specimen. The diffracted wave is phase shifted by an amount δ that depends on the RI of the medium (n_1), and the specimen (n_2) along with specimen thickness t. The optical path difference $(\Delta) = (n_1 - n_2) \cdot t$, the phase shift δ is

$$\delta = \frac{2\pi\Delta}{\lambda}. \tag{2.10}$$

The refractive index of a cell is usually ~1.36. The phase shift is introduced by the cell is nearly equal to $\lambda/4$. The spatially separated diffracted and un-diffracted wave from the object traverse through the objective. A phase plate introduced in the back focal plane of objective is used to modify the relative phase and amplitude of these two waves. The phase plate then changes the un-diffracted light's speed by $\lambda/4$, so that this wave is advanced or retarded by $\lambda/4$ with respect to the higher order diffracted waves. The total $\lambda/2$ phase difference introduced between the two waves [3]. Thus, when the two waves come to focus together on the image plane, they interfere destructively or constructively (Fig. 2.27).

There are two forms of phase contrast: positive and negative phase contrast (Fig. 2.29). They mainly differ by the phase plates used for illumination. In positive phase contrast the narrow area of the phase plate is optically thinner than the rest of the plate. The un-diffracted light passing through the narrow area of phase ring travels a shorter distance resulting its phase is advanced compared to diffracted light. This causes the details of the specimen to appear dark against a lighter background, and so is called positive or dark phase contrast. In negative phase contrast the ring phase shifter is thicker than the rest of the plate, the un-diffracted wave is retarded in phase. The image appears bright on a darker background for negative or bright contrast. This is much less frequently used.

The central component of a phase contrast microscope is the phase ring. Usually it is composed of a neutral density filter and a phase retardation plate. The portion of light that passed the specimen without experiencing diffraction passes the phase ring

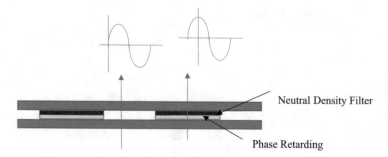

Fig. 2.28 Structure of phase ring

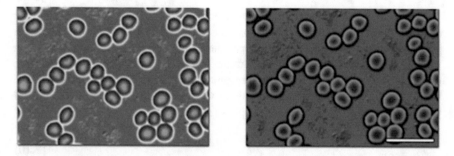

Fig. 2.29 Images of erythrocytes in positive and negative contrast optics (Source: D. B. Murphy and M. W. Davidson, Fundamentals of Light Microscopy and Electronic Imaging, 2 edition. Wiley-Blackwell, 2012)

(right arrow). The neutral density filter reduces the light intensity to avoid irradiation. The phase retardation plate retards the phase of the non-diffracted light to allow interference with the light waves that experienced phase shift and diffraction by passing the specimen (left arrow) (Fig. 2.28).

Limitations of Phase Contrast System
1. The phase contrast microscope, however, has some problems with its image quality. One is the so-called halo effect. This effect causes spurious bright areas around phase objects or reverse contrast images. Halos form because the low spatial frequency wave fronts, diffracted by the specimen traverse the phase ring as well. The absence of destructive interference between these diffracted wave fronts and un-diffracted light waves produces a localized contrast reversal (manifested by the halo) surrounding the specimen. These halos are optical artefacts and can make it hard to see the boundaries of details.
2. Another problem in phase contrast microscopy can be contrast inversion. If the objects are thick with very high refractive index, they will appear brighter instead of darker (for positive phase contrast). In such regions the phase shift is not the usual shift of $\lambda/4$ for biological specimens, and instead of destructive interference, constructive interference occurs (opposite for negative phase contrast).

Alignment of Phase Contrast Microscopy

1. Set up the Köhler illumination in microscope.
2. Install the phase ring in the condenser.
3. Remove one of the eyepieces and replace these with the phase contrast cantering telescope.
4. Put the phase contrast telescope into focus, so that the phase plate of objective and phase ring are in focus. Observe a sharp image of the phase ring in the back focal plane of objective.
5. Put the lowest magnification phase objective and corresponding phase annulus in place. For example, a 10× Ph1 objective with a Ph1 phase annulus (low magnification objectives have large diameter phase annuli (normally inscribed "Ph1" for "phase 1" and suitable for 5× or 10× objectives); intermediate magnification objectives have Ph2 annuli (e.g. 20× and 40× objectives) and the 60× or 100× objectives have the smallest diameter annuli, generally inscribed "Ph3").
6. Look at the phase plate and phase ring through the phase telescope.
7. Use the centering screws for the condenser inserts to centre the phase contrast ring, so that the bright ring overlaps the dark ring within the field of view (Fig. 2.30). If the phase ring and annulus are slightly misaligned (rotate the turret slightly), the background light intensity increases, and the quality of the phase contrast image falls.
8. Repeat the steps 5, 6, 7 for each phase and contrast ring set.
9. Once the centering operation is complete, remove the centering telescope and replace it with the eyepiece.

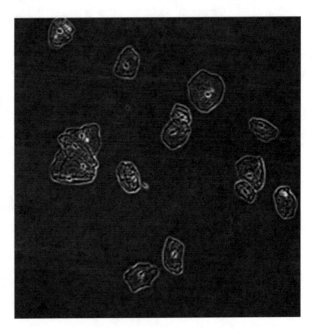

Fig. 2.30 Cheek cell image using phase contrast microscopy (Source: M.K. Kim, Digital Holographic Microscopy: Principles, Techniques, and Applications, Springer Series in Optical Sciences)

10. Focus the image with the fine focus of the microscope.
11. Widen the field iris diaphragm opening until the diaphragm image circumscribes the field of view.

2.8.3   Differential Interference Contrast (DIC) Microscope

Differential interference contrast (DIC) microscope uses the dual beam interference mode where the light beam from sample is replicated and sheared by the passage through specially designed Wollaston prism (or Nomarski prism). When these two identical and mutually coherent sample beams are made to overlap in the image plane with small shear between them, minute phase variations are visualized in white light illumination as graded and colourful intensity variations. The difference between DIC and phase contrast microscopy is discussed in Table 2.1.

The major advantage of DIC over phase contrast is that the full aperture of the microscope is used. In phase contrast the condenser's annular stop restricts the aperture, and therefore the resolution of the system. Compared with phase contrast images, differential interference contrast: (1) produces high-resolution images, (2) has better contrast, (3) can be used with thick specimens, and (4) lacks the distracting halo of phase contrast.

The main element in DIC microscope is Wollaston prism. Wollaston prism is a polarizing beam-splitter made of quartz or calcite (which are birefringent, or doubly-refracting materials). This device splits the light ray into two linearly polarized rays, and the resulting rays vibrate perpendicular to each other. One of the waves is designated the *ordinary* (*O*) wave and vibrates in a direction perpendicular to the optical axis of the prism, while the other is termed the *extraordinary* (*E*) wave with a vibration direction parallel to the prism optical axis.

Table 2.1 Difference between DIC and phase contrast technique

Study	DIC observation	Phase contrast observation
How contrast is added	Contrast added by gradients in sample thickness	Contrast added at sample borders or points
Image features	Bright/dark or colour contrast added, conveying a three-dimensional appearance Shadows added depending on orientation	Bright/dark contrast added Pronounced halo around thick samples
Contrast adjustment and selection	Fine adjustment of three-dimensional contrast possible	Choice of negative or positive contrast
Suitable sample	Capable of observing structures with sizes ranging from minimum to large. Sample thickness up to several 100 μm	Useful for observing minute structures Sample thickness up to 10 μm
Resolution	High	Poor compared to DIC

This is briefly how Nomarski DIC images are produced:

1. Light passes through a standard polarizer before entering the condenser, producing plane-polarized light at a 45-degree angle with respect to the optical axes.
2. This light enters a Wollaston prism situated in the front focal plane of the condenser. The two wavefronts, ordinary and extraordinary, are separated by a very small difference (less than the resolution of the system). A separation like this is called shearing and is one of the most important features of the system.
3. The two wavefronts pass through the specimen and are retarded to varying extents in doing so.
4. The light now enters a second Wollaston prism which recombines the wave fronts. If there has been a phase shift between the two rays as they pass through areas of different refractive index, then elliptically polarized light is the result.
5. Finally, the light enters a second polarizing filter, termed an analyser. The initial polarizer and this analyser form a crossed polarized light. The analyser will permit the passage of some of the elliptically polarized light to form the final image (Figs. 2.31 and 2.32).

Basic Components of DIC

Condenser: The condensers designed for DIC usually have a built-in polarizer. This can be slid out of the light path for bright field illumination. The polarizer can fully be rotated, but is marked to permit correct east-west orientation and a locking screw is provided. The main body of the condenser is the rotating, phase contrast type.

Wollaston Prism: In DIC each Wollaston prism consists of two precision made wedges of quartz, cemented together so that their axes of birefringence are at right angles to each other. The prism itself is mounted in a circular cell. These prisms are specific for the objectives to be used, so if DIC observation at $10\times$, $40\times$, and $100\times$ is required, then three matching prisms need to be installed.

Objectives: Theoretically any objectives can be used, but in practice higher grade objectives (fluorite and apochromatic types) are generally specified to benefit from the high-resolution potential. In many cases phase contrast fluorite objectives are chosen, permitting bright field, DIC, phase contrast, and fluorescence observation with a single set of objectives.

DIC slider: The second Wollaston prism arrangement is a slider fitted above the objectives but below the tube lens. In this case only one prism is required, and it is provided with a means of sliding it across the light path. The DIC slider is orientated northwest–southeast, i.e. diagonally in the light path.

Analyser: The output polarizer in the system, termed the analyser, is installed above the DIC slider. The polarizer and analyser need to be aligned so that their transmission axis is orthogonal to each other.

Alignment of DIC

1. In place of the condenser used for bright field observation, a "universal condenser" fitted with a built-in polarizer and a DIC prism are required.
2. A "DIC prism (DIC slider)" and an "analyser" are required below the objective.

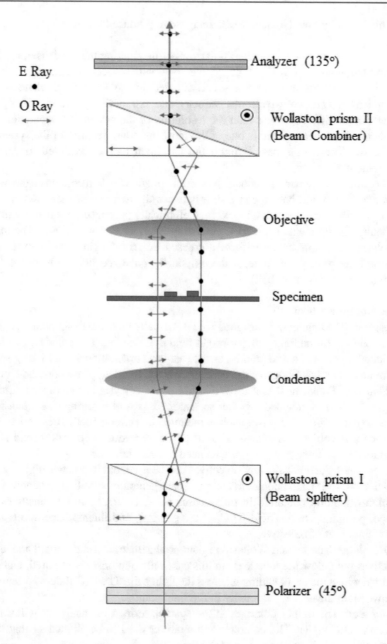

E Ray
•

O Ray
←——→

Analyzer (135°)

Wollaston prism II
(Beam Combiner)

Objective

Specimen

Condenser

Wollaston prism I
(Beam Splitter)

Polarizer (45°)

Fig. 2.31 Schemetic diagram of DIC

3. Focus on a blank sample plate using either a 4× or 10× objective in bright field mode.
4. Move the DIC slider with the analyser into the light path.

Fig. 2.32 Cheek cell image using DIC (Source: M.K. Kim, Digital Holographic Microscopy: Principles, Techniques, and Applications, Springer Series in Optical Sciences

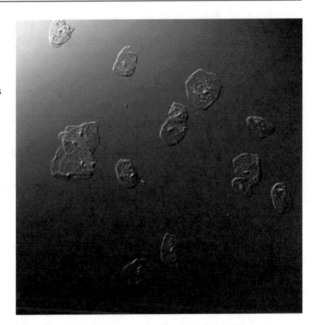

5. If using a trinocular head, remove one eyepiece and view the sample directly down the trinocular head.

6. (a) When using eyepieces: rotate the polarizer until there is a dark stripe through the centre of the field of view. This indicates that the transmission axis of polarizer and analyser is aligned at 90 degrees to each other. (b) When using a camera: rotate polarizer until the image is at its darkest.

7. If the condenser and objectives were removed, put them back in their position and also check the illumination condition.

8. Place the specimen on the stage and bring the specimen into focus by moving the objective up or down.

9. Adjust the field iris diaphragm so that its image circumscribes the field of view.

10. Adjust the aperture iris diaphragm to enhance the contrast.
    Move the prism movement knob of the DIC slider to select the interference colour that can provide the optimum contrast in accordance with the specimen.

2.8.4 Digital Holographic Microscopy for Quantitative Phase Measurement

The phase contrast microscope, DIC allowed only qualitative evaluation of phase which was sufficient to visualize the internal structure of living cells without the need of dyes. The alternative approach to phase imaging is through the use of interferometry where small phase variations of the light emerging from the specimen are rendered in intensity as shifts of the interference fringes. Interferometry provides the measurement of the defects of samples with resolutions of fractions of the wavelength of light. Based on interference phenomenon, digital holographic micros-copy (DHM) is developed for quantitative phase imaging (QPI) [17, 18]. The knowledge of this microscopic technique is important because it permits true three-dimension (3D) visualization and 3D phase display of any unstained specimen. A comparison study of different microscopes can be found in Table 2.2.

The basic DHM setup consists of an illumination source, an interferometer with microscopic imaging optics, a digitizing camera (CCD), and a computer to run the algorithms. A laser is used for illumination with the necessary coherence to produce interference. The common interferometer for DHM is Mach–Zehnder configuration as depicted in Fig. 2.33. The spatially filtered and collimated laser beam is amplitude divided by the cube beam-splitter (CBS1) into an object (O) and reference beam (R). The specimen or object is placed at the working distance of the microscope objective (MO1) and this MO1 collects the object wave transmitted through the transparent sample. After passing through CBS2 these beams interfere and hologram (interfer-ence pattern) is recorded by the CCD. The image is numerically reconstructed from the hologram. The reconstruction algorithm consists of two steps: (1) Multiplication of a reference wave with the hologram and (2) Convolution of the propagation transfer function with the digital hologram. The propagation transfer function is calculated by diffraction integral using Fresnel transform method or angular spec-trum method. Two images, real and virtual image, are formed from this digitally recorded hologram. The hologram (E_h), recorded by CCD, can be expressed as

$$E_h = |R + O|^2 = |R|^2 + |O|^2 + O^*R + O \cdot R^*. \tag{2.11}$$

In the above equation, the first term $|R|^2 + |O|^2$ is known as the dc term. Real and virtual image of the object are, respectively, given by the terms OR* and RO*. There is no significant difference between real and virtual image besides 180° rotation; both are known as the twin images.

The image reconstruction algorithm is depended on well-developed fast Fourier transforms (FFT). Reconstruction of image is described by the following equation:

$$E_I = \mathcal{F}^{-1}[\mathcal{F}(E_h \cdot R)\mathcal{F}(h)], \tag{2.12}$$

where R is the reference wave and h is the propagation transfer function which is calculated by Fresnel diffraction integral.

The Fourier transform of first two terms $|R|^2 + |O|^2$ in Eq. (2.10) being real, and in frequency plane their transform is centred at origin. The object can only be reconstructed from the last two terms. To improve the reconstruction quality, the dc and twin-image terms have to be eliminated. One of the methods to achieve this is to apply the spatial filtering operation. The spatial filtering method is used not only to suppress the DC term, but also to select one of the twin terms as well as to eliminate spurious spectral components due to parasitic reflections and interference.

Table 2.2 Main features and usage of different infinity-corrected microscopic techniques

Name of method	Features	Main area of use
Bright field microscopy	Commonest observation method Entire field of view illuminated by light source	Observation of stained specimen
Dark field microscopy	Zeroth-order un-diffracted light is rejected at the objective back focal plane and does not contribute to image formation. Only interference of higher order diffracted light contributes to image formation. Transparent specimens appear bright against a dark field of view	Suitable to the examination of minute refractile structures, which scatter light well. Observation of phase objects, such as the silica of the frustules (i.e. shells) of diatoms, bacteria, aquatic organisms, small inclusions in cells, and polymer materials.
Phase contrast microscopy	By using annular stop into the condenser, the zeroth-order un-diffracted lights will appear as a ring at the objective back focal plane, whereas the specimen diffracted light will be inside or outside this ring. Introduction of an annular quarter-wave ($\lambda/4$) plate at the objective back focal plane results in a total $\pm\lambda/2$ (90°) phase shift of diffracted light relative to un-diffracted light, as well as specific attenuation of the undiffracted light. At the image plane, interference of this "modified" diffracted and un-diffracted light leads to good image contrast without sacrificing resolution.	Observation of phase objects, such as bacteria, living cells. Does not work well with thick specimens
Differential interference microscopy	Wollaston prisms (one at the condenser aperture plane and the other very close to the objective back focal plane) are used to create two parallel and orthogonally polarized beams (O-rays and E-rays) out of every beam that would be incident upon the sample. Any phase difference between O-rays and E-rays is converted into elliptically polarized light when the rays are recombined. Specimen appears three dimensional.	Observation of phase objects, such as bacteria, living cells
Digital holographic microscopy	Two-step imaging process: Recording the hologram and numerical reconstruction of image Interferometric technique where object and reference wave interfere to generate the hologram. Reconstruction algorithm depends on optical configuration	Observation of phase objects, such as protozoa, bacteria, and plant cells, mammalian cells such as nerve cells, stem cells, tumour cells, bacterial-cell interactions, red blood cells or erythrocytes, etc. Quantitative depth measurement is possible.

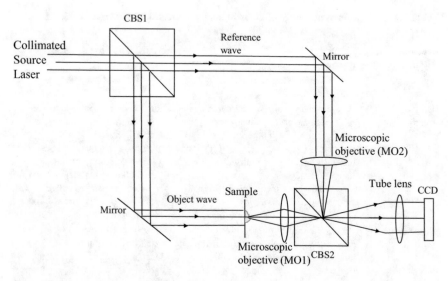

Fig. 2.33 Schematic diagram of a digital holographic in microscopic configuration

The intensity image is calculated from the complex amplitude distribution E_I and it is represented by following equation:

$$I = \text{Re}\,[E_I]^2 + \text{Im}[E_I]^2. \tag{2.13}$$

If n is the refractive index of the sample, then the sample thickness t is possible to calculate from the reconstructed phase information of the sample. The sample phase reconstruction is given by

$$[\delta(x,y)]_{\text{Sample phase}} = \tan^{-1}\left[\frac{\text{Im}(E_I)}{\text{Re}\,(E_I)}\right]. \tag{2.13}$$

$$\text{So}, t = \frac{\lambda[\delta(x,y)]_{\text{sample\_phase}}}{2n\pi}. \tag{2.14}$$

In DHM, the phase image is a quantitative representation of the object profile with subnanometre precision [18–20] (Fig. 2.34).

A well-known distinctive feature of holography is the reconstruction of image from the single hologram at various distances. Spatial resolution of DHM is limited by the wavelength of source, NA of objective, and pixel size of CCD. The interferometers may also include various apertures, attenuators, and polarization optics to control the reference and object intensity ratio. The polarization optics may also be used for the specific purpose of birefringence imaging. There are also low-coherence sources (LED) used in DHM for reducing speckle and spurious interference noise, or generating contour or tomographic images.

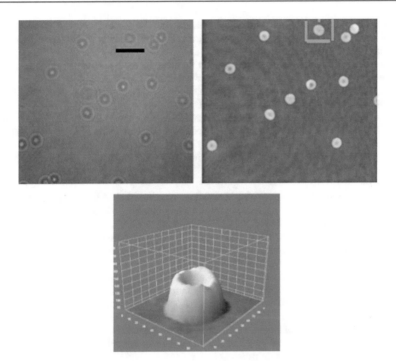

Fig. 2.34 (**a**) Recorded hologram of red blood cells (black bar is 500 μm), and (**b**) reconstructed phase image (green bar is 10 μm). (**c**) Phase reconstruction of single RBC cell (Source: [20])

Take Home Messages
The limiting resolution of all microscopes depends on the wavelength (λ) of the light used and the NA of the objective. Dirty or misaligned optics or vibration, or both, can reduce the achieved resolution. To reduce the aberration different types of infinity-corrected objectives are designed. Test resolution regularly, and especially pay attention to the iris setting and full illumination of the condenser aperture, to assure optimal performance of the microscope.

References

1. van Lommel ATL. From cells to organs: a histology textbook and atlas. Dordrecht: Springer; 2003.
2. Murphy DB, Davidson MW. Fundamentals of light microscopy and electronic imaging. 2nd ed. New York: Wiley-Blackwell; 2012.
3. Sanderson J. Understanding light microscopy. New York: Wiley; 2019.
4. Born M, Wolf E. Principles of optics, 60th anniversary edition. Cambridge: Cambridge University Press; 2019.
5. The Diffraction Barrier in Optical Microscopy. Nikon's MicroscopyU. https://www.microscopyu.com/techniques/super-resolution/the-diffraction-barrier-in-optical-microscopy

6. Oldenbourg R, Shribak M. Microscope, Chapter 28. In: Geometrical and physical optics, polarized light, components and instruments, vol. I. 3rd ed. New York: The McGraw-Hill Companies; 2010.

7. Chen X, Ren L, Qiu Y, Liu H. New method for determining the depth of field of microscope systems. Appl Opt. 2011;50(28):5524. https://doi.org/10.1364/AO.50.005524.

8. Sheppard CJR, Gu M. Aberration compensation in confocal microscopy. Appl Opt. 1991;30 (25):3563. https://doi.org/10.1364/AO.30.003563.

9. Kam Z, Agard DA, Sedat JW. Three-dimensional microscopy in thick biological samples: a fresh approach for adjusting focus and correcting spherical aberration. Bioimaging. 1997;5(1): 40–9. https://doi.org/10.1002/1361-6374(199703)5:1<40::AID-BIO4>3.0.CO;2-W.

10. Sheppard CJR, Gu M, Brain K, Zhou H. Influence of spherical aberration on axial imaging of confocal reflection microscopy. Appl Opt. 1994;33(4):616–24. https://doi.org/10.1364/AO.33. 000616.

11. Wan DS, Rajadhyaksha M, Webb RH. Analysis of spherical aberration of a water immersion objective: application to specimens with refractive indices 1.33-1.40. J Microsc. 2000;197 (Pt 3):274–84. https://doi.org/10.1046/j.1365-2818.2000.00635.x.

12. Botcherby EJ, Juskaitis R, Booth MJ, Wilson T. Aberration-free optical refocusing in high numerical aperture microscopy. Opt Lett. 2007;32(14):2007–9. https://doi.org/10.1364/OL.32. 002007.

13. Zhang Y, Gross H. Systematic design of microscope objectives. Part I: system review and analysis. Adv Opt Technol. 2019;8(5):313–47. https://doi.org/10.1515/aot-2019-0002.

14. Coverslip Correction. Nikon's MicroscopyU. https://www.microscopyu.com/microscopy-basics/coverslip-correction

15. Rottenfusser R. Proper alignment of the microscope. Methods Cell Biol. 2013;114:43–67.

16. Hammond C. A symmetrical representation of the geometrical optics of the light microscope. J Microsc. 1998;192(1):63–8. https://doi.org/10.1046/j.1365-2818.1998.00408.x.

17. Nadeau JL. Introduction to experimental biophysics: biological methods for physical scientists. 1st ed. Boca Raton, FL: CRC Press; 2011.

18. Kühn J, et al. Axial sub-nanometer accuracy in digital holographic microscopy. Meas Sci Technol. 2008;19(7):074007. https://doi.org/10.1088/0957-0233/19/7/074007.

19. Mann CJ, Yu L, Lo C-M, Kim MK. High-resolution quantitative phase-contrast microscopy by digital holography. Opt Express. 2005;13(22):8693–8. https://doi.org/10.1364/OPEX.13. 008693.

20. Jaferzadeh K, Sim M, Kim N, Moon I. Quantitative analysis of three-dimensional morphology and membrane dynamics of red blood cells during temperature elevation. Sci Rep. 2019;9(1): 14062. https://doi.org/10.1038/s41598-019-50640-z.

Epifluorescence Microscopy

3

Rezgui Rachid

Contents

What You Will Learn in This Chapter

In previous chapters, we have seen that the interaction of light with matter produces one or a combination of the following phenomena: transmission, absorption, reflection, scattering and diffraction, refraction and polarization, phase change and fluorescence emission [1]. Each one of these effects can be used to generate contrast and hence create an image. In this chapter, we will discuss the light-matter interaction that leads to the absorption of a photon and the subsequent emission of a photon with lower energy: Fluorescence. We will explore its principles, advantages over classic bright field techniques, limitations and some of its main applications in life and material sciences.

By providing technical analysis as well as a step-by-step protocol, the reader will be able to understand the concept of fluorescence microscopy, get an introduction to

R. Rachid (✉)

Core Technology Platforms - New York University Abu Dhabi, Abu Dhabi, United Arab Emirates

e-mail: rachid.rezgui@nyu.edu

V. Nechyporuk-Zloy (ed.), *Principles of Light Microscopy: From Basic to Advanced*, https://doi.org/10.1007/978-3-031-04477-9_3

labelling techniques, understand the components of a fluorescence microscope and learn how to design and set up experiments with the optimal compromise between Acquisition Speed, Signal-to-Noise Ratio and Resolution.

The chapter will be divided into four sections: theoretical aspects of fluorescence microscopy, microscope setup, sample preparation and key applications of widefield fluorescence microscopy.

3.1 Fluorescence

Fluorescence is a natural property of individual molecules. These molecules absorb photons of specific wavelengths and subsequently emit photons of red-shifted wavelengths. Considering red light has lower energy than blue light, the term red-shifted signifies that some of the energy absorbed is lost due to vibrational relaxation, dissipation or other intramolecular processes. It is worth noting that, since fluorescence covers the visible range, the term 'colour' and wavelength are used interchangeably.

3.1.1 Jablonski Diagram

In order to understand the mechanism behind fluorescence, we use a Jablonski diagram (Fig. 3.1). This diagram shows the different energy states of a particular

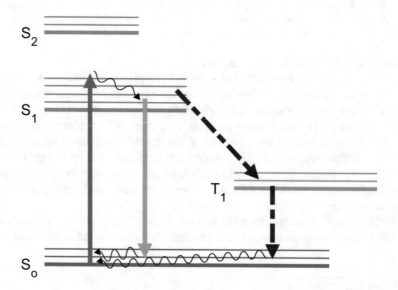

Fig. 3.1 A Jablonski diagram displaying the molecular singlet and triplet states and possible transitions for a fluorophore. Continuous arrows show radiative transitions (excitation and emission), dashed arrows show nonradiative transitions

molecule that govern the spectroscopic transitions of its electrons. Quantum mechanics dictates that the energy levels are distinct and that a transition between states occurs when so-called quanta with energies matching the energy gaps are absorbed or emitted. These transitions are either radiative or nonradiative. Radiative transitions indicate that a photon is absorbed or emitted. Nonradiative transitions do not involve photon emission. Depending on the electron spin state, they can happen in the same singlet state and are called internal conversions or between a singlet and a triplet state, where they are called intersystem crossings.

In its natural ground state G_0, a molecule is described as being in a singlet state. When a fluorescent molecule, a fluorophore, is exposed to light with high energy, photons are absorbed, and electrons get transitioned to the excited state S1 (Fig. 3.1). This absorption typically happens in the time scale of femtoseconds. The excited molecule can relax radiatively by emitting a photon with lower energy. This fluorescent de-excitation takes place in the time scale of nanoseconds [2]. Alternatively, the molecule can transition into a triplet state T_1 via intersystem crossing by flipping its spin. The triplet state is more stable than the singlet state, and the relaxation to the ground states happens in a matter of seconds to hours and is called phosphorescence.

3.1.2 Fluorescent Markers

Fluorescent molecules, or fluorophores, can be divided into three families: fluorescent proteins that can be natural or engineered, organic fluorophores, which are small molecules extracted from natural compounds and nanoparticles that are generally made from molecules like Cadmium, Zinc or rare-earth elements.

All fluorophores have five main properties:

1. Stokes shift (SS): The Stokes shift is defined as the difference between the absorption and emission peaks (Fig. 3.2) and corresponds to the average energy loss during the fluorescence process. The Stokes shift is the most important property of any fluorophore because it allows the separation between the excitation light and the emission light, significantly increasing the contrast. Most fluorophores have a Stokes shift of 30–50 nm. The higher this shift is, the better and easier the separation becomes. However, a high stokes shift reduces the total number of dyes that can be imaged together.
2. Fluorescence Lifetime (FL): This is the average time it takes a fluorophore between the absorption of a photon and the emission of a Stokes-shifted photon. The fluorescence lifetime ranges from a few nanoseconds to multiple seconds and can also be used to increase contrast in Time-Correlated-Single-Photon-Counting applications.
3. Quantum Yield or Efficiency (QE): It is the ratio between the number of absorbed photons and the number of emitted photons per unit of time and is used as an indicator of the efficiency of a particular fluorophore. QE is an important

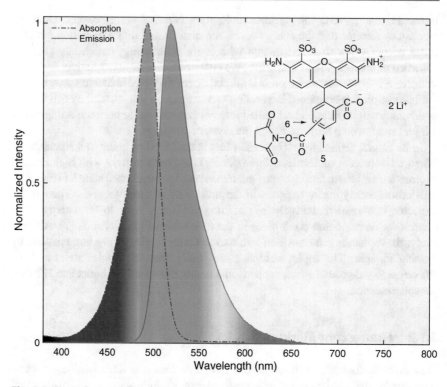

Fig. 3.2 Excitation, emission spectra and molecular structure of Alexa 488. Alexa 488 has an absorption maximum at 490 nm (blue) and an emission maximum at 525 nm (green)

parameter to be considered when choosing the right fluorophore since a high QE allows for low excitation powers.

4. Extinction Coefficient (EC): It is also known as the attenuation coefficient and describes the probability of absorption of a photon at a particular wavelength. It is usually measured at the absorption maximum.
5. Photon Yield (PY): It is the total number of photons a fluorophore emits before going extinct and varies between different fluorophore types. The Photon Yield is a critical parameter for single-molecule applications.

3.2 Fluorescence Microscope Setup

The fluorescence microscope consists of four parts: the excitation module, the emission module, the filter cube and the objective that focuses the excitation light on the sample and collects the emitted light (Fig. 3.3). The fluorescence microscope exists in two illumination forms: transmitted light illumination, also called *dia-*illumination and reflected light illumination better known as *epi*fluorescence

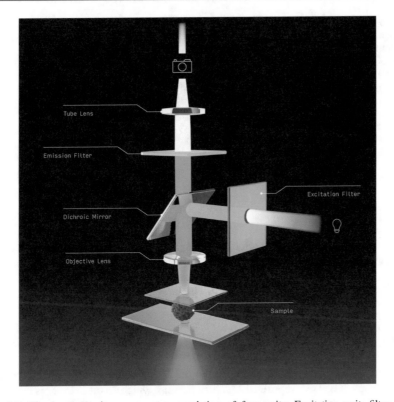

Fig. 3.3 Fluorescence microscopy setup consisting of four units: Excitation unit, filter unit (excitation filter, dichroic mirror and emission filter), magnifying unit (objective and tube lens) and detection unit (camera)

Table 3.1 Comparison of different excitation sources for fluorescence microscopy

	Mercury	Xenon	Solid State	Lasers
UV brightness	+	−	−	++
Blue-green brightness	+	+	+	++
Red brightness	−	++	−	++
Stability	−	+	++	++
Lifetime	−	+	++	++[b]
Safety	−[a]	+	++	−
Cost	++	+	−	−

[a]UV light, Heavy Metal and Ozone exposure
[b]valid for diode-based lasers

microscopy. In this chapter, we will focus on the epifluorescence setup as it is the most common type of microscope.

In epifluorescence, the excitation light (Table 3.1) is directed on a dichroic mirror that reflects it perpendicularly through the objective (Fig. 3.4). The objective focuses the light on the sample. Fluorescent light is emitted from the sample in all directions,

Fig. 3.4 Typical filter cube setup used in fluorescence microscopy. Green: Excitation Filter. Purple: Dichroic Mirror. Yellow: Emission Filter

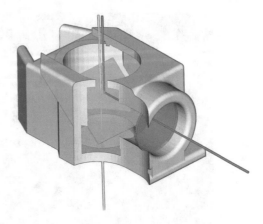

and only part of it will be collected through the same objective where it passes through the dichroic mirror, the emission filter as well as the tube lens to reach the detector. Since this type of illumination covers the whole field of view of the objective, a camera is used to collect the fluorescence.

3.2.1 Excitation Module

Mercury Lamps Mercury lamps are the oldest light sources used in fluorescence microscopy. They use vaporized Mercury through which an electric arc is created to produce light. They are cost-effective and have the highest UV irradiation among other white light sources. However, Mercury lamps are hazardous as they contain mercury, can generate ozone and have a short lifetime of 200 h. They also need to be aligned for optimal illumination of the sample.

Metal Halide Lamps Similar to mercury lamps, metal halide lamps are gas discharge lamps that use a mixture of mercury and metal halides like bromine or iodine. They have a continuous spectrum with peaks at different ranges of the visible spectrum. They have sufficient intensity in the UV as well as the far-red range. Compared to Mercury lamps, metal halide lamps are more energy-efficient and have a longer lifetime (2000 h).

Solid State Light Sources or Light Emitting Diodes (LED) LEDs are the newest light sources available for fluorescence microscopy. Previously mainly available in the blue, green and red ranges, their UV emission had improved in recent years. LEDs have three main advantages. They can be switched on and off in milliseconds, have a lifetime of more than 10,000 h, and their intensity can be electronically controlled. An example is shown in Fig. 3.5.

Lasers Lasers are usually reserved for high-end widefield microscopy techniques like Total Internal Reflection Fluorescence Microscopy (TIRF) or Stochastic Optical

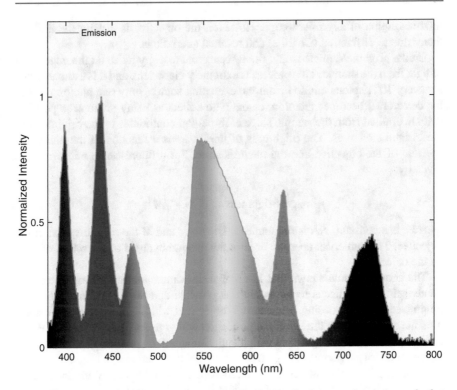

Fig. 3.5 Lumencor Spectra X emission spectrum. The light engine has several emission peaks that are matched to common fluorescence dyes like DAPI, FITC, TRITC or CY5

Reconstruction Microscopy (STORM) due to their higher cost and to laser safety regulations. They have the advantage of being: i) monochromatic, which removes the need for an excitation filter and ii) coherent, which gives the best sample illumination profile. However, this optimal sample illumination profile is highly dependent on the optical alignment requiring frequent servicing of the microscope.

3.2.2 Objective

In an epifluorescence setup, the objective has two functions: it is both the condenser that focuses the excitation light on the sample and the lens that collects the fluorescence signal, magnifies it and creates the image that will be projected on the detector. The objective is the heart of any microscope. Successful imaging requires careful consideration of the different specifications of any objective: magnification, numerical aperture, field number, immersion medium, working distance, contrast method (brightfield, phase contrast, differential interference contrast DIC). Many of these objective properties are interdependent and have been described in more detail in previous chapters. It is important to note that each part of the setup contributes to the final resolution of any microscope. However, the objective is usually the limiting factor due to diffraction (Chap. 2) and residual aberrations.

Unlike brightfield microscopy, fluorescence microscopy relies on the generation of light from the sample. This process is extremely inefficient, and it is estimated that for every 10^6 photons emitted from the excitation source, only one photon reaches the detector. It becomes therefore essential to collect as many photons as possible that are emitted from the sample since each photon contributes to the brightness of the imaged structures. The brightness of the image is a function of the numerical aperture of the objective and the magnification. For a fluorescence microscope is given by:

$$\text{Image Brightness} = c \cdot \text{NA}^4 / M^2,$$

where c is a constant, NA is the numerical aperture and M the magnification of the objective. The numerical aperture defines the maximum angle of collection of light by the objective.

The above formula shows that in order to maximize the image brightness, the numerical aperture needs to be as high as possible and the overall magnification minimized. When choosing between two objectives with identical numerical aperture, the objective with the lower magnification will give the brightest image. When choosing between two objectives with identical magnification, the objective with the highest numerical aperture will give the best resolution and image brightness.

Another part of the fluorescence microscope is the tube lens. With an infinity-corrected objective, its role is to focus the light on the detector and to correct for some of the optical aberrations. Some microscopes have more intermediate lenses that can be used to magnify or demagnify the image on the camera in order to match the sensor's pixel size.

3.2.3 Fluorescence Filters

Fluorescence filters are a key component in fluorescence microscopy. They separate the light emitted from the sample from that used to excite the fluorophores, thereby allowing the image to be seen or detected with sufficient contrast. In conjunction with advances in camera technology, fluorescence filters have contributed significantly to the success of fluorescence microscopy in life science research. They are usually mounted in a cube containing the excitation filter, the dichroic mirror and the emission filter (Fig. 3.4). The excitation and emission filters are usually bandpass filters that have been designed to transmit only a specific part of the spectrum that can be matched to the excitation and emission of a specific fluorophore.

Dichroic mirrors, on the other hand, are filters with a cut-off wavelength at which they switch from transmission to reflection (Fig. 3.6). Dichroic mirrors are optimized for an incident angle of 45°, while excitation and emission filters work best with an incident angle of 90°. All three components are made of a glass substrate coated with several layers of metals or metal oxides with different thicknesses. Depending on its

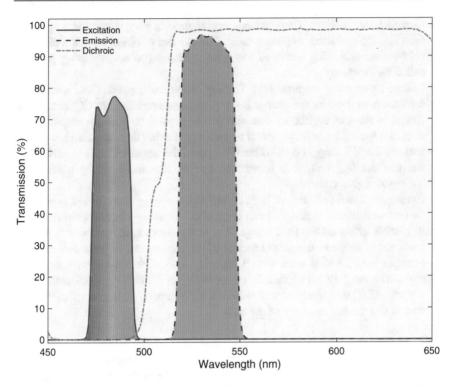

Fig. 3.6 Transmission curves for an Alexa 488 filter cube. Blue line: excitation filter. Green dashed line: emission filter. Red dash-dot line: dichroic mirror

incident angle, the light can either be reflected, transmitted or absorbed at the different layers giving the filters its optical properties. The visible spectrum has traditionally been divided into four primary colour ranges known as the blue, green, red and far-red ranges, with recent advances permitting a fifth colour in the near-infrared range.

3.2.4 Emission Module

Charge-Coupled Device CCD and Complementary Metal Oxide Semiconductors CMOS are the main two types of cameras used for widefield microscopy. CCDs and CMOS cameras are two-dimensional arrays of photosites (pixels) that convert the fluorescence light to a digital signal in a process involving:

1. Light to electron conversion. Here the photons hit the active surface of the chip and get converted to electrons through photo effect. The efficiency of this process is called Quantum Efficiency QE and varies across the spectral range reaching more than 95% in the green range for high-end cameras.

2. Charge accumulation: The electrons accumulate in each pixel for a specified duration, the so-called 'exposure time'. The number of electrons that a pixel can hold before becoming saturated and leaking charges to neighbouring pixels is called well capacity.

3. Charge transfer and amplification. The key difference between CCD and CMOS cameras is in the way the charge is converted and amplified. In CCD chips, the charge is transferred line by line and pixel by pixel to a single amplifier that amplifies the signal and converts it to a voltage making them slower than CMOS sensors. CMOS chips, on the other hand, have the amplifier built in each pixel, and the resulting voltage is transferred line by line and pixel by pixel to the analog-to-digital converter.

4. Analog-to-digital conversion: Here the amplified voltage is converted to an 8, 12, 14 or 16-bit digital signal that can be read by the computer and converted into brightness levels on the final image. It is important to note that there is a direct relationship between the output format of the camera and its well capacity. For example a CCD chip with a well capacity of 3000 electrons, will require a digitalization of 12 bit (4096). In comparison, an EMCCD camera with a well capacity of 60,000 electrons will require a digitalization of 16 bits (65,536) to be able to use the full capacity of the chip.

CCD or CMOS?

CCD and CMOS cameras have been around for decades. The traditional choice was CCD sensors for low light applications thanks to their higher signal-to-noise ratio and CMOS sensors for speed applications or in consumer electronics thanks to their smaller size and lower power consumption. However, recent developments of both technologies made this choice a little more complicated. For example EMCCDs have been developed in the early 2000s making video frame rates possible at full frame and reducing blooming effects. Meanwhile, CMOS cameras have seen an increase in their quantum efficiency as well as a decrease in noise levels, making them a valuable option for low light applications.

Due to the difference in sensor and pixel sizes, it is challenging to compare CCDs and CMOS camera performances. We have therefore summarized the main specifications of two similar high-end sCMOS and EMCCD cameras in Table 3.2.

Camera Noises

There are four primary noise sources: (i) The photon noise or shot noise is due to the Poissonian distribution of the fluorescence emission and is proportional to the square root of the number of emitted photons. (ii) The readout noise is the result of the charge/voltage transfer through the chip. This noise can be reduced by changing the readout speed at the cost of lower time resolution. (iii) The dark noise is an inherent noise to each chip and is associated with thermally generated charges on the photosites. The dark noise can be reduced by cooling the camera chip with forced air or liquid. (iv) The amplification noise is the noise created during the amplification process. This noise is the same across all pixels for CCDs as they have one amplifier

Table 3.2 Technical specifications of Andor Ixon 888 and Photometrics Prime 95B

	Photometrics Prime 95B	Andor Ixon888
Sensor type	sCMOS	EMCCD
Pixel size resolution	11 μm 1200 × 1200	13 μm 1024 × 1024
Sensor diagonale	18.7 mm	18.8 mm
Readout rate and noise	1.6 e RMS	<1e @ 30 MHz
Well capacity	80,000 e	80,000 e
Max frame rate	41 @ 16 bit 82 @ 12 bit	26 @ 16 bits
Quantum efficiency	Up to 95%	Up to 95%
Dark current		
Air	0.55 e	0.00025 e
Liquid	0.3 e	0.005 e

but varies with each pixel on CMOS cameras. It is hence given by its Root Mean Square (RMS) or Median Value.

Tips
- Match the camera's sensor (Fig. 3.7) to the Field Number (FN) of the microscope and objective. Modern CMOS cameras are available with a 25 mm diagonal chip size. When combined with an FN 25 objective, they offer 70% more area for the same magnification compared to the standard FN of 19 mm.
- Minimize over- or under-sampling by matching the objective's theoretical resolving power to the pixel size of the camera (see step-by-step protocol).
- Binning should be used when: (i) resolution is not essential and (ii) the sample has low fluorescence emission. Binning increases acquisition speed, especially in CCD cameras, improves the signal-to-noise ratio and also reduces the file size, which improves processing speeds. For example binning is recommended when performing long-term live-cell imaging with sCMOS cameras as it allows the user to reduce the exposure of the cells to high intensities of light, ensuring cell viability.
- Cool the camera with forced air or liquid to reduce the dark noise in low light level conditions and lower the readout speed when time resolution is not limiting to reduce the readout noise.

3.3 Basics of Sample Preparation for Fluorescence Microscopy

Sample preparation can be divided into two groups: live sample and fixed sample preparation. In live sample imaging, the specimen needs to be kept in its natural environment or the closest possible to it. This usually means using an incubator with temperature and gas control for cells or tissues and a perfusion system for organs like the mouse brain [3]. In order to fluorescently label the specimen, the fluorophores need to be small enough to penetrate the membrane and bind to the desired

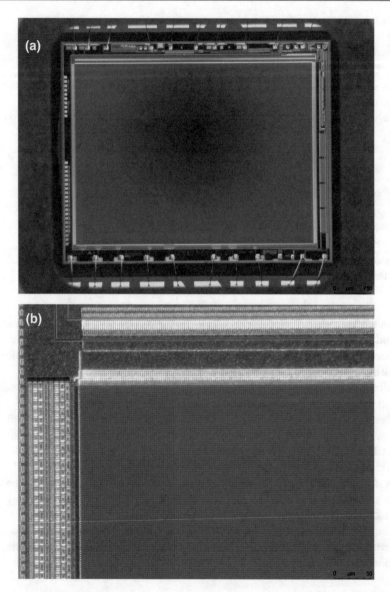

Fig. 3.7 (**a**) Image of a Sony ICX274 CCD chip with 1600*1200 pixels. (**b**) Zoom on the top left corner showing individual pixels

structures, or they have to be genetically engineered using fluorescent proteins vectors. For live sample imaging, the sample needs to be placed in a dish that is suitable for the type of microscope used. In case of upright microscopy, dipping objectives are generally used for high-resolution imaging and require dishes with diameters of 35 mm. For inverted microscopy, glass-bottom single or multiwell dishes can be used.

In the case of fixed samples, the specimen needs to be sliced to 2–4 μm slices for widefield microscopy, fixed using fixatives like paraformaldehyde or glutaraldehyde, permeabilized with detergents like Triton and then labelled using immunohistochemistry protocols. This includes blocking non-specific epitopes with blocking reagents like Bovine Serum Albumin (BSA), incubation with primary antibody and then with a secondary antibody with several wash steps in between.

Sample mounting is the last important step before imaging and is as essential as the previous steps. Mounting has two objectives: (i) avoiding physical damage or deformation and (ii) matching the refractive indices of mounting medium and sample. Mounting media with different refractive indices can be made by mixing water and glycerol. By changing their relative concentrations, the refractive index can be tuned to match that of the specimen of interest (organelles, cytosol, mitochondria, tissue type) [4]. To reduce photobleaching, antifade agents should be added to the solution. Alternatively, a multitude of mounting media is commercially available (Vectashield, Prolong series, Moviol etc...).

Sample preparation and mounting is a tedious process. Although a plethora of protocols can be found in the literature, each user needs to optimize the predefined protocols to obtain the best results. For the sake of conciseness, detailed labelling protocols are referenced at the end of this chapter [5–9].

Several companies (incl. Sigma Aldrich, Thermofisher Scientific, Spirochrome, Ibidi) have specialized in developing and commercializing fluorophores for both live and fixed sample preparation. Their products include antibodies, plasmids for cell transfection as well as stains for the most common cell organelles. They also provide labelling kits with organic or silicon rhodamine dyes conjugated to maleimides, NHS ester, carboxylic acid groups that can be used to label a variety of biomolecules.

Tips
- Use 0.17 mm coverslips (thickness 1.5) for magnifications above 20×. Most objectives are corrected for this type of coverslips. Some objectives have correction collars that can be used to adjust for different thicknesses.
- Try different staining protocols and vary the concentration of the fluorophores (organic fluorophores or primary and secondary antibodies). Use multiwell dishes for initial testing.
- High concentrations of markers lead to non-specific binding.
- There are 6.022×10^{23} molecules in 1 mol! When changing concentrations, use factors 10 and 100 to notice a change.
- Avoid mounting media that contains DAPI.
- Ensure proper sample mounting and seal the coverslip with nail polish to avoid contaminating the microscope.

3.4   Main Applications of Fluorescence Microscopy

Fluorescence microscopy covers a wide range of applications in life sciences research and diagnostics. Classically used as a structural analysis tool, fluorescence is nowadays a key component in many research areas. Modern fluorescence microscopes offer not only high resolution and, in some cases, super-resolution capabilities, but also high-speed functional imaging capabilities. Moreover, other tools like flow cytometers and multimode readers have emerged with high-speed imaging capabilities. With the advances in camera technology, it is possible today to image with more than 100 frames per seconds with relatively high resolution.

Most modern high-end methods are based on standard fluorescence microscopes. Drawing a sharp line between widefield microscopy and other imaging modalities like TIRF, STORM or even spinning disk confocal microscopy is a difficult task. Although all these techniques provide unique features and advantages, they are essentially an improvement of the standard fluorescence microscopy setup. This setup is in contrast to another technique in microscopy, Laser Scanning Microscopy (LSM), that is based on sample scanning for imaging. The specimen is illuminated point-by-point with one or more lasers, and the fluorescence emitted from the sample is collected and associated with one pixel in the final image. A raster scan reconstructs the full image.

Several Laser scanning techniques like confocal microscopy or multi-photon microscopy will be discussed in the next chapters. Table 3.3 summarizes the most common techniques in microscopy.

3.4.1   Structural Imaging

Being an imaging technique, microscopy is fundamentally a structural analysis tool. The microscope not only magnifies an object but also resolves the spatial distribution of individual molecules of interest within that object. The key success of fluorescence microscopy comes from the fact that only labelled structures are visible. This

Table 3.3 Fluorescence microscopy techniques separated by image acquisition method

Widefield techniques	Laser scanning techniques
– Epifluorescence Microscopy	– Confocal Microscopy
– Total internal reflection fluorescence Microscopy TIRF	– Multi-photon Microscopy
– Photoactivated localization Microscopy PALM	– Fluorescence (cross) correlation spectroscopy FCS and FCCS
– Stochastic optical reconstruction Microscopy STORM	– Raster image correlation spectroscopy RICS
– Light-sheet Microscopy	– Fluorescence recovery after Photobleaching FRAP
– Spinning disk Microscopy	– Stimulated emission depletion Microscopy STED
– Structured illumination Microscopy	– Minimal photon flux MINFLUX
– Fluorescence recovery after Photobleaching FRAP	– Fluorescence lifetime imaging Microscopy FLIM

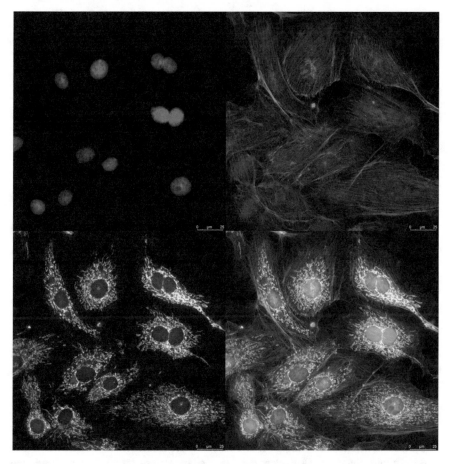

Fig. 3.8 Fluorescence image of bovine pulmonary artery endothelial (BPAE) cells stained with three colours. Nuclei are shown in Red, F-actin in green and Mitochondria in white

means that any molecule that can be labelled with a fluorophore can be made visible. By targeting specific lipids, peptides, amino acids or proteins with fluorophores, subcellular structures like the nucleus, the mitochondria, the actin filaments or any other molecule of interest can be imaged with high levels of detail.

Another advantage of fluorescence microscopy is its capability to use the full spectrum of visible light. A fluorophore's excitation and emission usually extend around 50 nm. This makes it possible to divide the visible spectrum into 4–5 channels (colours) which can be imaged simultaneously or sequentially (Fig. 3.8). This type of multiplexing has made colocalization studies possible: Instead of imaging only one element at a time, several molecules can be labelled and imaged during the same experiment. Here, the co-distribution of these biomolecules is analyzed, and their interaction is assessed.

3.4.2 Functional Imaging

Functional imaging essentially adds a time dimension to the imaging process. Here, structural information, in two or three dimensions, is combined with time information. Cellular dynamics can be resolved with nanometre and millisecond precision. These so-called time-lapse experiments can be performed to study molecular processes like replication, transcription or DNA repair as well as cellular processes like proliferation, immune reactions or migration. Other applications include Fluorescence Recovery After Photobleaching (FRAP) to investigate kinetics processes and Förster Resonance Energy Transfer (FRET) to study protein-protein interactions.

Protocol for Successful Imaging
Step-by-Step Protocol
In this protocol, a step-by-step guide is provided, and methods are presented on how to navigate iron triangle (Resolution–Speed–Signal to noise ratio). This protocol is optimized for 0.17 mm coverslips (thickness 1.5).
Before You Start:

1. Start by defining the purpose of the microscopy experiment: Is it structural analysis, colocalization studies, nuclei phenotyping, tracking of dynamic processes or others?
2. Estimate the number of images that will be needed in the study: for example, structural studies do not require multiple images when the resolution is sufficient, tracking dynamic processes needs a high number of images to achieve great statistical significance.
3. Identify the type of images to be acquired. XY (2D), XYZ (3D), XYT (time-lapse), XYZT(4D) and the number of colours that will be used. Modern microscopes can easily image four colours (three fluorescence channels and one brightfield channel). However, the more channels are required, the bigger the bleed-through between channels and the more control experiments will be required. It is advised to use a brightfield channel, especially for live-cell imaging, as it shows the overall condition of the sample.
4. Estimate the resolution required. This is a critical step as it can vastly decrease the duration of the experiment. A common pitfall is to go with the highest resolution simply because the images look beautiful. In other words, do not try to resolve details that will not be used in post-processing. For example if subcellular structures need to be resolved, then the feature size will be limited by diffraction, and the resolution will be equal to roughly 200 nm. Alternatively, if cell nuclei need to be imaged for counting or phenotyping, then a resolution of 1 μm will be sufficient.
5. Match the desired resolution to the microscope camera to find out which objective to use. You need to know the pixel size of your camera for this calculation. This can be easily found out online and is available on the software interface in modern commercial microscopes. It is usually between 5 and 15 μm. According to Nyquist's Sampling Criterion, the pixel size should be twice to three times

smaller than the size of the features that need to be resolved. Divide this pixel size by the desired resolution to get the magnification. Here it is assumed that the camera diagonal is matched to the microscope field number, and no intermediate magnification lens is used.

Example: a feature size of 600 nm corresponds to a pixel size of 200–300 nm in the image plane as per Nyquist's Sampling Criterion. With a physical pixel size of 10 µm, a 20× or 30× objective will be required. 30× objectives are uncommon, so a 40× is more appropriate. Note: magnification is in each dimension. A 20× objective will have a field of view that is four times bigger than a 40× objective. Considering that a plan-apochromat 20× objective has a NA of 1.0 and an equivalent 40× objective a NA of 1.4, the latter will bring 40% more resolution at the cost of losing 75% of the imaged area.

6. Choose the appropriate mounting medium/immersion medium combination [10]. For optimal resolution, the refractive index of the immersion medium should match the refractive index of the sample medium. Where possible, use water, glycerol or silicon oil immersion objectives for live-cell imaging and oil immersion for fixed samples with hardening mounting media like Prolong Gold or Prolong Glass.

On the Microscope:

1. Place the sample on the holder and fix it with clamps or tape.
2. Make sure the stage is levelled and that the inserts are properly placed. This is important to avoid unnecessary reflections from the coverglass. Some stages have a spirit level built in.
3. Use brightfield microscopy to focus on the sample. This greatly reduces photobleaching. In some cases, like thin sections, it is hard to find the sample. Here, use phase contrast or DIC contrast to locate them. Alternatively, look for the reflection of the covergalss or close the aperture diaphragm. Use fluorescence only when brightfield microscopy does not work.
4. Start with the objective that has the lowest magnification (5× or 10× objective). This objective will have the largest depth of focus and will make it easy to focus on the sample.
5. Use the binoculars to focus on the sample. Modern microscopes have an autofocus feature that can focus on the sample. This autofocus has improved over the years, but a trained eye is still better and faster.
6. Find the area of interest and centre it. Use the crosshairs on the binoculars, if available. Many commercial companies offer them as an option.
7. Set up the imaging channels in the software. Use multi-band filters only when acquisition speed is limiting. Otherwise, use single band filters that reduce cross excitation and bleed-through.
8. Set up the camera exposure and excitation power in a way that utilizes the whole dynamic range of the camera. Avoid saturation as it reduces resolution. We recommend starting with a few milliseconds of exposure for brightfield imaging and a few hundreds of milliseconds for fluorescence imaging and adjusting

either exposure or light attenuation until some of the pixels get saturated. Use a high-low colourmap/look-up table, if available, as it shows saturated pixels in a different colour (blue or red).

9. Set up the other dimensions (time, Z-stack) depending on the experiment.
10. Start your experiment.

Tips for multi-position long term live-cell imaging

- Turn the incubator on well before the start of the experiment (1–2 h) and make sure the temperature sensor is placed as close as possible to the sample.
- Make sure the stage is levelled, and the sample dish/slide is fixed on the stage. Use putty-like adhesive (Blu-tack) if necessary.
- Use hardware focus control systems, if available (Leica Adaptive Focus Control, Olympus Z-Drift Compensation, Nikon Perfect Focus System, Zeiss Definite Focus). Alternatively, make sure the microscope is in a location with minimal vibration (basement) or placed on an anti-vibration table.
- For suspension cells, wait 30 min or 1 h before starting your experiment. This will ensure that the cells reach the dish bottom.
- Use a slow stage speed (100–250 µm/s).
- Use the highest gain and lowest exposure possible.
- Add the individual positions in spiral-like shape to minimize stage movement.
- Use Z-stacks only when necessary. This will greatly increase the viability of the cells. Most cells are small, and a 2D image will deliver most of the required information.
- Make sure the shutter goes off after each image to avoid unnecessary exposure.

Take-Home Message
- Epifluorescence microscopy is a widefield modality of microscopy that uses fluorescence contrast to generate images.
- Epifluorescence microscopy is the most common type of imaging in life science research.
- The main advantages of epifluorescence microscopy are:
  - A dark background (high contrast) thanks to the separation of excitation and emission.
  - The labelling specificity: only the desired structures are visible on the image.
  - Its versatility: the possibility to combine multiple colours.
- The main disadvantages of epifluorescence microscopy are:
  - The indirect nature of the technique: The fluorescence is emitted from the fluorophore and not from the molecule of interest.
  - Photobleaching: the emission of fluorescence is finite in time. Photobleaching can be critical for long-term experiments.
  - Bleed-through from fluorophores with spectral overlap. Bleed-through is critical for colocalization studies.

- For optimal imaging results, excitation, emission and sample have to be matched and optimized for each other.
- Many high-end techniques are based on the epifluorescence setup.

Acknowledgments This work was supported by the microscopy core facility at New York University Abu Dhabi. The author thanks Jumaanah Al Hashemi and Dr. Oraib Al Ketan for the creation of some of the figures.

References

1. Bradbury S, Evennett PJ. Contrast techniques in light microscopy. 1st ed. New York: Routeledge; 1996.
2. Diaspro A, Pratim Mondal P. Fundamentals of fluorescence microscopy - exploring life with light. New York: Springer; 2014.
3. Sun N, Malide D, Liu J, Rovira I, Combs CA, Finkel T. A fluorescence-based imaging method to measure in vitro and in vivo mitophagy using mt-Keima. Nat Protoc. 2017;12:1576–1587.
4. Takamura K, Fischer H, Morrow NR. Physical properties of aqueous glycerol solutions. J Petrol Sci Eng. 2012;98–99:50–60. https://doi.org/10.1016/j.petrol.2012.09.003.
5. Streiblová E, Hašek J. Light microscopy methods. Yeast protocols. Methods Mol Biol. 1996;53:383–90.
6. Zaglia T, Di Bona A, Chioato T, Basso C, Ausoni S, Mongillo M. Optimized protocol for immunostaining of experimental GFP-expressing and human hearts. Histochem Cell Biol. 2016 Oct;146(4):407–19. https://doi.org/10.1007/s00418-016-1456-1.
7. Rezanejad H, Lock JH, Sullivan BA, Bonner-Weir S. Generation of pancreatic ductal organoids and whole-mount immunostaining of intact organoids. Curr Protoc Cell Biol. 2019 Jun;83(1): e82. https://doi.org/10.1002/cpcb.82.
8. Wu J, Luo L. A protocol for dissecting Drosophila melanogaster brains for live imaging or immunostaining. Nat Protoc. 2006;1:2110–5. https://doi.org/10.1038/nprot.2006.336.
9. Manning L, Doe C. Immunofluorescent antibody staining of intact Drosophila larvae. Nat Protoc. 2017;12:1–14. https://doi.org/10.1038/nprot.2016.162.
10. Jonkman J, Brown CM, Wright GD, Anderson KI, North AJ. Tutorial: guidance for quantitative confocal microscopy. Nat Protoc. 2020;5:1585–611. https://doi.org/10.1038/s41596-020-0313-9.

Further Reading

Kubitscheck U. Fluorescence microscopy: from principles to biological applications. 2nd ed. New York: Wiley-VCH; 2019.
Litchman J, Conchello JA. Fluorescence microscopy. Nat Methods. 2005;2:910–9. https://doi.org/10.1038/nmeth817.
Papkovsky DB. Live cell imaging. Totowa, NJ: Humana Press; 2010.
Pawley BP. Handbook of biological confocal microscopy. 3rd ed. New York: Springer; 2006.
Sanderson J. Understanding light microscopy. 1st ed. New York: Wiley; 2019.

Basic Digital Image Acquisition, Design, Processing, Analysis, Management, and Presentation

4

Rocco D'Antuono

Contents

What You Will Learn in This Chapter

Understand what is a single image or what a more complex multidimensional dataset represents; identify the technique used for the acquisition and read the metadata; consider the limits deriving from the imaging technique; be able to visualize and

R. D'Antuono (✉)
Crick Advanced Light Microscopy STP, The Francis Crick Institute, London, UK

Biomedical Engineering, School of Biological Sciences, University of Reading, Reading, UK
e-mail: rocco.dantuono@crick.ac.uk

© The Author(s) 2022
V. Nechyporuk-Zloy (ed.), *Principles of Light Microscopy: From Basic to Advanced*, https://doi.org/10.1007/978-3-031-04477-9_4

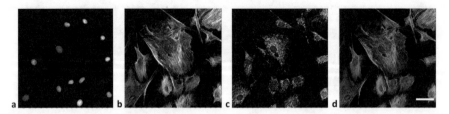

Fig. 4.1 BPAE cells. (**a**) Nucleus (DAPI), (**b**) actin (AlexaFluor 488), (**c**) mitochondria (MitoTracker Red), (**d**) merge of the channels after the assignment of different color scales to the original grayscale images (**a–c**). Scale bar: 50 µm

render the dataset using different software tools; apply basic image analysis workflows to get data out of images.

Present images and data analysis results in an unbiased way.

4.1 Image Acquisition and Analysis: Why they Are Needed

4.1.1 Image Relevance in Science from Biology to Astronomy

The use of the image as a vector of information is not a feature that uniquely belongs to modern science. From the first attempts made by Renaissance artists to accurately describe the human body anatomy to the representation of the lunar topography done by Galileo Galilei, the common surprising feature is the effort to render the image as representative as possible of the investigated object.

Modern science can, fortunately, rely on detectors to capture images of the reality (thanks mainly to the theoretical formulation and experiments of those "analogical" geniuses such as Galilei, Newton, and Einstein[1]).

A variety of imaging techniques, some more invasive than others, can be used to study the human anatomy or functional aspects of cellular organelles at a different scale of resolution. For example, Fig. 4.1 shows BPAE cells properly stained to characterize different cellular compartments: nucleus, cytoplasm, and mitochondria.

Therefore, the image, considered as the intensity of detected radiation, is not simply a qualitative description of an object or a biological sample, but has the potentiality to convey useful and unexpected information, often about very different scientific aspects, and so it becomes a "measurable entity."[2]

---

[1] Among the astonishing scientific contributions, regarding the optics: Galilei improved the telescope design for his observations of celestial bodies, Newton gave fundamental insights on the nature of light, and Einstein explained the photoelectric effect and formulated the theory of laser emission.

[2] Here we consider a measurement as any comparison with a standard object, known physical quantity, or constant.

Examples of scientific applications of imaging can come from extremely different disciplines:

- in astronomy, the radiation from the cosmos is detected to measure which wavelengths of the light spectrum are missing (absorption bands); it is then possible to determine the atomic composition of stars which are billions of kilometers distant from the Earth [1].
- in biology, the use of fluorescence resonant energy transfer can be used to obtain the stoichiometry of interacting molecules in living cells [2], where typical dimensions are in the order of nanometers.

4.1.2 Need for Quantitative Methods Is Extended to Light Microscopy

Light microscopy allows wonderful discoveries about the shape and functions of biological samples, but beyond the descriptive power, it is necessary to apply reproducible acquisition settings and quantitative methods to analyze the resulting images. This is not only to eliminate the subjective bias of the observer but also to take into account the possible variability between different imaging sessions:

- *Experimental conditions and image acquisition parameters.* The image itself is determined by sample preparation protocols and acquisition conditions that not always can be set in a reproducible way from session to session. A few examples of conditions that can change during image acquisition in microscopy are illumination level and evenness, detectors gain, optics alignment, image format (bit depth, type), etc. Figure 4.2 shows how images of the same mouse tissue can differ if acquired with (a) low, (b) intermediate, and (c) high gain.
- *Data handling, processing, and visualization settings.* The image can be visualized with settings that might induce the observer to draw wrong conclusions or can be easily altered by unattentive manipulation, or even worse, by mischievous practices. In Fig. 4.2 the same image is visualized with (a) low and (d) high contrast. Increasing the contrast might render similar to the eye two images acquired with hugely different settings (in this case Fig. 4.2d and the image in Fig. 4.2c). The real intensity distributions can be comprehended by using the histogram, where the pixel values are plotted: in Fig. 4.2e the histogram of the image Fig. 4.2a shows an overall low signal and in f) the histogram of the image Fig. 4.2c shows saturation (peak at the maximum value of 255).

The examples reported in Fig. 4.2 highlight the need for precise and quantitative methods in light microscopy, in order to achieve a more reproducible approach at each step of the experiment, from sample preparation to image acquisition and analysis, to ensure more reliable data.

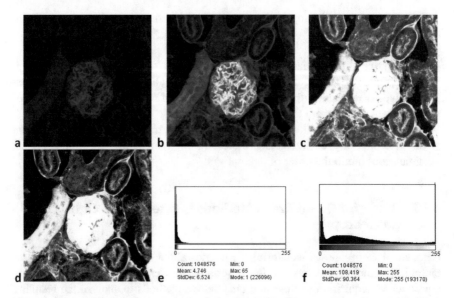

Fig. 4.2 Mouse kidney section acquired and visualized with different settings. Same field of view acquired with relatively (**a**) low (500 V), (**b**) intermediate (600 V), and (**c**) high gain (700 V). (**d**) Same image contained in (**a**) is visualized with high contrast, showing how visualization settings can give a deceptive image similar to (**c**). (**e**) The histogram of image (**d**) shows the overall low signal, compared to the (**f**) histogram of image (**c**), which covers the entire dynamic range and exhibits saturation (accumulation of counts at the maximum value 255). More info in Sect. 4.2.2.7 "Digitization"

4.2   Representation of Reality: Image Acquisition

4.2.1   What Is the Difference Between a Material Object and a Representation of It?

During imaging and analysis of the resulting image-type datasets, it should never be forgotten that images are a "representation" of reality. Indeed, in the field of theoretical physics there has been an evolution of the concept of reality: from Galileo's famous sentence "(the universe) it is written with mathematical language" [3] to the most modern interpretations, according to which "the universe we live in is itself a mathematical structure" and more importantly is computable [4].

Nevertheless, the computational power to compute all the aspects of reality is not always available and we do imaging to get a *representation* of the reality itself. The information about an object of interest is obtained collecting radiation from the region of space where the object sits, making use of detectors that may have different design; for example, with positron emission tomography (PET) imaging is possible to detect the radiation emitted by decaying isotopes contained in the drug administered to the patient, or with magnetic resonance imaging (MRI) neurosurgeons can obtain a scan of the brain before an operation [5].

In a way that resembles painters' grid technique, image acquisition can be summarized as the association of a number with each cell of a grid: the well-known *pixels* (picture elements).

The same region of space where our sample sits is hit by a number of different radiations (i.e., cosmic gamma rays, solar neutrinos, thermal radiation emitted by any warm body, visible and ultraviolet light from the sun rays). A curious example is the application of particle physics to archaeology, in the search for secret pyramid rooms with the measurement of muon (cosmic particles) flux [6].

So what simplifies the reality of a beautiful landscape painting or a "nice" cell image? The blindness of the used detector to other forms of signal: human eyes of a genius painter do not get disturbed by neutrinos and microscope cameras are "shielded" by filters to read only a certain range of the visible light spectrum.

Depending on the imaging technique selected we can capture just one aspect of our sample, but not obtain a complete description of it. This is a warning about the care we should have when we state that a sample has a certain feature (healthy versus diseased, higher versus lower protein concentration, etc.). The attributes that we tend to "easily" assign are relative to the state of the imaging system, acquisition parameters, visualization settings, and of course subject to the physical measurement variability.

Detectors used in light microscopy count photons (discrete light units) coming from the sample environment and have a sensitivity usually limited over a certain range of wavelengths, which for the most common applications is ~400–800 nm (Fig. 4.3).

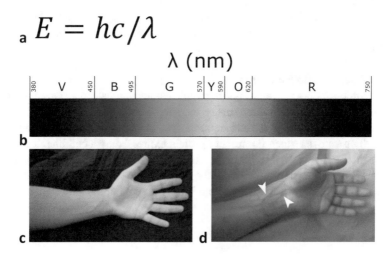

Fig. 4.3 Light wavelength and photon energy. (**a**) Planck equation: photon energy depends on light wavelength. (**b**) Electromagnetic spectrum region used in light microscopy. Commonly used cameras exclude infrared (IR) radiation above ~700 nm, to acquire picture similar to what human eye sees. The figure shows how undetected light carries useful information: blood vessels visibility of a human arm imaged in a similar pose (**c**) with and (**d**) without IR filter

4.2.2 Light Microscopy Key Concepts

4.2.2.1 What Is Light Microscopy?

Light microscopy is the study of small objects making use of the visible part (VIS) of the electromagnetic spectrum (Fig. 4.3b). As the technology improves and more fluorescent tools become available, the area of applications of light microscopy is expanding toward the ultraviolet (UV) and infrared (IR) regions [7] of the electromagnetic spectrum.[3]

4.2.2.2 Optical Resolution

The observation of objects under any lens or complex optical system is often seen with wonder for the aspect of magnification (the ratio between the size of the object obtained in the image plane and the real size of the object in the sample plane [8]), although magnification does not result in informative images if is not obtained with correction of the optical aberrations (like the chromatic and spherical) and enough resolving power. Therefore, an object can be seen as magnified, but the image could be still blurred and containing aberrations.

Every point-like source of light imaged through a microscope results in an image or structure (in 3D) that is "distorted" by the lens: this blob is referred to as point-spread function (PSF) (Fig. 4.4a–c). The size of PSF influences the *resolving power*, which is defined as the smallest distance at which two objects can be *separated*, i.e., identified as distinct [9].

The equation that describes the resolution, the famous "Abbe law" (Fig. 4.4g) [10], states that the resolution improves by shortening the wavelength of the used light and increasing the numerical aperture (N.A.) of the lens. Figure 4.4d, e show cell organelle structures (red) that can be better resolved by using an objective with higher N.A. (Fig. 4.4f). Concisely, the lens N.A. is more relevant than its magnification.

4.2.2.3 Microscope Elements

Any light microscope, as well as photographic cameras, relies on a *source of light* to illuminate the sample; the source can be the simple brightfield lamp to illuminate a macroscopic object under a stereomicroscope, or a fluorescence illuminator to specifically excite the green fluorescent protein expressed by a cell line. The illumination light is conveyed by *optics*, which include optical fibers that allow the light to travel flexibly and at long distance on the optical table with reduced loss, lenses to spread or focus the light beam, dichroic mirrors to reflect or transmit light according

---

[3] The availability of pulsed laser in the UV region makes possible fluorescence lifetime imaging microscopy (FLIM) experiments also at shorter wavelengths. On the opposite side, devices such as optical parametric oscillator (OPO) allows deeper sample penetration in intravital microscopy and third harmonic generation. For info about these techniques, see the additional references at the end of the chapter.

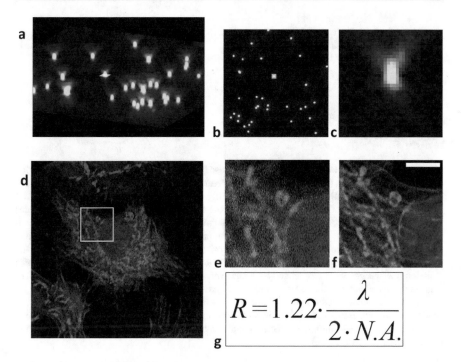

Fig. 4.4 Optical resolution in fluorescence microscopy. The point-spread function (PSF) describes how a point-like object is reconstructed through a microscope. The obtainable resolution depends on the wavelength of the light and the numerical aperture (N.A.) of the lens. Sub-diffraction limit fluorescent beads (0.1 um size) are shown as (**a**) 3D reconstruction, (**b**) 2D slice middle plane (a bead is annotated with a white square). (**c**) Bead orthogonally resliced. (**d**) BPAE cells described in Fig. 4.1 acquired with a 20×/0.8 N.A. objective. (**e**) Detail cropped from the 20×/0.8 N.A. image. (**f**) Same field of view of (**e**) acquired with a 63×/1.4 N.A., where better details of mitochondria are visible; scale bar: 5 um. (**g**) Abbe law for optical resolution

to the wavelength, prisms to separate different wavelengths, and crystals which respond to the applied electrical tension with variable optical properties. One of the most common microscope configurations is epifluorescence, where the objective has the dual role of focusing the illumination light that arrives at its back aperture and collect the fluorescence emitted by the sample.[4] Fluorescence is then sent back toward the *detectors*, eventually spatially filtered with a pinhole (confocal microscopes; see Sect. 4.2.3) to get optical sectioning, and then spectrally filtered in order to read signals in different channels.

The light falling on a target can be absorbed, reflected, or scattered by any substance along its path [11, 12], while a fraction of it travels through the sample

---

[4]In some optical design, such as light-sheet microscopes, an objective is used to focus the illumination beam whereas other ones are responsible to collect the light from the sample.

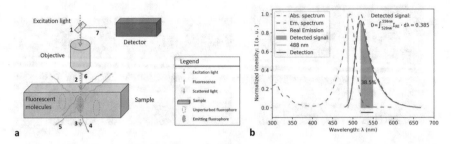

Fig. 4.5 Light path for transmitted light and fluorescence microscopy. (**a**) Excitation light (dotted blue) passes through the (1) dichroic mirror (dichromatic beam splitter) and (2) is focused on the sample by the objective. Depending on light wavelength, sample type, and thickness, some light can pass through and be observed as (3) transmitted or (4) scattered light (blue). Fluorescent molecules in the samples can absorb some excitation light and emit part of the energy as fluorescence (green) (5) in multiple directions. (6) Fluorescence traveling backward can be collected with the objective and (7) sent to the detector by the dichroic mirror. (**b**) Light of any wavelength at which the fluorophore absorption spectrum is nonzero (dashed blue) can be converted into fluorescence emission (dashed green), characterized by longer wavelengths (Stokes shift) and intensity dependent on the wavelength of the excitation light. The example is relative to the Alexa Fluor 488: by exciting with monochromatic light at 488 nm (cyan line), the emission intensity is lower (green curve) than the maximum obtainable hitting the absorption peak. If fluorescence is read between 520 and 550 nm (black line), the detected signal (lime area) is only about 38% of the total (area under the dashed green curve)

and can be detected on the other side. These processes are the basis of fluorescence and transmitted light detection that are described in Sects. 4.2.2.4 and 4.2.2.5.

4.2.2.4 Fluorescence

The first and principal use of fluorescence is the measurement of the intensity and location, after the collection of signal in a "*reflected geometry*": the excitation light passes through a dichroic mirror (Fig. 4.5a.1) and after being focused by the objective hits the sample (Fig. 4.5a.2). Depending on the light wavelength and the sample composition, the light can pass through (Fig. 4.5a.3), be scattered (Fig. 4.5a.4), or eventually excite fluorophores that emit light at a longer wavelength in multiple directions (Fig. 4.5a.5). A reduced amount of fluorescence light traveling backward can be collected with the same objective (Fig. 4.5a.6), separated from the excitation light using the dichroic mirror and sent to the detector (Fig. 4.5a.7).

Absorbed light can either be dissipated as thermal vibrations, induce chemical changes, or eventually cause electronic energetic transitions. Different molecular mechanisms induce energy losses which bring the electron from a higher to the lowest energy level of the excited state: the starting point of fluorescence decay. Therefore, only a part of the incoming photon energy is converted into fluorescence [13]. Due to the inverse proportionality between energy and wavelength of the light (Fig. 4.3a), the fluorescence photons possess a longer wavelength than the excitation ones ("Stokes shift") (Fig. 4.5b).

The absorption and emission spectra determine the amount of signal that can be obtained by fluorophores in the sample, depending on the choice of illumination and detection filter sets (Fig. 4.5b).

In samples where multiple fluorophores are present, the cross talk of signals can constitute a problem.

To overcome this mixing of signals, it is fundamental to use specific excitation wavelengths, limited detection ranges for each channel, and a sequential acquisition (see experiment design in Sect. 4.2.4).

Fluorescence can additionally be characterized by the lifetime of the excited state (property used in a technique called FLIM), the polarization of the emission (used to study molecular rotational time and complex formation), resonant energy transfer from a donor fluorophore to an acceptor (FRET), and other properties exploited to answer different biological questions [2, 14, 15].

4.2.2.5 Transmitted Light

When a sample is studied using the absorbed light, a transmitted light path can be adopted by positioning the illumination source[5] on the opposite side of the sample, with respect to the position of the objective (in Fig. 4.5a light would travel in the opposite direction, from the point 3 to 7).

Figure 4.6 shows how the same tissue slice appears using different approaches that employ both the transmitted light path and a "reflected" geometry (epifluorescence).

The transmitted light path is used for simple brightfield illumination acquired with a color camera (similar to the eye vision, if an equivalent magnification could be achievable) (Fig. 4.6a), observation of immuno-histochemistry staining (dyes like hematoxylin and eosin), or with contrast techniques such as phase contrast (Fig. 4.6b) or DIC [16].

The "reflected" geometry is instead used to detect the fluorescence in different channels (the blue, green, and red regions of the visible light spectrum) (Fig. 4.6d–f).

4.2.2.6 Detectors: Cameras and PMT/Hybrid

The detectors used in microscopy can be divided into two main classes: cameras and PMTs.

– Cameras are used on systems where the signal is expected to be emitted in a short period of time from the entire field of view, mainly for wide-field microscopy (epifluorescence, DIC, phase contrast, stereo), selective illumination systems (light-sheet, TIRF), localization microscopy (e.g., STORM), and spinning disk

---

[5] For transmitted light, a low power illuminator can be used. For example, a filament lamp used for the brightfield illumination can have a nominal power consumption as low as 6 W, while a fluorescence light bulb (arc lamp or halogen) consumes around 100 W. The higher power required is to compensate the energy dissipation of the fluorophores.

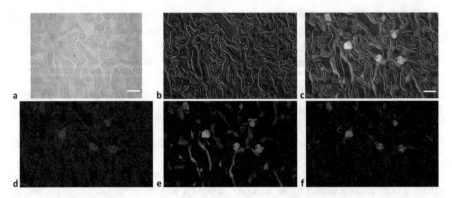

Fig. 4.6 Same sample imaged with transmitted light and fluorescence. Tissue slice imaged on a wide-field microscope with (**a**) brightfield (transmitted light, color camera), (**b**) phase contrast (transmitted light, monochrome camera), and (**c**) merge of phase contrast and fluorescence channels (monochrome camera) of (**d**) DAPI, (**e**) AF488 WGA, and (**f**) AF568 phalloidin. Scale bar is 10 um for both (**a, c**)

confocal (where illumination of the sample and light rejection is operated by multiple microlenses/pinholes).

The sensor, divided into a grid of physical pixels, is manufactured with chemical doping of a silicon chip, to allow the formation of hole–electron pairs every time that some light with a specific wavelength illuminates the device surface.

Once simultaneously excited by the light, the pixels release photoelectrons in the conduction band. The charge is then read, eventually amplified accordingly to a chosen gain, and converted into a digital signal. Relevant parameters for the signal to noise are the exposure time and the binning factor [17, 18].

– PMT (photo-multiplier tube) detectors are built to have a single detection area and are instead used on laser scanning confocal microscopes. The original design includes a photocathode plate that emits electrons when hit by light; these charged particles are then accelerated under an applied voltage (variable changing the gain) so that they can hit secondary plates (dynodes), which in turn emit more electrons. This results in a multiplication of charges, which are collected by an anode at the end of the tube, and the electrical current is converted to a digital value. Improved versions of PMT detectors are using a combination of design and materials to increase the quantum yield (ratio between the detected and total received light). Examples of improved PMTs are GaAs, GaAsP, and hybrid detectors; each type results in better performance in specific regions of the spectrum [18, 19].

4.2.2.7 Digitization

The digitization of an optical image (Fig. 4.7a) is obtained by the analog to digital converter (ADC) of the detector, which converts the fluorescence or transmitted light signal from electrical current into a digital pixel intensity (Fig. 4.7b). The intensity values are assigned over an interval of possible measurable signal called "dynamic

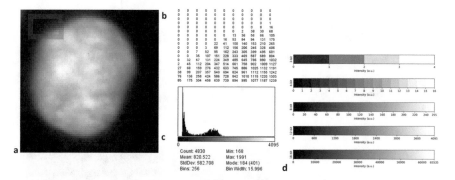

Fig. 4.7 Images and numbers. HeLa cell nucleus stained with (**a**) DAPI. A small region of the image (red square) is inspected to show (**b**) the pixel values (sample has been acquired on a 12-bit monochrome camera). (**c**) Image histogram. (**d**) Pixel intensities range obtained with different "bit depth": 2-bit, 4-bit, 8-bit, 12-bit, and 16-bit

range" (Fig. 4.7c). The storage of each pixel information requires an adequate number of bits ("bit depth") to render the digital data format capable of representing all the divisions of the dynamic range (Fig. 4.7d). The image bit depth depends on the type of detector and/or the software settings used during the acquisition. A common choice is a digitization at 8-bit, corresponding to $2^8 = 256$ different intensity values (from 0 to 255), 12-bit ($2^{12} = 4096$ total values), or 16-bit ($2^{16} = 65,536$ total values). The choice of a higher bit-depth digitization makes available more possible intensity values, improving the capability of comparison among very similar samples. The downside of high bit-depth datasets is in the image handling, given the larger storage space and computational resources required, and the limitations of most software algorithms designed around the 8-bit format.

Images acquired with a color camera are digitized assigning a triplet of values per each pixel, considered the red, green, and blue (RGB) channels, usually each varying over an 8-bit range, to obtain all the other colors by additive RGB composition.

4.2.3   Examples of Microscope Systems: Simplified Light Path for Wide-Field and Confocal Laser Scanning Microscopy

Every microscope system needs a source of light to illuminate the target sample, lens to convey the light appropriately toward and from the sample area, and detectors to read the signals.

The choice and arrangement of optical components diversify the systems on the basis of their capability to image different samples. Here are described two different types of commonly adopted microscopes, with a very essential representation of their components (actual microscopes are more complex and technological refinements often add more lenses, filters, and devices along the optical path, in addition to an expensive aberration-corrected objective). Figure 4.8a shows the

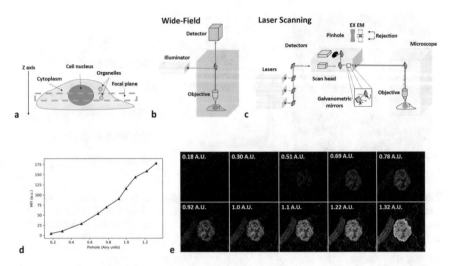

Fig. 4.8 Comparison of wide-field and confocal imaging modalities, and optical sectioning. (**a**) A cell consist of multiple compartments of different sizes distributed across the cell body. The image acquired on any microscope system, for a sample thicker than the depth of focus of the lens, contains light arriving both from the focal plane and out of it. (**b**) A wide-field microscope illuminates the sample extensively across the Z-axis; the fluorescence received from the sample derives from both the focal plane and the out-of-focus regions. The resulting image does not allow us to attribute correctly the light to the originating Z layer. (**c**) Differently from a wide-field, a confocal microscope is equipped with a pinhole that operates the rejection of out-of-focus light, yielding the optical sectioning. (**d**) The pinhole size influences the amount of signal that is received due to the level of light rejection. The plot represents the mean fluorescence intensity (MFI) measured as a function of pinhole size in (**e**) a mouse kidney tissue section acquired on a confocal microscope, at different pinhole sizes. A region of interest (ROI in yellow) has been measured in all the images to obtain the plot in (**d**)

cartoon of a cell and some of its compartments that are usually investigated in life sciences (nucleus, cytoplasm, organelles). Beyond the resolution, one of the most addressed aspects of light microscopy is the *optical sectioning*, which consists in the acquisition of information coming only from the focal plane or from a limited section of the whole sample, without the need to physically slice it. Figure 4.8b, c represents the essential microscope body and describes the salient parts of the most diffused microscope configurations: wide-field and confocal microscope.

A *wide-field* microscope (Fig. 4.8b) allows us to illuminate and observe a sample along the entire Z-axis, given that light can penetrate through it enough to either excite fluorophores (fluorescence detection) or simply pass through and be detected on the other side (transmitted light detection). As an illuminator, it often includes gas-filled or filament light bulbs, otherwise specific LED sets; the choice of the illumination source should ensure enough excitation power for the sample at specific wavelengths (illuminators show different spectral profile of the power curve[6]). The image can then be detected with a camera.

A *confocal* microscope (Fig. 4.8c) uses lasers to illuminate the sample in a subregion of the field of view (FOV). This allows the use of monochromatic light to specifically excite the fluorophore of interest. The most widely adopted confocals are laser scanning systems, which means that the laser spot is moved by a scanner along the FOV to serially illuminate points and contextually read the signal along lines in XY directions. More correctly these points should be thought as *spots,* since the size of the laser beam is not infinitesimally small in XY and the excitation occurs also along the Z-axis. The higher acquisition time due to the use of a serial illumination/read-out is compensated by the optical sectioning obtained with the use of a pinhole. The role of the pinhole in detection is to allow the passage of light originated at the focal plane while excluding the out-of-focus one. This light rejection yields the optical sectioning.

In most of the modern confocals, the pinhole size can be controlled and optimized for the specific objective, in order to get enough optical sectioning while still reading a sufficient signal. In general, the light rejection favors the optical sectioning, but reduces the signal read from the sample (Fig. 4.8d, e).

4.2.4 Experiment Design

When a sample includes multiple fluorophores, the design of a multichannel acquisition should be optimized to answer the salient questions posed in the experiment. Unfortunately, in microscopy it is not possible to have a combination of settings that allow the simultaneous optimization of different aspects such as the signal-to-noise ratio, acquisition speed, spatial resolution, and sample viability.

Therefore, the acquisition parameters are chosen to address possibly one aspect at the time, while often worsening the others.

For example, a possible experiment might be imaging fixed cells to determine the fine structure of a specific organelle type, in the attempt to optimize the aspect of spatial resolution. In this case the choice could be made among some super-resolution techniques, such as the photo-activation localization microscopy (PALM) [20]. However, PALM requires particular sample preparation protocols and long acquisition time compared to other techniques.

Conversely, in a live cell imaging experiment the acquisition speed is a relevant parameter while a suboptimal resolution can be accepted; then the use of a spinning disk confocal is more suitable [21].

---

[6]The choice of the right light source has been historically driven by the technological development so nowadays it is shifting from dangerous mercury-filled bulbs to power efficient and electronically controllable LEDs.

Table 4.1 Selected acquisition settings used to capture the image in Fig. 4.4d

Sample and staining	BPAE cells stained with: DAPI, Phalloidin-AF488, MitoTracker Red
Excitation	405 nm @ 2% 488 nm @ 10% 561 nm @ 2%
Optics and scanner options	Plan Apochromat 20×/0.8 M27 Format: 1024 × 1024 Pixel size: 0.073 um Dwell time: 0.244 us Line average: 2
Detection	Ch #1: 417 nm–459 nm, gain: 700 V Ch #2: 499 nm–543 nm, gain: 800 V Ch #3: 573 nm–630 nm, gain: 700 V

When designing an experiment, in order to get sufficiently good data sets that will facilitate the image analysis, it is fundamental to consider a variety of aspects such as optical resolution and digital sampling, signal to noise, acquisition speed, photobleaching, sample viability, and cross talk of signals.

Table 4.1 contains an excerpt of parameters used to acquire the image in Fig. 4.4d, as an example of a multichannel experiment aimed to obtain a good signal to noise and an optimal spectral separation on a confocal system.

4.2.5 Questions and Answers

Questions

4.2.5.1 Can I obtain all the properties of an apple simply looking at its picture?

4.2.5.2 How the image of a sample (and the scientific content of it) is influenced by the acquisition technique and image handling?

Answers

4.2.5.1 No, the image of a sample is obtainable by the detection of some radiation. Light microscopy uses the visible part of the electromagnetic spectrum to create images that are a "representation" of the reality: most of the properties of the objects are ignored and cannot be computed.

4.2.5.2 The information contained in an image depends on the adopted imaging technique and acquisition settings. If the technique is not appropriate or the settings are not well tuned, any analysis done on the dataset is subject to failure.

4.3   Image Visualization Methods

Datasets obtained through different imaging modalities need to be inspected before starting any analysis. The purpose is to get an overall idea of the dimensionality (FOV size, channels, slices, frames, positions). Additionally, any spark of intuition about the experimental results can start by visual inspection of the acquired images. It is therefore fundamental to adopt visualization methods that are not prone to error and fairly render the scientific information carried by the data, along with a robust quantitative analysis.

If no reference to the acquisition parameters is available, the first step is to identify the file format and seek for a software able to open the dataset. In some cases, the microscope system saves the images in file formats that are interpretable only by proprietary software. Fortunately, this has become a rarer occurrence, mainly thanks to the efforts of the "Bio-formats" project, which aims to make readable all the formats of image-type data [22]. The "Bio-formats" plugin is available in the most commonly adopted open-source software such as *ImageJ* [23], *FIJI* [24], *CellProfiler* [25], *Icy* [26], or commercial ones. When the dataset is opened in the correct way, its dimensionality (X, Y, Z, positions, time points, etc.) can be easily assigned and metadata are available for inspection. Metadata are additional data, attached to images, which contain useful info regarding the acquisition parameters, can be checked to establish whether the samples have been acquired in the same way, and additionally support data analysis, especially if they are presented in an open and standardized format [22]. An example of metadata extraction is reported in Table 4.1.

A single image in grayscale can be visualized with a look-up table (LUT), a map between pixel values and different hues of a color (e.g. green, red, etc., as in Fig. 4.6d–f) or a combination of colors (to highlight some features as intensity saturation or outstanding structures, as in Fig. 4.2c).

Multichannel images are composed of individual grayscale ones. Assigning the LUT to each channel makes then possible the visualization of the merge (as in Fig. 4.6c).

Z-stack or timelapse experiments can be shown in a montage (as in Fig. 4.8e), to compare slices/frames.

A relevant issue is the choice of visualization settings, which can be hugely changed until the point where it is no longer possible to fairly compare images. To be safe, the main references remain the pixel intensity, the adoption of the same range for brightness and contrast (B&C),[7] and metadata inspection to evaluate the acquisition conditions.

---

[7]Differently from their optical counterparts, in the context of image visualization, the definition of brightness and contrast can be given in terms of minimum and maximum visualization values, which are represented with the first and last color of the chosen LUT. For instance, by acquiring an image, the contrast can be increased if the min and max are chosen to be very close values; the brightness can be increased moving the min/max range to lower values.

The following protocols require the use of different software and describe how to run commands either in the graphical user interface (GUI) or in the script editor. Conventionally, the name of a software is in *italic and bold* (e.g. *CellProfiler* is a software), while the menu structure of the commands to be executed is in "*italic*" (a file can be opened clicking in succession "*File/Open*"). Lines of code are instead reported using another `font` type: to show the value of the variable called "pixelSize" the command `print(pixelSize)` should be used.

Almost all the operations described as image visualization (Sect. 4.3), analysis (Sect. 4.4), or data presentation (Sect. 4.5) can be equivalently executed by using different approaches, ranging from an exclusively visual in the GUI, to a strictly code-only one in the script editor. Notably, each presented protocol could be executed in a specific software only, to pursue the easiest approach. However, a comparison of different methods to perform the same operations can highlight what are the advantages of a GUI or the flexibility of a script. The levels of difficulty are defined as: "basic", "intermediate", "advanced". The presented protocols are available in the following repository: https://github.com/RoccoDAnt/Basic-digital-imaging_protocols.

Protocol 4.3.A: Visualization in *ImageJ*/*FIJI*—Level: Basic
FIJI [24] is a distribution of *ImageJ* [23] software: it includes several useful plugins for bioimage analysis and supports further developments of *Java* libraries for image analysis.

1. Download *FIJI* software through the website: https://fiji.sc/.
2. Unzip and open the folder: launch *ImageJ*/**FIJI**.[8]
3. **ImageJ** allows us to open images/datasets through different methods, some of them are:
   (i) Drag and drop on the software bar.
   (ii) Open specific format or sequence of images with "*File/Import/*".
   (iii) Import bioimage datasets through "*Bio-Formats*" plugin: "*Plugins/Bio-Formats/Bio-Formats importer*".

The acquisition software can be set to attach metadata to the saved images; this info can simplify the import of the dataset, making no longer necessary specifications such as the pixel size, number of channels, and Z-stack slices.

Use of method 3 (i) can result in loss of some metadata, while method 3 (iii) is usually successful in the recognition of the acquisition metadata (thanks to the work of the Bio-formats developers [22] and the microscopy community, which constantly demands the inclusion of more data formats).

---

[8]On Win10 the application is called *ImageJ-win64.exe*; on Mac it is visible as *FIJI*.

4. The properties of the image are visible in different places:
    (i) The image frame contains usually the title and a text area below specifying the number of pixels, the bit depth, and the size of the file; for example, *image 1.czi* and *1024x1024; 8-bit; 3 MB.*
    (ii) The command *"Image/Properties..."* shows the dimensions (channels, slices, frames) and the voxel size (pixel width/height and depth).
    (iii) Metadata is visible with *"Image/Show Info..."*
5. The multidimensional dataset (obtained through 3) iii) with the option "View stack with: Hyperstack") can be browsed using the sliders in the bottom part of the image window, and the merge of channels can be seen with *"Image/Color/ Channels Tool.../Composite."*
6. The brightness and contrast of the single channels can be changed individually, moving the *"C"* slider in different position and using *"Image/Adjust/Brightness/ Contrast...."*
7. Pixel values of the image can be obtained with the tool available in the main bar "Pixel Inspection Tool."

Protocol 4.3.B: Import Images in *CellProfiler*—**Level: Basic (Propaedeutical for 4.4.B)**

CellProfiler is a software for batch image analysis [25]. It has a GUI that allows us to build an image analysis pipeline in a really easy way, by adding modules to be executed in series. Each module performs one operation, such as the identification of nuclei (*"IdentifyPrimaryObjects"*) according to the size and the fluorescence intensity or other measurements of the detected cells (for instance, *"MeasureObjectIntensity"* will produce statistics about the average intensity, standard deviation, etc.). The pipeline efficacy can be checked through each module step by step in "Test Mode," with the option of enabling/disabling and hiding the results of a particular module.

This first protocol includes two modules to import images and visualize them, in order to show how the pipeline can be built using the visual inspection on test images. The aim is to assess which are the parameters to use in the analysis (e.g. intensity threshold and size range).

1. Download *CellProfiler* software (version 3.1.9) through the website: https:// cellprofiler.org/.
2. Install and launch *CellProfiler*.
3. *CellProfiler* has two alternative methods to set up the import of images:
    (i) The legacy input module *"LoadImages,"* which can be added with *"Edit/Add module/File Processing/LoadImages,"* or simply clicking on the "+" symbol in the bottom left part of the GUI. This opens the window *"Add modules"* that lists all the available operations that can be included in the pipeline (equivalently called *"project"*). The *"input"* folder where the "LoadImages" module will find the picture, as well as the *"output"* folder where we want to save the results, can both be specified by clicking *"View output settings"* (bottom left of the GUI).

(ii) Use the default method called *"Images"*, where special filter can be set to choose specific substrings contained in the filename: for example, to only import the image of the DAPI from the *"input"* folder containing also a GFP channel, the string "DAPI" should be specified.

4. For the simple inspection of the two channels, the option 3 (i) can be used, adding the individual pictures called DAPI and GFP in the *"LoadImages"* tab. The module *"GrayToColor"* allows us to merge multiple channels into an RGB image. Choosing *"Start Test Mode"* allows the execution of one step at a time and see the results. Once executed *"GrayToColor,"* a window containing also the original image can be used to inspect the pixel intensities by moving the pointer (values are shown in the bottom part of the frame).

5. With right click it is possible to call *"Show image histogram."*

Protocol 4.3.C: Use Coding to Open/Visualize Image—Level: Intermediate (Propaedeutical for 4.4.C)

Few lines of code in *python* to import and visualize a single image.

Python is a really flexible programming language and there are numerous packages supporting image analysis. To execute a python script, a development environment called *JupyterLab* (https://jupyter.org/) can be used. It presents a really clean interface that includes a file browser and visualization of running terminals and kernels, together with a support for autocompletion and syntax highlighting in the text editor. Code is organized inside this development environment in the form of *notebooks*, which are documents composed of different independent cells, containing pure text, code, or markdown[9] text.

JupyterLab can be installed and launched within *Anaconda Navigator* (see below), in order to make the management of packages and environments much easier. The 3D rendering and orthogonal views of an HREM[10] dataset, obtained with *napari* [27], are shown in Fig. 4.9.

1. Download *Anaconda Navigator* software through the website: https://www. anaconda.com/products/individual .

2. Install and launch *Anaconda Navigator*.

3. Install and launch *JupyterLab*.

4. *"File/New/Notebook"* choosing "Python 3" for the new kernel. This is a new ". ipynb" file that we can edit.

5. The usefulness of notebooks consists in the possibility to run chunks of code in independent cells ("Ctrl+Enter"). A simple test can be run "17 + 13" in a cell. The protocol code is in the Table 4.2.

6. It is advisable to use the terminal *Anaconda Prompt* to create a specific environment for *napari* (simply run the command contained in cell 1).

7. Activate the new environment and install *napari* (copy and run also the second cell in *Anaconda Prompt*).

---

[9] A language for quick formatting of the text.

[10] Courtesy of Fabrice Prin, HREM platform manager, The Francis Crick Institute.

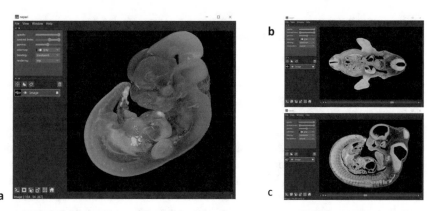

Fig. 4.9 3D rendering and reslicing in napari. A mouse embryo imaged with high-resolution episcopic microscopy (HREM) is visualized with napari software as (**a**) 3D rendering, (**b**) XY slice view, (**c**) YZ reslicing. The use of a few lines of code for simple image visualization is worth to simplify operations such as reslicing and 3D rendering. Additionally, *napari* supports the use of different layers, allows manual annotations, and includes an iPython terminal for integrated image analysis of the dataset. Annotations and results of segmentation can be added as new layers

Table 4.2 Python code to visualize a dataset with napari. The code is divided into cells that have to be run in the terminal called *Anaconda Prompt* (cells 1 and 2) and cells that can be run in a notebook (.ipynb), inside the *JupyterLab* editor. Cell 1) A special environment can be created with *Anaconda Prompt* to host *napari* software. Cell 2) The environment has to be activated before the installation. Cell 3) *napari* utilizes "QT" for the graphical user interface; this and the following cells have to be run in *JupyterLab*. Cell 4) The method "io" is imported from "skimage" library, to read the "Mouse_embryo_HREM.tif" dataset; finally, a viewer to render in 3D is created and the image stack is added as a layer, ready to be visualized, annotated or further processed (see Fig. 4.9)

Cell 1	`conda create -y -n napari-env python=3.8`
Cell 2	`conda activate napari-env` `pip install napari[all]`
Cell 3	`%gui qt`

Cell 4	`from skimage import io`
	`myImage =`
	`io.imread('Documents/data/Mouse_embryo_HREM.tif')`
	`import napari`
	`viewer = napari.Viewer(ndisplay=3)`
	`viewer.add_image(myImage, rgb=False)`

8. Choose the newly created environment *napari-env* in the GUI *Anaconda Navigator*, and restart *JupyterLab*. The command in the third cell and the following ones have to be run inside a new ".ipynb" notebook. Cell 3 is the command to choose the visualization library.

9. The fourth cell contains 5 lines executing the following:
    (i) Import the method "io" from the library "skimage" to read images.
    (ii) Import the embryo dataset (or any other local image z-stack) and store it in a variable called "myImage."

(iii) Import *napari* package.
(iv) Create a viewer to render in 3D.
(v) Add the dataset "myImage" to the viewer.

4.4   Image Analysis

Data acquired on any microscopy platform often contain a level of information that is not comprehensible by simple visual inspection. Human perception is additionally subject to bias due to its variability among subjects (different light sensitivity, color blindness, or eye disorders). Furthermore, it depends on color representation (human eye has developed around peak emission of solar light), visualization settings (B&C, gamma, screen brightness, etc.), and personal beliefs (agreement with previous data or published literature). It is therefore a good practice, and nowadays a quite non-dismissible need, to sustain hypotheses regarding data interpretation with a reliable image analysis.

Image analysis makes use of a variety of software tools identifiable as *components,* which can be combined to build the entire *workflow*. Frequently, the use of a single software to complete the whole process is not possible and it turns advantageous to become confident with different approaches and tools, spanning from GUI-based ones to scripting [28].

Well-defined algorithms allow to analyze datasets in a robust way with an enormous advantage in automatizing the image analysis workflow. Furthermore, the development of machine and deep learning makes possible the extraction of scientific information, otherwise inaccessible with other methods. Additionally, the classification and processing of humongous volumes of data, which would be extremely time-consuming for human operators, can be executed in a much shorter time [29].

Despite these recent developments, the limits of image analysis should not be forgotten: image analysis cannot "show" what is not contained in the data, often because of the inappropriate biological sample preparation or image acquisition. Even more, the scientific hypotheses may simply be wrong.

Data should be interrogated to obtain information to validate hypotheses, instead of being forced to sustain biased positions developed only with the visual inspection.

To promote the knowledge of image analysis, a wise use of bioimage analysis tools and the mutual interchange between biology and computer science, initiatives such as NEUBIAS training schools and community meetings have been promoted in the last few years.[11] The performances and results obtained by the use of different algorithms can be compared with *benchmarking*, whose complete approach is

---

[11] Network of European BioImage Analysts (NEUBIAS): more info at https://eubias.org/NEUBIAS/

available through the BIAFLOWS project: a platform to deploy and fairly compare image analysis workflows [30].

One of the main goals of image analysis is the object *segmentation*, which consists of the identification of structures of interest such as cell nuclei, organelles, etc. Common workflows are aimed at operations that include simple cell counting in an FOV, intensity measurements (like MFI or value distribution), size, and shape determination (occupied area, elongation, convexity, etc.)

The protocols 4.4.A-B-C show how those tasks can be run with different software tools.

4.4.1 Questions and Answers

Questions

4.4.1.1 The results of image analysis are susceptible to the choices I make. Is this science? For example, let us talk about something that should be simple like setting a threshold.

4.4.1.2 Do I need all the possible measurements or which ones to choose?

Answers

4.4.1.1 Image analysis results are definitely influenced by the choice of parameters we use in the workflow; the threshold is a clear example of how choosing a too low value might result in *undersegmentation* (e.g., more cells identified as a single clump), while setting it too high might break the objects into multiple ones and cut part of them. For the benefit of science it is important to use reliable algorithms, properly selected with benchmarking.

4.4.1.2 Additional measurements can help to better discriminate the different samples (such as positive control and treated cells). However, starting with *high dimensionality,* as first approach, can slow down the development of the workflow; in addition, multiple features might be correlated. The suggestion is to start with simple operations, such as object counting, MFI and diameter measurements, and check of positiveness in the different fluorescence channels.

Protocol 4.4.A: Measure Areas and Count Objects in *ImageJ/FIJI*— Level: Basic

The protocol uses a fluorescence image of cell nuclei ("HeLa-20Xdry-DAPI_300ms. tif") and shows how to apply a threshold and get a binary image, to count and measure cell nuclei.

1. Open the image with *FIJI* (protocol 4.3.A)
2. Check if pixel size is already calibrated: "*Image/Properties...*"; use the image specific pixel size (1.95 um in this case).
3. Duplicate the current image; the copy will be used to apply the threshold: "*Image/ Duplicate...*"

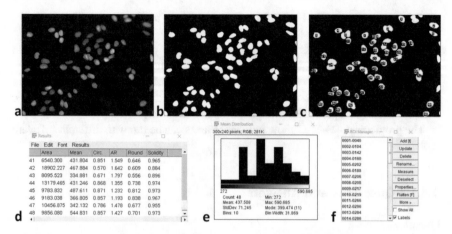

Fig. 4.10 Nuclei detection and measurements with FIJI (**a**) Original image of cell nuclei. (**b**) Applied threshold on an image copy originates a binary image. (**c**) Object detected by the "*Analyze Particles…*" command. (**d**) Measurements are shown in "Results" window. (**e**) Distribution of MFI obtained from "Results" table. (**f**) Identified regions are listed in the "ROI Manager."

4. Select the copy and apply a threshold: "*Image/Adjust/Threshold…*". Can choose Min = 250. Click on "Apply". The one obtained is a binary image: pixels have value 0 for background, 255 for foreground (objects of interest).
5. Segmentation can be improved by splitting touching nuclei: "*Process/Binary/Watershed.*"
6. The measurements can be chosen with: "*Analyze/Set Measurements…*". Choose: "Area," "mean gray value," and "shape descriptors." Select the original fluorescence image name in "Redirect to."
7. Identification of individual nuclei is run calling "*Analyze/Analyze Particles…*" on the binary image. The measurements will instead be redirected to the original image. In the panel that pops up choose the options: "Display results", "Add to Manager", "Exclude on edges", "Include holes".
8. The measurements are displayed in the window "Results", the identified regions in the "ROI Manager". It is possible to get the distribution of every column of the "Results" table: "Results/Distribution…".

The protocol should lead to the content of Fig. 4.10.

Protocol 4.4.B: Count Cell Nuclei and Related Vesicles with *CellProfiler*—Level: Basic

The advantage of using *CellProfiler* for cell counting and measurements is the possibility to easily build a pipeline that can also detect subcellular vesicles and analyze several images automatically. The protocol analyzes a folder containing 3-channels images: the cell nucleus, a cytoplasmic staining, and vesicles.

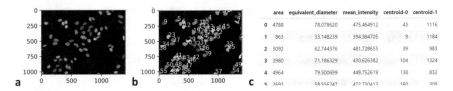

Fig. 4.11 Counting nuclei and measuring properties in python. (**a**) Cell nuclei in a fluorescence image are (**b**) segmented and labeled, (**c**) measured finding the area, MFI, and position with a *python* script

1. Launch *CellProfiler* and set the input/output folder (explained in protocol 4.3.B); set *LoadImages*.
2. Add *IdentifyPrimaryObjects* to find the nuclei in the first channel: start with a typical diameter between 80 and 150, do not discard objects outside this range, and set a manual threshold of 0.005 in the advanced options.
3. Add *IdentifySecondaryObjects* to get the cytoplasm by propagation in the second channel (using primary objects' name as "input").
4. Add *IdentifyPrimaryObjects* to find the vesicles in the third channel: choose typical diameter between 3 and 40, and use a manual threshold of 0.0025 in the advanced options. Comment: the choice of good parameters for vesicle segmentation might result a bit tricky, due to the diversity of structure size and intensity in the same sample; the indicated parameters work to count the smaller and brighter ones. Opportune retuning allows the detection of bigger ones.
5. Add *RelateObjects* to find "Childs" vesicles per each "Parent" Nucleus.
6. Add *MeasureObjectsIntensity* to get statistics of the vesicles (use the third channel image).
7. Add *ExportToSpreadSheet* to save all the results to a .csv file.

Protocol 4.4.C: Use Coding to Count and Measure Cells—Level: Advanced
The approach described in this protocol may look more difficult compared to the use of a GUI, but represents the beginning of a walk along "the road to freedom". Learning new coding skills opens much more possibilities than having several software GUIs available. The following script includes a minimal number of lines in *python* to: apply a threshold, find connected components, and print their properties (intensity and size). It can be run in a *JupyterLab* notebook, and uses the packages: *scikit-image* [31], *pandas* [32], *matplotlib* [33]. The outputs of the script should be similar to Fig. 4.11.

Image processing	```from skimage import io, measure```
Data Analysis	```import pandas as pd```
Plotting	```import matplotlib.pyplot as plt```
Import image	```myImage=io.imread('Documents/images/HeLa-20Xdry-DAPI_300ms.tif')```
Threshold	```binary=myImage>250.0```
Object Detection	```labelled=measure.label(binary)```
Measurements	```properties=skimage.measure.regionprops_table(labelled,```
	``` myImage,properties=['area','equivalent_diameter','mean_intensity','centroid'])```
Properties in a table	```properties_df=pd.DataFrame(properties)```
Create a figure	```fig, (ax1, ax2) = plt.subplots(1, 2)```
	```plt.subplots_adjust(wspace = 0.3)```
Show the original image	```ax1.imshow(myImage, cmap='gray')```
Set a color map	```cmap = plt.cm.prism```
to	```cmap.set_under(color='black')```
show the labelled image	```ax2.imshow(labelled, cmap, vmin=0.1)```
Loop over the objects	```for index in properties_df.index:```
to print the labels	``` ax2.text(properties_df['centroid-1'][index],properties_df['centroid-0']```
	``` [index], index, color='white')```
Show properties table	```properties_df```

4.4.2   Questions and Answers

Questions

4.4.2.1 Are there segmentation "rules" other than threshold?

4.4.2.2 What if I do not know the properties of my objects/cells of interest?

Answers

4.4.2.1 Yes, threshold is the most immediate method, but also patterns and statistics can be employed.

4.4.2.2 Use dimensionality reduction methods such as PCA or machine learning (unsupervised classification).

4.5   Publication of Images and Data

The presentation of images for the purpose of publication requires careful dataset handling, which should not be processed by changing pixel intensities or by altering proportions (both spatial and signal related).

The most diffused graphics software allows the inadvertent use of operations such as smoothing, change of gamma, resizing, and compression to 8-bit from higher bit depth. All these manipulations may result in compromised data that contain altered scientific information.

Available guidelines for safe image visualization can avoid the spread of wrong image manipulation, scientific misconduct, or trivial errors. These include the obvious avoidance of copy-paste procedures, spatial transformations, use of "lossy" compressed saving formats such as ".jpeg", the limitation to simple linear transformation (B&C levels and not the nonlinear gamma), and mainly the inclusion of references for image interpretation, such as calibration bars and history log [34, 35].

All the guidelines constitute the essence of the "Image data integrity."[12]

Additionally, image analysis should always accompany any claim regarding the formulation of scientific hypotheses based on image data. Results can be easily presented with scatterplots or histograms with specific color combinations that help readers to be unbiased (e.g. using color-blind friendly representations[13]).

Figure preparation can be done using several tools like *FIJI*, *Inkscape*, *Adobe Illustrator*, or *OMERO Figure*.

The protocol 4.5.A describes how to assemble a panel as in Fig. 4.1.

4.5.1   Questions and Answers

Questions

4.5.1.1 Do I need to use the same visualization settings for all the images? Is it always possible?

4.5.1.2 Which method should be used in order to be consistent in the analysis and figure preparation that include multiple images?

Answers

5.1.1 Yes, in the case of sample comparison. If images have been captured from samples that are not comparable, then different B&C can be used, but the visualization range used in both cases should be stated and/or shown with a calibration bar.

5.1.2 Log every step (for example if the software plugin is recordable like in *FIJI* macro recorder) or use scripting.

Protocol 4.5.A: Figure Preparation in ImageJ/FIJI—Level: Basic

1. Import the multichannel image in *FIJI* using "Bio-Formats," as hyperstack and composite.
2. Adjust B&C in every channel, as "Composite": *"Image/Adjust/Brightness/Contrast..."*.
3. Get an RGB image showing the composite: *"Image/Type/RGB Color"*. Show the scalebar with (*"Analyze/Tools/Scale Bar..."*); use "overlay" option if unsure about the scalebar aesthetics.
4. Convert to RGB the original stack in single "Color" view and concatenate it with the composite *"Image/Stacks/Tools/Concatenate..."*.
5. Make a montage: *"Image/Stacks/Make Montage..."*.

---

[12]More on "Image data integrity" in the webinar held by Kota Miura at NEUBIAS Academy (additional references).

[13]Color-blind friendly palette for example: see additional references.

Take-Home Message

- Image acquisition is possible by using a variety of techniques, which determine the type of datasets obtained and the extracted scientific content.
- Image analysis should be done consistently across the data and with benchmarked workflows.
- Science requires a fair presentation of data and openness about the outcome of image-based experiments. It is incredibly important to log every step of the process: from acquisition, through image analysis, until results presentation.

Acknowledgments Being surrounded by experts is a luxury and helps improving ourselves every day: my acknowledgments go to the Crick Advanced Light Microscopy (CALM STP, the Francis Crick Institute, London) as a stimulating place where to work and in particular to Dr. Kurt Anderson (CALM head) for supporting personal development and encouraging community-addressed initiatives. We live in and are fed by the inputs of the community, so I would like to thank NEUBIAS for the promotion of the knowledge on bioimage analysis. Passion for your work can arise from internal drives, but is much promoted by the trust and the teaching received: thanks to Dario Parazzoli (Imaging Facility, IFOM, Milan).

Lastly, I am grateful to Dr. Giuseppina Pisignano (University of Bath) for the suggestions to improve this chapter.

This work was supported by the Francis Crick Institute which receives its core funding from Cancer Research UK (CC0199), the UK Medical Research Council (CC0199), and the Wellcome Trust (CC0199).

References

1. Cox AN. Allen's astrophysical quantities. New York, NY: Springer; 2002. https://doi.org/10.1007/978-1-4612-1186-0.
2. Hoppe A, Christensen K, Swanson JA. Fluorescence resonance energy transfer-based stoichiometry in living cells. Biophys J. 2002;83(6):3652–64.
3. Galilei G. Il saggiatore, 1623. Text version: IntraText (V89), Èulogos [2007 05 21]
4. Tegmark M. The mathematical universe. Found Phys. 2008;38:101–50. https://doi.org/10.1007/s10701-007-9186-9.
5. Maier A, et al. Medical imaging systems. New York: Springer; 2018.
6. Morishima K, Kuno M, Nishio A, et al. Discovery of a big void in Khufu's Pyramid by observation of cosmic-ray muons. Nature. 2017;552:386–90. https://doi.org/10.1038/nature24647.
7. Wang T, Ouzounov DG, Wu C, et al. Three-photon imaging of mouse brain structure and function through the intact skull. Nat Methods. 2018;15:789–92.
8. Fowles GR. Introduction to modern optics. 2nd ed. New York: Dover; 1989.
9. Borne M, Wolf E. Principles of optics. 6th ed. Oxford: Pergamon Press; 1980.
10. Abbe E. Beiträge zur Theorie des Mikroskops und der mikroskopischen Wahrnehmung. Arch Für Mikrosk Anat. 1873;9:413–68.
11. Berne BJ, Pecora R. Dynamic light scattering. Mineola: Courier Dover Publications; 2000.
12. Keiser G. Biophotonics. Singapore: Springer; 2016. https://doi.org/10.1007/978-981-10-0945-7.
13. Lichtman J, Conchello JA. Fluorescence microscopy. Nat Methods. 2005;2:910–9. https://doi.org/10.1038/nmeth817.
14. Bajar BT, Wang ES, et al. A guide to fluorescent protein FRET Pairs. Sensors (Basel). 2016;16(9):1488.
15. Lakowicz J. Principles of fluorescence spectroscopy. 3rd ed. New York: Springer; 2006.

16. Allen RD, David GB, Nomarski G. The zeiss-Nomarski differential interference equipment for transmitted-light microscopy. Z Wiss Mikrosk. 1969;69(4):193–221.
17. Stuurman N, Vale RD. Impact of new camera technologies on discoveries in cell biology. Biol Bull. 2016;231(1):5–13.
18. Pawley JB. Handbook of biological confocal microscopy. 3rd ed. New York: Springer; 2006.
19. Hamamatsu. Handbook photomultiplier tubes – basics and application. 3rd ed. Hamamatsu: Hamamatsu Photonics K.K.; 2007. https://www.hamamatsu.com/resources/pdf/etd/PMT_handbook_v3aE.pdf
20. Gould TJ, Verkhusha VV, Hess ST. Imaging biological structures with fluorescence photoactivation localization microscopy. Nat Protoc. 2009;4(3):291–308. https://doi.org/10.1038/nprot.2008.246.
21. Gräf R, Rietdorf J, Zimmermann T. Live cell spinning disk microscopy. Adv Biochem Eng Biotechnol. 2005;95:57–75. https://doi.org/10.1007/b102210.
22. Linkert M, et al. Metadata matters: access to image data in the real world. J Cell Biol. 2010;189 (5):777–82. https://doi.org/10.1083/jcb.201004104.
23. Schneider CA, Rasband WS, Eliceiri KW. NIH Image to ImageJ: 25 years of image analysis. Nat Methods. 2012;9(7):671–5.
24. Schindelin J, Arganda-Carreras I, Frise E, Kaynig V, Longair M, Pietzsch T, Preibisch S, Rueden C, Saalfeld S, Schmid B, Tinevez JY, White DJ, Hartenstein V, Eliceiri K, Tomancak P, Cardona A. Fiji: an open-source platform for biological-image analysis. Nat Methods. 2012;9 (7):676–82. https://doi.org/10.1038/nmeth.2019. PMID: 22743772; PMCID: PMC3855844.
25. McQuin C, Goodman A, Chernyshev V, Kamentsky L, Cimini BA, Karhohs KW, Doan M, Ding L, Rafelski SM, Thirstrup D, Wiegraebe W, Singh S, Becker T, Caicedo JC, Carpenter AE. CellProfiler 3.0: next-generation image processing for biology. PLoS Biol. 2018;16(7): e2005970.
26. de Chaumont F, et al. Icy: an open bioimage informatics platform for extended reproducible research. Nat Methods. 2012;9:690–6.
27. napari contributors. napari: a multi-dimensional image viewer for python. Cambridge: Cambridge University Press; 2019. https://doi.org/10.5281/zenodo.3555620.
28. Miura K, Paul-Gilloteaux P, Tosi S, Colombelli J. Workflows and components of bioimage analysis. In: Miura K, Sladoje N, editors. Bioimage data analysis workflows. Learning materials in biosciences. Cham: Springer; 2020.
29. Skansi S. Introduction to deep learning. Cham: Springer; 2018. https://doi.org/10.1007/978-3-319-73004-2.
30. Rubens U, et al. BIAFLOWS: a collaborative framework to reproducibly deploy and benchmark bioimage Analysis Workflows. Patterns. 2020;1(3):100040.
31. van der Walt S, et al. scikit-image: image processing in Python. PeerJ. 2014;2(e453):10.7717/peerj.453.
32. McKinney W. Data structures for statistical computing in python. In: Proceedings of the 9th Python in Science Confernce (SCIPY 2010)
33. Hunter JD. Matplotlib: a 2D graphics environment. Comput Sci Eng. 2007;9(3):90–5.
34. Cromey DW. Avoiding twisted pixels: ethical guidelines for the appropriate use and manipulation of scientific digital images. Sci Eng Ethics. 2010;16(4):639–67. https://doi.org/10.1007/s11948-010-9201-y.
35. Rossner M, Yamada KM. What's in a picture? The temptation of image manipulation. J Cell Biol. 2004;166(1):11–5.

Additional References

AF488 spectrum has been downloaded from "fluorophore.org": http://www.fluorophores.tugraz.at/
Anaconda Navigator. Computer software. Vers. 1.10.0. Anaconda, 2016. https://anaconda.com
Bio-formats. https://www.openmicroscopy.org/bio-formats/

Color blindness awareness. https://www.colourblindawareness.org/ - https://venngage.com/blog/color-blind-friendly-palette/

FLIM. https://www.picoquant.com/applications/category/life-science/fluorescence-lifetime-imaging-flim

HREM technique. https://dmdd.org.uk/hrem/

Image data integrity: In Defense of Image Data & Analysis Integrity - [NEUBIASAcademy@Home] Webinar. https://www.youtube.com/watch?v=c_Oi2HKom_Y

Machine Learning and Deep Learning: Intro to Machine Learning-DeepLearning-DeepimageJ - [NEUBIASAcademy@Home] Webinar. https://www.youtube.com/watch?v=0vTbsO8Vnuo

Measurement in Science - Stanford Encyclopedia of Philosophy - Eran Tal. https://plato.stanford.edu/entries/measurement-science/

Visible spectrum image has been modified by https://upload.wikimedia.org/wikipedia/commons/d/d9/Linear_visible_spectrum.svg

Confocal Microscopy

5

Jeremy Sanderson

Contents

What You Will Learn in This Chapter

Why optical sectioning is required to image thick fluorescent specimens

How microscopists manage fluorescence blurring

What we mean by 'optical sectioning'

Why the confocal microscope is used to obtain blur-free images by optical sectioning

How the confocal microscope works

J. Sanderson (✉)

Bioimaging Facility, Mary Lyon Centre, MRC Harwell, Oxfordshire, UK

e-mail: j.sanderson@har.mrc.ac.uk

© The Author(s), under exclusive license to Springer Nature Switzerland AG 2022

V. Nechyporuk-Zloy (ed.), *Principles of Light Microscopy: From Basic to Advanced*, https://doi.org/10.1007/978-3-031-04477-9_5

When to select the confocal microscope to image fluorescent specimens

How to operate a single beam-scanning confocal microscope

How to improve signal-to-noise ratios (SNR) in your images

What alternatives are available to using the confocal microscope for optical sectioning

What to consider when using the confocal microscope as a quantitative instrument

How to keep your confocal microscope functioning optimally

5.1   Introduction

The confocal microscope is key to successfully viewing and analysing thick fluorescent specimens. Whilst fluorescence is a very versatile contrast enhancement technique that—under the right conditions—is capable of producing very high contrast images, it suffers from three inherent disadvantages: bleaching, bleed-through and blurring. The scientific worker and microscopist must be aware of these three phenomena and manage them to acquire scientifically rigorous images. The first is bleaching, which has been covered in the chap. 3 of this volume on fluorescence microscopy. Light and free oxygen radicals bleach fluorescent molecules so that they don't forever emit light when illuminated. Think of bleaching as an 'ageing' process. The second disadvantage of fluorescence is bleed-through. It occurs when a sample is labelled with more than one fluorophore. The third disadvantage is blurring, and this necessitates using 'optical sectioning' in one form or another to acquire a blur-free image. The multi-photon microscope is used to image very thick tissue samples, discussed elsewhere in this book. In this chapter we discuss optical sectioning stratagems, in particular the function and operation of the single beam-scanning confocal microscope to obviate fluorescent blurring.

5.2   Fluorescence Blurring

5.2.1   The Advantages and Disadvantages of Fluorescence

Fluorophores have been used for many years to label cells and tissues [1]. Biological systems are highly dynamic environments with molecules diffusing around the cell or undergoing active transport along protein or nucleic acid tracks. Even when fixed, cells and tissues still represent complex environments worthy of study. Inorganic fluorophores, quantum dots, conjugate tags and fluorescent proteins are suited to marking and tracking proteins and other cell moieties [2–4] allowing us to see cells, and parts of cells, that would not be possible to see otherwise. Reverting to fundamentals, a microscope must provide:

1. Resolving power to carry fine detail in the specimen to the image, with
2. sufficient contrast to show differences between image features and the background, at
3. sufficient magnification to present the resolved detail to the eye or digital detector.

Whilst the primary function of the microscope is to resolve fine detail, this cannot be achieved satisfactorily unless sufficient contrast is present in the image. It is more difficult to resolve details in a pellucid image arising from an unstained object than one that is stained. Resolving power and contrast are therefore linked [5]. Of all the contrast enhancing techniques, if properly applied, fluorescent markers give the highest contrast with a dark background and also high signal-to-noise ratio. Fluorescent labels are advantageous in that they are highly sensitive even at low concentration and, generally, non-destructive to the target cell or tissue.

To summarise, fluorescent markers offer the following advantages to marking cells and tissues: fluorophores are self-luminous and under the right conditions exhibit very high contrast; they are very sensitive markers, detecting small concentrations of proteins and cell moieties; they are very specific markers. With proper control of background and blocking of non-specific signal, fluorescent markers generally give good discrimination of the tissues, cells, organelles and proteins they are employed to mark. A wide range of fluorescent probes are available, and these can be used simultaneously to label multiple targets in cells and tissues. Endogenous fluorescent proteins are very versatile and can also be used to label multiple cell components. Finally, fluorophores are relatively cheap and easy to apply and the high quality CCD, CMOS and PMT detectors available are able to collect weak signals easily. These are the advantages afforded by using fluorescent probes.

Against these important advantages there are three disadvantages that must be considered and managed in order to achieve meaningful and scientifically rigorous data with high signal-to-noise ratios. These are: bleaching, blurring and bleed-through. Unless the specimen is very thin, blurring in the image is probably the most serious of the three disadvantages that must be addressed. At present, we shall discuss blurring, for bleaching is addressed in the chapter on fluorescence and we shall briefly address bleed-through when covering the operation of the confocal microscope.

5.2.2 Why Blurring Occurs

Fluorophores are self-luminous; when illuminated with light of sufficient energy they emit light (for an excellent review, see [6]). The entire thickness of the fluorescently-labelled cell or tissue fluoresces, as shown in Fig. 5.1. Moreover, the depth of field of the high numerical aperture microscope objectives used to acquire the image is very small. The depth of field is the axial depth of the space on both sides of the object plane within which the object can be moved without detectable loss of sharpness in the image, and within which features of the object appear acceptably sharp in the image while the position of the image plane is maintained.

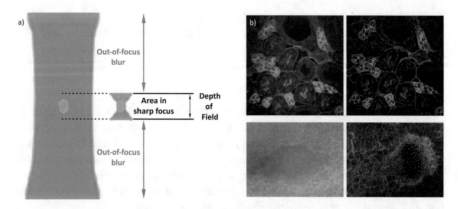

Fig. 5.1 Since the entire fluorescently-stained tissue, section or cell emits signal when illuminated with high energy light, and because the thickness of the fluorescently-stained sample is greater than the depth of field of the microscope objective (**a**) blurring will occur. The effect is shown in (**b**) with two different samples: kidney (upper panels) and cilia in the node of a mouse embryo (lower panels). The left-hand panels (upper and lower) show the fluorescent blurring that occurs, whilst the right-hand panels show the improvement from acquiring a single optical section with the confocal microscope. Compare this figure with Fig. 5.11 showing a maximum intensity projection of an entire z-stack. Mouse images courtesy of Dr. J Keynton. Reproduced from *Understanding Light Microscopy* with permission from John Wiley & Sons Ltd. Image copyright, author: J Sanderson

Typically, the objective depth of field is less than 1 μm, and values of 200–400 nm are common. Even flattened adherent cells grown onto a coverslip are much thicker than this, approximately 3–5 μm [7, 8]. The result will be a low-contrast image, composed of a bright but very blurred background upon which is superimposed the much dimmer in-focus information (Fig. 5.1). That part of the image which is in focus will be degraded by light that is emitted or scattered by the tissue outside the narrow plane of focus.

If an adherent cell is, say, 4 μm thick and the depth of field of the microscope objective is 400 nm, then 90% of the image seen will be blurred because most of the light is contributed by regions that are not exactly in focus. The contribution of the blurred background light from the out-of-focus regions is superimposed over the weaker (less-intense) in-focus image. This will reduce the signal-to-background ratio, and thus the contrast, of the image. This out-of-focus haze will also reduce the resolution of detail in the image. Signal-to-noise ratios (SNR) and signal-to-background ratios (SBR) are different. The SBR is a measure of contrast in an image whereas SNR describes the variability in photon intensity in time at a single sampling point, or pixel, in the image [9]. A confocal microscope generates good SBRs, but is still inherently noisy, for reasons we shall discuss later.

For non-adherent cells (cells generally round up when undergoing division) and tissues or tissue slices, the situation is worse. With cells and tissues of a thickness greater than about 10 μm, scattering of light from the refractive index differences between the cell, its constituents and the local environment also begin to degrade image quality [10, 11]. This scattering occurs because curved surfaces of moieties in the cell act as microlenses scattering and reflecting light randomly. Light may be scattered whilst propagating down both the illumination and imaging pathways [12]. The reason the multi-photon microscope (Chap. 9) is successful at viewing very thick tissues (those that are too thick to be imaged by the confocal microscope) is because non-linear multi-photon microscopy is insensitive to scattering of the emitted signal; the longer-wavelength (red; infrared) illumination scatters minimally; the optical train is simpler, does not require descanning and the excitation volume at the plane of focus is exceedingly small [13, 14].

In order to acquire a thin blur-free section, or an in-focus blur-free z-stack of images, we must employ the so-called optical sectioning in one form or another. If the cells or tissue section that you are investigating is suitably thin, then use a widefield fluorescence microscope to acquire your images. The flow chart in Fig. 5.2 will guide you in the choice of which microscope to use.

5.2.3 Optical Sectioning

The confocal approach to optical sectioning scans the illumination across the image sequentially, the single-beam point-scanning confocal being the most widespread and popular of all the optical sectioning fluorescence microscopes. Several designs are available on the market, but they all do one thing: optical sectioning is achieved by point illumination and point detection. By restricting the illumination to a pinhole, rather than illuminating the entire field of view at once it becomes possible to build up a blur-free image in three dimensions (Fig. 5.3). The entire image is built up point by point in each frame as the diffraction-limited spot of the illuminating beam rasters across the frame, like reading words in lines along a page of script in a book. To take the analogy further, each two-dimensional section is like a page, and a 3-D so-called z-stack (a series of multiple images taken at different lateral focal planes to provide a composite image with a greater depth of field) is equivalent to a chapter or entire book. The single-beam laser-scanning microscope does not form a real image, as a widefield microscope does, but builds it up pointwise.

Marvin Minsky originated the concept of optical sectioning in 1955 to produce blur-free images, motivated by the need to see and understand how neural networks were connected. He reasoned that the blurring and scattering 'would be gone if we could only illuminate one specimen point at a time'. In a delightful memoir well worth reading [15] he goes on to say 'the price of single-point illumination is being able to measure only one point at a time. This is why the confocal microscope must scan the specimen point by point and that can take a long time'. Minsky also commented 'In retrospect, it occurs to me that this concern for real time speed may have been what delayed the use of that scheme [confocal microscopy] for almost thirty years'. Figure 3 of Minsky's patent application in 1961 is remarkably similar to the configuration of the modern single-beam point-scanning confocal (see also Fig. 1 in [16]). Practical implementation of Minsky's design was impeded firstly by the lack of sufficiently bright illumination, secondly from a means to display and capture the image and thirdly because scanning was implemented by moving the

Which type of microscope to choose

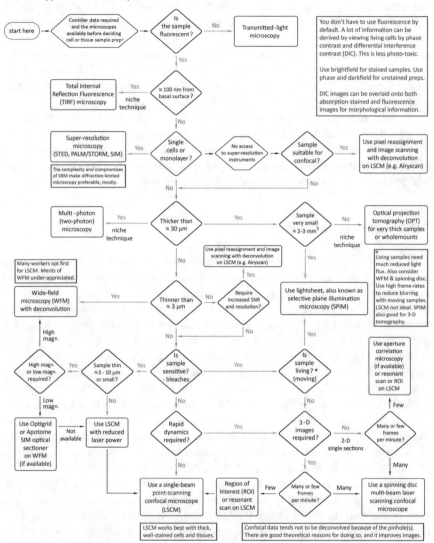

Fig. 5.2 Flow chart to help choose which type of optical sectioning microscope to use. Author's artwork, first reproduced in *Current Protocols in Mouse Biology* published by John Wiley & Sons Ltd. Image copyright, author: J Sanderson

stage rather than the illuminating beam. Stage-scanning confocal microscopes are mainly used now for inspection in the microchip industry and in materials science where specimens are very much larger and flatter than biological tissues. Beam scanning does not vibrate or insult delicate biological tissues, is not limited in the raster scan by the weight of the stage, does not suffer loss of resolving power from

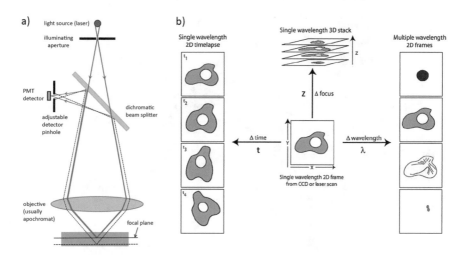

Fig. 5.3 The ray path of a single-beam, laser-scanning confocal microscope, showing how (**a**) only the in-focus signal passes through the pinhole to the PMT detector. The versatility of the confocal microscope allows data to be collected in up to five dimensions (x, y, z, t and λ) as shown in (**b**). Panel (**a**) reproduced with permission from Springer. Panel (**b**) reproduced from *Understanding Light Microscopy* with permission from John Wiley & Sons Ltd

mechanical vibration or geometrical distortion in the image and is potentially much faster.

Mojmír Petráň developed the reflected-light tandem scanning confocal microscope in 1967, a pocket-sized instrument with multiple pinholes that was notoriously difficult to align, but the image could be seen directly in real time. The Nipkow disc of Petráň's design attenuated the illumination and signal to such an extent that only the brightest-stained samples could be viewed, and the advantage of its fast frame rate was lost. In 1981, Wilson and Sheppard (for a historical summary, see [17]) proposed a theoretical solution to combine the resolving power and depth discrimination of a conventional fluorescence microscope. The first practical single-beam laser-scanning design of confocal microscope, in which the illumination was scanned rather than moving the stage, was developed by William 'Brad' Amos, John White, Mick Fordham and Richard Durbin at the Medical Research Council laboratory in Cambridge in 1986 [18, 19] followed very shortly thereafter by a Swedish group (Carlssen and Aslund 1987). The breakthrough [20] making this possible was not only due to bringing together suitable lasers, galvanometer mirrors and computers to display the image, but achieving accurate raster scanning at high speed. When considering the development of practical confocal microscopy, the important development of the dichromatic beam-splitter by JS 'Bas' Ploem [21] is almost always overlooked.

The *confocal* microscope is so-called because the projection of the illumination pinhole, the plane of focus within the specimen and the back-projection of the detector pinhole are all situated at *con*jugate *focal* planes [22]. Every laser-scanning

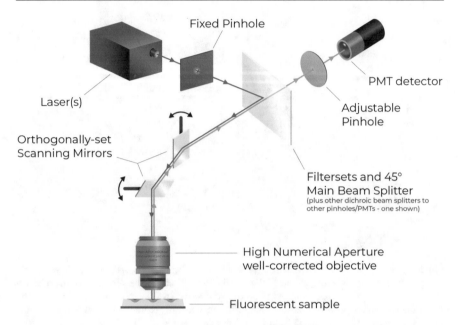

Fig. 5.4 The common parts found on a confocal microscope. See the text for details. Image copyright, author: J Sanderson. Redrawn by Gareth Clarke, MRC Harwell Institute

confocal (a generic design is shown in Fig. 5.4) has the following features in common:

- One or more gas or diode laser(s) providing illumination of specific wavelengths
- Fluorescence filters and a dichroic mirror as a filter set to direct the illumination onto the sample and direct the emitted fluorescence of specific bandwidth towards the detector
- A mirror and galvanometer-based raster scanning mechanism
- One, or more, pinhole apertures
- Photomultiplier tube (PMT) detectors for each channel

Because the confocal microscope depends upon point-scanning and a pinhole aperture to exclude out-of-focus light, lasers are used because they have sufficient collimation and power to provide the necessary illumination flux. Lasers produce very intense, coherent beams of very narrow wavelength that are ideal for illumination in a microscope where 90% or more of the emitted signal may (intentionally, due to the pinhole) not reach the detector. It is essential to be able to adjust the laser intensity so that fluorophores are not saturated and bleached. Lasers are generally used at fractions of a per cent of their maximum output power unless, of course, they are intentionally used to bleach the sample [23]. The lasers can be switched very rapidly, and adjusted in intensity, by the use of an Acousto-Optic Tuneable Filter

(AOTF) or neutral density filters. Generally, AOTFs are used; they are more versatile and also allow specific regions of interest to be scanned in the field of view.

The beam is raster scanned using two oscillating mirrors driven electromagnetically by a mechanism similar to a moving-coil galvanometer. As such they are colloquially referred to as 'x- and y-galvos'. The sawtooth duty cycle is demanding on power, with the x-galvo working harder than the y-galvo. Higher frame rates can be achieved by driving the mirrors at resonant frequency. Although confocal microscopes have been made without mirrors, they endure because of their achromatic behaviour.

Photomultiplier tubes (PMTs) are used as detectors in point-scanning confocal microscopes because these enhance the weak signal and can also very rapidly collect single points of light emitted as signal. Using a CMOS or CCD detector would be far too slow. The PMT does not 'see' the entire image as a CCD detector does, but very rapidly produces a voltage equivalent to the photon intensity emitted at each sampling point in the object, which is then digitised to form a corresponding pixel in the image.

With only a single point illuminated (rather than the entire field of view as in 'widefield' mode) the illumination intensity rapidly falls off above and below the plane of focus as the beam converges to a point and then diverges. This reduces excitation of fluorescence of these objects situated out of the focal plane, improving depth discrimination. The pinhole in front of the PMT detector excludes out-of-focus light giving a sharp, blur-free image. The confocal microscope differs from widefield fluorescence microscopy in that the microscope configuration for widefield (both brightfield and fluorescence) microscopy is designed to be used with Köhler illumination where the excitation illumination is maximally out of focus at the specimen plane (i.e. the lamp filament is not imaged at the specimen plane, but a different conjugate plane: the entire raison d'être of Köhler illumination). In the confocal microscope the illumination *is* focused onto the specimen plane as a diffraction-limited point. For an explanation of Köhler illumination, see Chap. 9 in Sanderson [24].

5.2.4   Structural Illumination Microscopy (SIM)

The confocal microscope remains the most popular optical sectioning instrument, but there are two other approaches: structural illumination microscopy and deconvolution. If we regard confocal microscopy as the optical solution to optical sectioning, then deconvolution is a purely mathematical approach, sometimes called computational optical sectioning microscopy (COSM) whilst SIM is a mixture of the two. In COSM a 3-D dataset is collected as a z-stack of a series of 2-D images, each with the microscope focused at a different plane through the specimen.

In SIM a grid pattern is superimposed upon the specimen at the focal plane, and subsequently shifted to allow the out-of-focus signal to be subtracted from the in-focus information to yield an optical section. The Ronchi grid pattern introduces an artificial high frequency spatial modulation which rapidly attenuates either side of the focal plane. The optical sectioning efficiency depends upon the pitch and contrast of the grating; in practice, different gratings are used with particular objectives. The grating is held in a custom-built slider which is inserted into a slot in the microscope body conjugate with the field diaphragm, to project the grid image onto the plane of focus. A plane-parallel plate is automatically moved and tilted, by piezoelectric motor, to shift the image of the grid above and below focus. The microscope hardware and software then acquires at least three images: one in-focus and two out of focus, to generate an optical section. This method has been called 'the poor man's confocal' but this does no justice to the elegance of the technique. The advantage of deconvolution is that it is software-based and can be used on any microscope, whereas SIM requires a specialist upgrade or third-party add-on to the microscope. The advantage of SIM is that raster scanning of the sample by intense laser illumination is not required.

Optical sectioning SIM typically uses coarse gratings and incoherent light. High resolution SIM, which can improve resolving power by a factor of two over the Abbe limit, uses finer gratings, superimposing a moiré pattern onto the sample [25–27]. SIM is significantly faster than conventional point-scanning microscopic techniques, which are inherently limited in speed by illumination intensity, fluorophore saturation, and raster scanning. With SIM one needs to acquire several frames for a single reconstruction; thus, the speed of the method scales with the availability of fast camera technology.

The aperture-correlation microscope [28] uses a similar technique, but has a higher frame rate akin to the spinning disc microscope. In aperture-correlation microscopy, the final image is calculated in three steps: first, the two images have to be extracted from the side-by-side view and one image is mirrored to match the image orientations. The second step involves a registration of both images to ensure that the overlay is precise on a pixel-by-pixel basis. In the registration step, distortions as mapped in a previous calibration step are corrected between the two imaging beam paths. The third step is the actual calculation of the optical section itself. A scaled subtraction of both images will yield the optical section. Both SIM and aperture correlation microscopy are well suited for thin to medium thickness specimens, where the SBR is low. Sometimes artifactual stripes occur [29] due to absorption and scattering in the illumination path but these can be overcome [30, 31].

5.2.5 Deconvolution

We will only consider the basic principles of deconvolution here. A short and excellent primer which I recommend to my students is Shaw [32]; another good explanation of deconvolution is Biggs [33]. All imaging systems are imperfect. When any image is formed—or convolved—from an object it suffers degradation. Convolution is a formal mathematical operation, just like multiplication, addition and integration—which in our case is applied to image formation. A convolution operation takes two signals and produces a third signal, in our case the input signal (from the object) is convolved with the second signal (the PSF) to produce the output signal (the image), see [34].

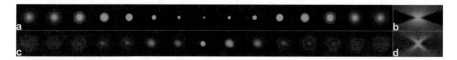

Fig. 5.5 A point spread function shown laterally in *x*, *y* dimensions in panels (**a**) and (**c**) with the corresponding *z*-stack projected in *x*, *z* in panels (**b**, **d**) respectively. A calculated PSF is shown in the upper pair of panels, while an experimentally-collected PSF from a fluorescent bead sample is shown in the lower two panels. The objective used to acquire these data-sets was a 63× NA 1.4 plan-apochromat with a prepared sub-resolution 0.1 μm bead slide, having a refractive index of 1.518—as close to homogeneous immersion as possible. The calculated PSF was generated with the PSF Generator plugin for Imagej/Fiji developed by the Biomedical Imaging Group at the École Polytechnique Fédérale de Lausanne (EPFL). Image copyright, author: J Sanderson

If a lens behaved perfectly, points in the object would not be smeared into a point spread function (PSF) and a defocused image of beads would appear black. As it is, a sub-resolution fluorescent bead forms the image seen in Fig. 5.5. A bright point is seen at the focal plane; either side of the focal plane, the focal spot is transformed into a disc that becomes both larger and dimmer in intensity. Ideally, the defocused image should be the same either side of the focal plane, but again this is usually not the case—compare the calculated theoretical PSF with the empirical PSF collected from a sub-resolution fluorescent bead. In three dimensions this image is seen as an hour-glass shape and describes the PSF characteristic of that particular microscope. The point spread function is the *signature* of the microscope. The confocal PSF is smaller than the widefield PSF (Fig. 5.8), as explained in Sect. 5.3.2 below.

Image degradation occurs not only because of fluorescence blurring (see Sect. 5.2 above), but also from the effects of diffraction and aberrations of the objective. This is inescapable. Because (from an image formation perspective) self-luminous fluorophores and self-luminous stars are similar, image restoration and sharpening by deconvolution is a useful technique in microscopy that was originally borrowed from astronomy [35].

Mathematically, convolution involves replacing each point in the specimen during the process of image formation to form a (blurred) point in the image. This direct convolution operation may appear complex, and indeed it is very computationally time-consuming. However the mathematical operations described above in 'real space' can be simplified by computation in the frequency domain, or 'Fourier space' using Fourier transforms, which describe images mathematically in terms of their sine and cosine components. The convolution operation reduces to: $FT_{microscope\ image} = FT_{object} \times FT_{PSF}$ Knowing the PSF, deconvolution can be performed in reverse ($FT_{object} = FT_{microscope\ image}/FT_{PSF}$) to sharpen the image as a more faithful representation of the object.

There are two main types of deconvolution algorithm: deblurring and restorative (Fig. 5.6). Restoration algorithms are more accurate and can be used for quantitative purposes because they keep the relative intensity relationships seen in the data forming the original image. Most restoration algorithms are iterative. With iterative deconvolution, the ideal 'model' image is compared with the results of computation

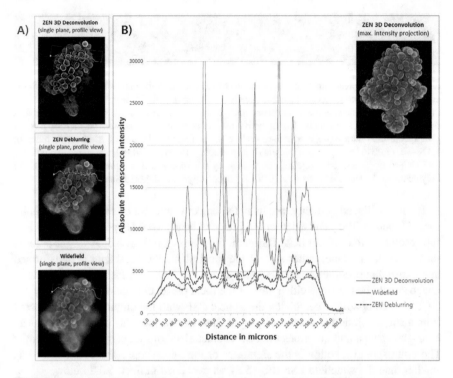

Fig. 5.6 Image improvement with deconvolution, showing the use of deblurring and restorative (constrained iterative) algorithms on a tissue sample. A widefield image and a maximum intensity projection (MIP) taken with a confocal are also shown for comparison. Further details may be found on page 9 of the Carl Zeiss Technical Note '*How to Get Better Images With Your Widefield Microscope*' (Stickler et al., March 2020). Reproduced with the kind permission of Carl Zeiss and René Buschow, Max Planck Institute for Molecular Genetics, Berlin. Image: Copyright ZEISS Microscopy

successively allowing for noise and signal integration. Blind deconvolution (more properly referred to as adaptive blind deconvolution) is an alternative restorative method that extracts the PSF *directly* from the image data. This may seem counter-intuitive, since the algorithm is trying to compute a solution for both the image and the PSF, but it is quick and proponents argue that the calculated PSF best fits the data. For those wishing to use freeware rather than a commercial software package, try DeconvolutionLab2 ([36]; EPFL Lab, Lausanne) in Fiji/ImageJ.

In order to acquire a meaningful scientifically rigorous deconvolved dataset, the correct algorithm must be selected for the task or application in hand *and* the correct point spread function, whether calculated or experimentally acquired, must also be used. Without the most appropriate algorithm or correct PSF, there is little, or no, benefit to post-processing the image, and the old computer adage holds true: 'garbage in equals garbage out'. Clearly, a PSF generated from sub-resolution fluorescent beads give more precise results than a calculated PSF, but acquiring datasets from the former is more labour intensive and a calculated PSF may be all that is needed. Deconvolution will not make bad data good; it will only make good data better. The key to successful deconvolution is acquiring an accurate PSF. Cannell et al. [37] report that measured PSFs are usually 20% larger than calculated PSFs and nearly always are non-symmetrical due to the presence of spherical aberration. Practical details for collecting PSFs from sub-resolution fluorescent beads are given in [38]. Most deconvolution software has dialogue boxes to plug in values to generate a calculated PSF.

Deconvolution should improve the contrast and SBR with reduced background haze [39]. The edges of objects should be sharper, and the intensity of features in the image enhanced. The artifactual elongation of features in the z-axis should also be reduced. Although fibre optics make this artifact less common, flickering lamps and a change in illumination intensity during acquisition can lead to stripes being seen in the image. If the deconvolution algorithm is unsuitable or else applied too aggressively, artifactual rings and points can appear in the image. Rings may occur from sampling irregularities; it is important to sample according to the Nyquist-Shannon sampling criterion. Clearly, if the incorrect PSF is applied, the image will probably be skewed or asymmetric in some form. This may also occur if refractive index mismatches occur in or between the specimen and its environment. Indeed, where these mismatches occur and—for an immersion objective—homogeneous immersion generally does not occur, then the firmly-held idea that optical sectioning occurs because light is excluded by the pinhole is also upset. You should be asking yourself whether the confocal plane of focus is 'an actual geometrical plane resembling a mechanical section, or merely the surface described by an array of points at which the rastered laser beam happens to reach best focus' [40]. If the specimen is too thick, deconvolution algorithms will fail to deblur the image or properly to reassign the out-of-focus light. Deconvolution is no longer useful when the image contrast between the signal and the background noise falls to <1% as seen through the entire thickness of the cell or tissue [41].

Should deconvolution be used with confocal data? Some people reason that the pinhole has already 'sampled' the data, rendering post-acquisition deconvolution less important than with widefield microscopy. Nevertheless, there are very sound reasons why datasets from confocal and other optical sectioning microscopes should be deconvolved, as Jim Pawley consistently advocated [42]. A laser scanning confocal microscope (LSCM) is inefficient at collecting light because not only is it noisier than widefield microscopes, it throws away out-of-focus light rather than reassigning it as a deconvolution system does. Low intensity signals contain a lot of high frequency Poisson, or shot, noise which can be erroneously sampled and digitised into non-existent single-pixel 'features' that could never have been resolved by the microscope. Secondly, deconvolution averages the signal over the many voxels in the image needed to sample the signal from a single point in the object sampled at Nyquist frequency. This is a very effective way of reducing the overall noise and boosting the SNR. To achieve the same result with the confocal microscope would require Kalman averaging the signal anywhere between 80 and 130 times. This is not feasible and would bleach the specimen. Thirdly, the spread of the image dataset in the z-axis is reduced. Fourthly, the Nyquist-Shannon sampling theorem stipulates that the bandwidth of the (output) display device must be equal to that of the (input) digitising device. Deconvolution is the correct way to implement this (often forgotten) condition of Nyquist sampling. Each objective has an optimal lateral and axial sampling rate. The values given by Biggs [33] are dependent upon NA, wavelength and refractive index: $d_{x,y} = 0.25\lambda/\text{NA}$ and $d_z = 0.5\lambda/(\text{RI} - \sqrt{[\text{RI}^2 - \text{NA}^2]})$. When you deconvolve confocal data, the contrast drops and the image doesn't look so sharp. Those 'sharp' objects were noise artifact and the contrast can always be raised to match the characteristics of the display monitor.

5.3   The Confocal Microscope

5.3.1   How the Confocal Microscope Works

The modern single beam-scanning laser confocal is offered by the big four microscope manufacturers, and also independent companies, as a turn-key system. This makes the instrument easy to use. The generic design and configuration has already been described above. In this section we consider practical functional aspects of the lasers, the pinhole and PMT detectors.

Traditionally the lasers in a confocal microscope were Argon-Krypton and Helium-Neon gas lasers, which offered a range of discrete wavelengths. These are still used, but dye lasers, diode and diode-pumped solid state lasers are increasingly common due to their size, low heat dissipation, convenience of use and longer operational lifetimes and supercontinuum white-light lasers [43] are now available. The air-cooled argon-ion laser is still widely employed as a light source for confocal microscopy because of its brightness level, small size, excellent beam geometry and the suitability of its spectral lines for green fluorophores. Typical laser line values are as follows: 351, 364, 405, 430, 458, 477, 488, 497, 514, 532, 543, 561, 568, 594 and 633 nm, the bandwidth being 1 nm. Not all these lines will be present in one instrument, a typical combination of lasers on a high specification instrument being: 405, 458, 488, 514, 561 and 633 nm. For further information, see Table 18.1 in Sanderson [24].

Different laser lines emanate from different power lasers, from 1–2 mW up to 10–15 mW. A typical working output value (usually selected automatically in software) will be 2%. The brightness of the illumination cannot be increased indefinitely. In fact, there is a very small window of opportunity regarding laser power. Irrespective of photo-damage to living cells and tissues, the photophysics of fluorescence limits the useful intensity of illumination that may be used. Above a certain threshold the fluorophore saturates and molecules remain in the excitation orbital rather than returning to the dark ground state to absorb further photons. The result is fluorescence signal intensity falling with further increases in laser power.

During normal operation of single point-scanning confocal microscopes, the beam dwells on each pixel for 1–20 µs. Averaged over 4 µs the laser intensity fluctuates by 2–3%, but fluctuations up to 10–15% are observed with sub-microsecond dwell times. This introduces significant noise [9], which may drown out weakly-fluorescent structures; it is not due to poor resolving power. When averaged over a much longer time (ms to s), the amplitude of the fluctuation is small, in the range of 0.5–1%.

Confocal microscopes are set up to receive laser light via a single-mode fibre connector. Coupling these is extremely challenging, requiring adjustments in six degrees of freedom (three rotational and three linear parameters) to control the alignment of the laser beam into the objective. This procedure can take hours to optimise and, for virtually all microscope users, it is simply not worth the effort. It is always advisable to maintain instruments under a service contract, and alignment is best left to the service engineer.

Once the different fluorophores have been set up in software, they are usually assigned to individual 'channels' for sequential acquisition. Two commercially-available specimens suitable for training are the Fluocell slides from ThermoFisher. Slide #1 (F36924) comprises adherent bovine pulmonary arterial endothelial cells stained with DAPI (nuclei), MitoTracker Red CMXRos (mitochondria) in the live cells. Following fixation, the preparation is further stained with Alexa Fluor 488 phalloidin (F-actin). Slide #3 (F24630) is a 16 µm thick cryostat section of mouse kidney, ideal for demonstrating optical sectioning and z-stack acquisition. It is stained with two lectins, Alexa Fluor 488 wheat germ agglutinin (glomeruli; convoluted tubules) plus Alexa Fluor 568 phalloidin (F-actin in glomeruli & brush border) and DAPI. Using slide #1, each fluorophore is assigned to a channel: red, green and blue. Sometimes fluorophores in different channels, whose emission spectra do not overlap significantly, may be combined in a single 'track' to allow faster acquisition rates providing significant bleed-through does not occur. For example, the decision might be made to collect the signals from DAPI (channel 1) and MitoTracker Red (channel 3) simultaneously into one track.

The pinhole in front of each PMT detector associated with each channel or track is set to collect an optical section. Altering the pinhole size will alter the thickness of the optical section: a wider diameter, open pinhole collects a thicker optical section with brighter signal and a higher SNR, but consequently more blurring. Closing the pinhole will increase the SBR, increase the resolution slightly but attenuate the signal and make the SNR worse. There will be a '1 AU' setting for the pinhole. This is a normalised value to allow the signal from the central Airy disc, excluding the diffraction rings, to enter the detector. Setting the pinhole to 1 Airy Unit isolates 80% of the photon intensity distribution for the point object from the central maximum to the first minimum (zero) either side of the PSF Gaussian intensity distribution: the central Airy disc. The Airy disc, named after Sir George Biddell Airy who studied diffraction patterns and image formation in stars, describes the *central* intensity maximum of the two-dimensional pattern of the point spread function, the PSF. If pinhole is opened above 1.35 AU, the increase in brightness is due to collecting unwanted extra-focal light—which was intended to be removed by investing in a confocal microscope! Below 0.6 AU a slight improvement in resolving power occurs from having a smaller PSF, but with a concomitant huge intensity loss.

The setting for the one Airy Unit value is dependent upon the wavelength of illumination and the numerical aperture of the objective. Clearly, the former can change by a factor of almost two, whilst the latter is constant. Most systems now have separate pinholes in front of each detector, and this will mean altering the pinhole diameter and normalised AU value to ensure optical sections of equivalent thickness are collected from each channel. Adjusting the pinhole to the optimum one Airy Unit reduces the background in the image from out-of-focus light by approximately 1000-fold relative to widefield microscopy [44]. Reducing the pinhole diameter below 1 AU will improve resolution, but at the expense of signal intensity. A recommendation is not to close the pinhole below 0.7 AU, conferring optimal axial resolving power.

The contrast and SBR are determined by the gain and offset settings of the PMT detectors. These should be adjusted so that the signal emitted from each fluorophore falls within the dynamic range of the PMT. Confocal images are usually displayed as 8-bit or 12-bit images in three-colour RGB format. The default is 8-bit (2^8 or 256 grey levels), but 12-bit (4096 grey levels) should be used when collecting images for quantitative analysis. Setting the offset (black level) and the gain (signal level) requires a false-colour look-up table to colourise undersaturated and oversaturated pixels in the image. This is because our eyes are approximately 6-bit devices (dynamic range of ≈ 64 grey levels) and we cannot discriminate the limits of the 8-bit or 12-bit dynamic range of the PMT detector. It is also worth mentioning that some users of confocal microscopes may be colour blind. For further information see Chap. 31 in Sanderson [24] and Note 14 in [1].

When sampling the analogue signal and converting the intensity value into individual pixel values, the detector must work extremely fast. The default raster on a confocal microscope is 512×512. That is, 512 horizontal lines, each composed of 512 individual sampling points, upon which the laser 'dwells' as it raster-scans across the specimen to produce an image of 512^2, or 262,144 pixels. To acquire such an image normally takes 1–2 s, thus the dwell time for sampling each point in the sample to create the corresponding pixel in the image is approximately 4 µs. Although a CCD or CMOS detector is more efficient at sampling each photon of signal and creating photo-electrons to convert to a pixel intensity value, neither quantise the analogue signal fast enough. That is why a photomultiplier detector is used. The associated downside is that its quantum efficiency (QE) is much lower than either a CCD or CMOS camera, which is suitable for widefield microscopy. A high quality CCD camera will have a QE of 80–90%, converting nine out of ten photons to photo-electrons, but a PMT has a QE of about 15%. Gallium arsenide phosphide is a semi-conductor gallium alloy material with an extended sensitivity into the red end of the spectrum. Even the latest GaAsP, avalanche photo-diode and hybrid detectors have QE values of no more than 40%. This means that at low signal levels the gain control must be increased, and the resulting image will show high shot (Poisson) noise unless measures are taken to improve SNR. Figure 5.7 shows the differing image quality from a PMT detector and CCD camera.

Because the single-beam laser-scanning confocal microscope rasters the laser across the sample with two galvanometer driven mirrors under computer control, this

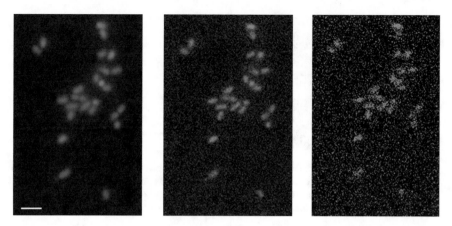

Fig. 5.7 Comparison of the quantum efficiency of PMT and CCD camera detectors. The widefield image is the left-hand panel, the single-scan confocal image on the right. The middle panel shows the confocal image taken with 16 average sans—the best possible. The sample is GFP-expressing Haemophilus influenzae bacteria and the scalebar is 2 μm. Reproduced from *Understanding Light Microscopy* with permission from John Wiley & Sons Ltd. Image copyright, author: J Sanderson

means that any scan zoom factor can be applied for image acquisition. All digital images are collected according to the Nyquist-Shannon sampling criterion [45, 46] such that an analogue photon signal must be sampled at least twice the highest frequency of the wave. In practice, the wave should be sampled at slightly higher than 2f because the image isn't formed until the sampled information is reconstructed, which always takes place through a 'filter' (in our case, the eye or display monitor) limiting the image to lower frequencies. Since the microscope objective forms the image, an optimal zoom setting yields pixel dimensions— equivalent to just under half the resolving power of the objective—sufficiently small to satisfy the Nyquist criterion, but still large enough to avoid over-sampling. Be aware that as the scan zoom increases, the laser power is concentrated into a smaller area, so the specimen will bleach more readily.

5.3.2 Resolving Power of the Confocal Microscope

Although primarily used for an improvement in SBR and optical sectioning, the confocal microscope also offers slight improvement in lateral and axial resolving power over widefield fluorescence microscopy [47]. This condition assumes using a perfectly clean highly-corrected objective and well-stained samples. Because the confocal utilises point illumination and point detection, the total PSF is the *product* of both the excitation illumination (λ_{ex}) PSF and the emitted signal (λ_{em}) PSF, improving resolving power by a factor of $\sqrt{2}$. This is seen in practice by the improved shape of the PSF in the confocal microscope as compared to a widefield PSF (Fig. 5.8).

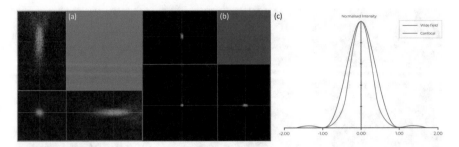

Fig. 5.8 Comparison of the PSF of a sub-resolution fluorescent bead taken with a widefield (**a**) and confocal (**b**) microscope respectively. The graph (**c**) represents the improvement in resolving power of a confocal over a widefield fluorescence microscope. Image copyright, author: J Sanderson. Graph redrawn by Gareth Clarke, MRC Harwell Institute

The classical equation for lateral resolving power becomes $d_{x,y} \approx 0.44\lambda/\text{NA}$, whilst in the axial direction $d_z \approx 1.4\lambda n/\text{NA}^2$. However, this assumes a pinhole diameter of much less than 0.5 AU, tending to zero, which is very rarely encountered because of rejection of most of the signal. Therefore for a pinhole of 1 AU, or greater, the following revised values apply: $d_{x,y} \approx 0.51\lambda_{\text{ex}}/\text{NA}$ and $d_z \approx 0.88\lambda_{\text{ex}}/[n - \sqrt{(n^2 - \text{NA}^2)}]$ for NA values greater than NA 0.5, whilst for NA values less than NA 0.5, $d_{x,y} \approx 0.37\lambda_{\text{ex}}/\text{NA}$ and $d_z \approx 1.28\lambda n/\text{NA}^2$ [48].

Since the minimum intensity at the first zero of the Airy pattern is hard to measure in practice, resolving power is usually calculated as the Full Width Half Maximum (FWHM) of the PSF of a sub-resolution fluorescent bead. From a z-stack, the FWHM is calculated from an intensity line profile of the image of the bead. In this case, the $\text{FWHM}_{\text{axial}} = 0.64\lambda_{\text{ex}}/[n - \sqrt{(n^2 - \text{NA}^2)}]$. For a mathematical explanation of the improvement in resolving power by a factor of 1.4, see Diaspro et al. [14] and the references cited therein. Various experimental stratagems have been adopted to improve the resolving power of the confocal microscope (e.g. focal modulation microscopy; [49]) and divided-aperture microscopy [50] but unfortunately these applications are not widespread.

5.3.3   Advantages and Disadvantages of the Single-Beam Confocal Scanning Microscope

Advantages
- Non-invasive optical sectioning—well-defined, sharp, optical sections
- Good reduction of background—increased SBR
- Laser switching allows good control of cross-excitation and bleed-through
- Magnification zoom can be adjusted electronically with ease
- PMT not limited by the matrix of a CCD detector—easy to draw regions of interest
- Can image several fluorophore markers simultaneously

- LSCM good for spectral unmixing fluorescent proteins with overlapping emission spectra
- Smaller PSF—slightly better lateral and axial resolution than widefield
- Can use deconvolution algorithms to further improve image quality

Disadvantages
- Real-time collection difficult—need to wait to collect a low-noise signal
- Use of laser illumination bleaches faster than widefield or spinning disc confocal
- PMT detectors have a low quantum efficiency—samples must be well-stained
- Scan speed and fluorescence saturation impose a frame-rate limit: slower than widefield or spinning disc
- Laser power noise and fibre optic coupler leads to artifactual pixel-pixel fluctuations
- Less sensitive than using the same objective on a fluorescence widefield microscope
- Unlike multi-photon, a relatively large specimen volume is still illuminated
- Light scattering/refractive index mismatches limits depth penetration to 100–200 µm

Despite the disadvantages listed, the single-beam laser-scanning confocal remains a versatile instrument with several significant advantages over alternative microscope options for collecting images from fluorescently-labelled cells and tissues. It is usually the optical sectioning microscope of choice and is ubiquitous in laboratories and research institutions worldwide. The majority of emitted photons are not detected: so a well-stained specimen, giving a low-noise signal, is required. The pixel dwell time cannot be too small (to increase acquisition speed) otherwise the light flux obtained from small volume of fluorophore contained within the focus of the scanned beam (about a cubic micron) is too small and affected by random fluctuations from shot noise. Image acquisition speeds can be increased using spinning-disc and line-scanning confocal techniques, but there will always be some loss of image resolution with these designs.

5.3.4   Line-Scanners and Array-Scanning Confocal Microscopes

The necessity to scan the illumination pointwise means confocal optical sectioning is not as fast as widefield fluorescence microscopy, but various stratagems are employed to increase image acquisition speed, by resonance scanning, programmable array designs or by line-scanning.

Line-scanning and programmable array microscopes use either slit apertures and by so doing sacrifice some confocality, or else use spatial grids or resonant scanning devices. In a single-beam point-scanning confocal, the x-galvanometer mirror works faster and is exposed to greater mechanical stress than the y-galvo because it must raster the point of the laser beam across the entire line scan, whereas the y-galvo merely drops the beam to the next line. If the mirror of the x-galvo is replaced by a slit, each line can be scanned at once. Furthermore a linear array CCD detector is used, with consequently greater quantum efficiency. Another arrangement is to sweep the illumination across the sample with a single galvanometer mirror, and the emitted signal is descanned using the reverse side of the same mirror. In another configuration the galvanometer mirror is replaced by an acousto-optical device which has no moving parts. The emitted signal cannot be descanned, as in the single-beam laser-scanning confocal, so is detected with a slit rather than a pinhole. With all these designs, the price paid for increased acquisition speed and frame-rate for live-imaging is partial loss of confocality in one direction.

An alternative method of increasing temporal resolution is to retain the single-beam point-scanning design and to increase the rate at which the galvanometer mirrors are driven. Unlike the spinning disc, swept-field or line-scanning confocals, resonant scanning single-beam point-scanning confocal microscopes are able to alter magnification without changing objectives by retaining the versatile confocal zoom functionality. The repetitive, shorter, exposures usually lead to brighter images, albeit with slightly worse signal-to-noise ratio but with much less bleaching per scan. The speed of the resonant scanner is fixed, usually at 8000 Hz, and it can be difficult to define regions of interest, except along horizontal lines.

5.4 Step by Step Protocol

5.4.1 Acquiring 2-D Sections and 3-D Z-Stack Images

This section gives guidance on how to view a specimen with a confocal microscope, set it up to collect in-focus optical sections and to acquire z-stacks. Like driving a car, once the basic principles are understood and have been learnt, it is possible to use different makes of confocal microscope and to obtain images of sufficient quality for publication. First and foremost, plan your experiment in advance. Don't blindly follow a previous protocol (although this can help prevent re-inventing the wheel) but plan your own experiment and consider how the images collected will support your working hypothesis, because microscope images are not merely pretty pictures but rather photon intensity datasets which, if they have been properly acquired in the first place, have scientific merit.

Before imaging on the confocal, check the cells/tissue and fluorescent signal on a widefield microscope, so that you are familiar with the specimen, the density of cells or features in the tissue section, and will therefore spend a minimum of time locating the desired field of view and plane of focus in the confocal microscope. You will, in any case, first use the confocal in non-laser widefield mode to locate and focus the specimen before switching to laser-illuminated confocal mode. This is done using an LED, metal-halide, short-arc mercury or xenon light source. This may be obvious, but is worth stating nevertheless: you *absolutely cannot* use laser light to view the specimen by eye down the microscope. Most lasers used for microscopy are rated class 3B or class 4. You must be aware of the dangers posed by laser illumination, and of your legal responsibilities using laser-illuminated confocal microscopes. If in doubt, speak to the laser protection supervisor in your workplace (for further details on this particular topic, see Chap. 18 in [24]).

Think about what spatial resolution is required. Users seldom consider this, instead asking facility staff about magnification. Also consider the field of view required—a related issue. There is little point in using a high-NA immersion objective with small working distance if all you require is a dry objective of lower NA with a consequent longer working distance, for it is easier and quicker to use than a high-NA objective with a very small field of view. Try to image only the volume you require. Don't scan large areas or depths, since this is time-consuming and bleaches tissue. Equally, image only for the minimum time required to collect the dataset needed. Use the lowest laser power possible.

For most tasks, plan-apochromat objectives are used, for they possess high numerical apertures [51]. The numerical aperture determines not only the resolving power but also the light gathering capacity of the microscope. When the NA is doubled, the light flux gathered is quadrupled—crucial for capturing fluorescent signal. Also as the NA increases, so also does the irradiance of the laser which means that the specimen will bleach more rapidly with a high-NA objective, particularly if a scan zoom is also applied when acquiring an image. Figure 5.9, reproduced from [52], compares three different types of objective. Some beginners assume—without checking with core facility staff first—that they can use a confocal to image cells in a multi-well plate. This is not the case unless a long working distance objective is used, in which case an inverted widefield fluorescence microscope is generally a better choice. Trying to image through a multi-well plate with a short working distance objective will mean that you won't be able to focus on the specimen. Figure 5.9a shows that much useful confocal optical sectioning microscopy can be done with a reasonably low magnification objective with a high NA. In this case, the advantage lies in the good working distance and larger field of view than using the highest NA objective available. Therefore, select the objective you require for the job in hand; don't waste photons.

Although manufacturers design objectives such that spherical aberration is minimised, it is possible for these corrections to be upset by the microscopist, and thus reintroduce spherical aberration into the image. Spherical aberration manifests itself as a loss of signal intensity and unsharp images [53]. Using the wrong thickness of coverslip, or not allowing for the thickness of mounting medium if the specimen is mounted on the slide is one common cause. The other is refractive index mismatch [10] from preparing the specimen, and which may be unavoidable. Use a water-immersion objective for aqueous samples and if possible a multi-immersion objective for specimens contained in a glycerol-based mountant. Opening the confocal pinhole may help, as does also using a lower NA objective.

Most samples prepared for confocal microscopy will be sandwiched between a glass slide and coverslip. It is important to use the correct thickness of coverslip (No. 1.5H—high tolerance 0.17 mm) and also to take into account the thickness of any mounting medium if the sample is attached to the slide rather than the undersurface of the coverslip. Where living samples in culture medium are prepared for imaging, use petri-dishes with coverslips on the base (e.g. Ibidi, MatTek, Willco

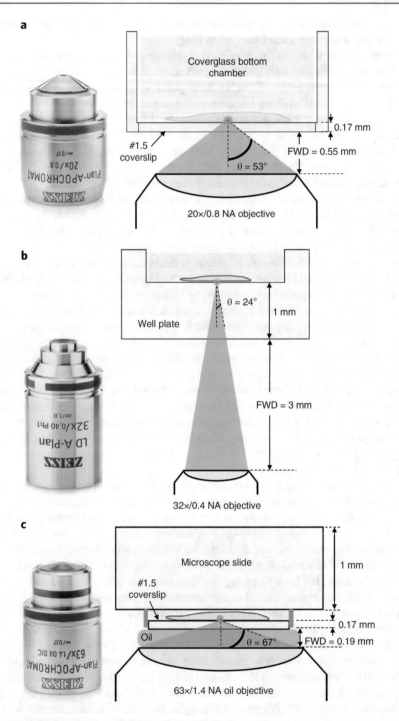

Fig. 5.9 Different objectives for confocal microscopy. For most work use the objective with the highest NA giving the shallowest depth of field for good optical sectioning. The 20×/NA 0.8 objective is good for low- to mid-magnification work (use a zoom value <1 to maintain light-

Wells). Proper fixation of excised tissue or dead cells is key; for other considerations regarding specimen preparation, see [52]).

Brief Generic Operating Protocol

Every point-scanning confocal microscope is operated in broadly the same fashion, differing only in minor points imposed by software design. For your particular confocal microscope, consult the manufacturer's specific operating instructions. Further details are given in Sanderson [24].

1. It is good practice—and essential for any form of quantitative analysis of fluorescence—to switch on the lasers at least 30 min beforehand in order for them to warm up sufficiently so as to put out a consistent illumination flux.
2. Also check that the stage insert, holding the slide or glass-bottomed petri dish, is inserted correctly and the specimen is level. Trying to image with a loose stage insert is self-defeating.
3. Image the brightest sample first—a positive control is ideal—to set a baseline for the other slides in the experiment and to ensure the gain settings chosen are sufficient to image all specimens at the same instrument settings.
4. From a Smart Setup or Experiment Configuration dialogue box, select the required fluorophores, appropriate laser lines and filter combinations. Activate the acousto-optical tuneable filter (AOTF) and select the correct primary dichromatic mirror acousto-optical beam-splitter (AOBS) for directing the emitted signal to the PMT detectors.
5. Selecting a configuration which assigns one PMT to collect each fluorophore independently in its own channel, sequentially from the longest to the shortest wavelength, will prevent bleed-through.
6. Adjust the pinhole value in each channel or track to give the optical section required. If one fluorophore is much weaker in intensity, or photo-bleaches more than another, try where possible to assign the larger pinhole to the PMT collecting this weaker or more photo-labile fluorophore, and adjust the others accordingly.
7. Check that the correct raster scan is selected for optimum resolution. There may be a radio button in the software for this to set the Nyquist sampling limit. You can generally manage with under-sampling in the x-y direction, to allow priority

◄───

Fig. 5.9 (continued) gathering capacity and resolving power) whereas the 63×/NA 1.4 plan-apo is the workhorse of most confocal work. The long-working distance objective (32×/NA 0.4) is ideal for tissue culture, but with a low NA is not optimised for optical sectioning. The dry 20× objective still has a reasonable working distance, but has only approximately half the resolving power and a quarter of the light-gathering capacity of the oil-immersion of the 63× objective. To achieve this ultimate resolving power and high sensitivity, the working distance of the 63× objective is very small, restricting observation to very near the underside of the coverslip. As such it is good for collecting high resolution images of adherent cells. Reproduced with permission from Springer Nature AG

for more important acquisition parameters which allow for better SNR and less photo-bleaching.

8. Select each PMT in turn, switching any other PMT off. Using the continual 'Fast Scan' or 'Live', scan the sample continuously. Work quickly (to avoid unnecessary bleaching) but efficiently, to first adjust the black level via the 'Offset' control on the PMT, and then the signal level via the 'Gain' control. The key points are (1) not to bleach the sample and (2) not to over-saturate the signal, otherwise details in the specimen are not recorded in the image. The photomultiplier will only record tones of grey, not colour. The colour applied to the display is a false pseudo-colour to help us recognise each fluorophore.

9. With all channels set with correct PMT offset and gain levels, a single image can be collected. Normally frame-by-frame collection is used, but if you are collecting moving objects collect line-by-line to minimise any blur in the captured image due to movement of the specimen.

10. Optimise the laser intensity, often this is left at a fixed value. At fluorophore saturation, the signal intensity no longer increases in proportion to the laser power: avoid this happening.

11. Image quality can be improved by either collecting a series of scans (averaging) to improve the signal-to-noise ratio or by reducing the scan speed so that the laser dwells on each sampling point for longer, allowing more emitted signal to be collected, but be aware (Fig. 5.10) of the potential for photo-bleaching.

12. The SNR may also be improved by opening the pinhole, increasing the laser power and increasing the scan zoom, but these stratagems will also increase bleaching.

13. To collect a z-stack, open the appropriate dialogue box, then manually focus through the sample and mark the first and last points in the stack. In most software, it is possible to perform a rapid x-z scan through the sample to help set these top and bottom limits. Take care, when setting these limits, to avoid collecting sections displaying no data and concomitantly bleaching the sample. Consider whether time may be saved by applying a region of interest to the z-stack (the ROI control can also be used to advantage when collecting single 2-D optical sections).

14. Set the optical section thickness to satisfy the Nyquist sampling criterion. There will be a radio button in software to do this to the minimum value of $2\times$ Nyquist. If required, manually select a thinner optical section to over-sample to some degree. This depends upon the sample.

15. When collecting a very large z-stack and imaging deep into tissue, it may be necessary to either alter the laser power, gain or offset values to maintain signal intensity when focusing through tissue that scatters both the illumination and the emitted signal.

16. The z-stack dataset contains a series of in-focus optical sections. This is normally viewed as an orthogonal presentation. The 3-D dataset can be rotated and saved as a movie, or it can be 'collapsed' into a 2-D dataset for publication (Fig. 5.11). There are several algorithms that will do this, but the most common is to render the image as a maximum intensity projection (sometimes referred to

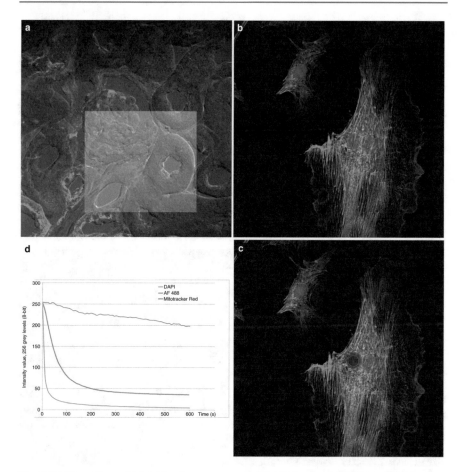

Fig. 5.10 The effects of bleaching from over-exposure to light. Panel (**a**) shows that bleaching only destroys the fluorophore, not the tissue (photo-toxicity to living cells is a separate issue). Here a square raster has bleached the fluorescent marker leaving the kidney tissue intact. In panels (**b**) before, and (**c**) after, the rates of bleaching of different fluorophores is seen, following irradiation of the nucleus and cytoplasm in the circular region of interest. The results are seen in the graph (**d**). The nuclear DAPI marker is much more resistant to bleaching than either the Alexafluor 488 labelling the tubulin, or the Mitotracker Red labelling the mitochondria. Image copyright, author: J Sanderson. Graph kindly drawn by Derek Storey

as an MIP). The brightest pixel is used for each location, regardless of which focal plane in the z-stack it originated from. An MIP is not suitable for colocalisation studies, because of loss of spatial information along the z-axis.

The confocal microscope can be used to collect a transmission image using phase-contrast or Differential Interference Contrast (DIC) in non-confocal mode with laser illumination, using a dedicated transmission PMT detector. When setting

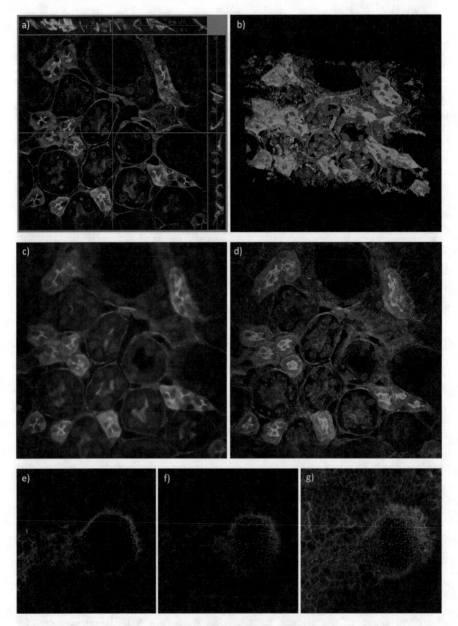

Fig. 5.11 Panel (**a**) shows an orthogonal view of a 3D z-stack of a 16 μm triple-labelled mouse kidney cryosection (ThermoFisher, Slide #3 F24630) which is useful for training. Panel (**b**) shows a 3-D rendering of the z-stack. Panel (**c**) shows the original blurred image of the entire tissue section seen in non-confocal 'widefield' mode whilst panel (**d**) shows the 2-D maximum intensity projection of the z-stack (compare this figure to Fig. 5.1). Panels (**e, f**) show two separate non-contiguous optical sections from the z-stack of mouse embryonic node, and (**g**) the maximum intensity projection. Image copyright, author: J Sanderson

up this detector, it may be necessary manually to switch the illumination path on the microscope. Failure to check and do this can catch even experienced users unawares.

What To Do If You Cannot See a Confocal Image

1. Stop scanning—so that you don't bleach the sample (do not panic and increase the laser illumination intensity in order to try and search for the specimen).
2. Check that the sample is actually in focus by viewing through the microscope. Check that the cells or tissues have actually been stained, and check the quality of the staining, that the specimen is not bleached. The sample slide should not be placed incorrectly 'upside-down' on the stage.
3. Check the laser interlocks are switched to scanning from observation mode, and that the laser bean is exiting from the objective.
4. Check that the laser beam is actually exiting from the objective (maybe an incorrect filter or dichroic beam-splitter has been selected. This is a separate issue from the laser interlock).
5. Check that the channel has not been switched off in the display software (easily overlooked).
6. Check that sufficient PMT gain has been applied to form an image on the monitor.
7. Open the pinhole to allow sufficient signal through to the detector.
   (also refer to Chap. 18 in Sanderson [24] and Box 3 in [52])

5.4.2 Spectral Unmixing

If you have only one target to investigate, a single fluorophore marker is all that is required. The goal with multiple labelling is to choose each fluorescent marker with sufficient separation between the emission spectra so that the signal arising from each fluorophore is collected into its own individual detection channel. Bleed-through occurs when emission from a short wavelength fluorophore overlaps the excitation filter or laser line illuminating subsequent longer-wavelength fluorophores. The re-excited fluorescence is collected erroneously by the detector assigned to the second longer wavelength fluorophore, rather than the first. Fluorophore emission profiles increase sharply in intensity towards the peak value, but exhibit a longer profile towards the red end of the spectrum that tails off much less sharply, increasing the likelihood of bleed-through. Ideally, fluorophores would exhibit long Stokes shifts and also have both narrow excitation and emission spectra that did not overlap with other fluorophores on the spectrum, but this is rarely the case. With such a wide choice of fluorescent markers, it ought to be possible to select those with distinct spectra, and so avoid bleed-through. Sometimes it is not that simple. Your lab may possess a limited choice of conjugated fluorophores, you may be using fluorescent proteins whose spectra lie close together or overlap significantly, or else the available laser lines on the microscope restrict your choice of fluorophores. Effective separation of signals that would otherwise bleed-through is often a compromise between using narrow excitation and emission filters that discard signal and collecting sufficient signal from samples that are not intensely stained.

To be able to implement linear unmixing, a 'lambda stack' (x, y, λ) is collected using a spectral dispersion element, that is usually a prism or grating [54] or a filterset. Although used mainly on confocal microscopes, with their ability to switch rapidly between laser excitation wavelengths, a lambda scan can also be collected on a widefield microscope [55]. A region of interest (ROI) emitting the fluorophore signal of interest is scanned at various intervals along the wavelength axis to create a spectrum. Each fluorophore has a unique spectral signature that is independent of any overlap with other fluorophores. For example, a spectral stack may comprise a dataset across 10 nm bandwidths spanning the visible spectrum from 380 to 720 nm, represented as 32 separate measurements. This unique spectral profile enables reliable discrimination of one fluorophore from another (including separation of autofluorescence as an independent 'hue'). A reference spectrum is taken; for accuracy, this is best taken from specimens containing only the signal of interest, under the same conditions as the imaging experiment. If no separately-stained sample is available, a reference spectrum can be taken from the multiply-labelled experimental sample, provided an ROI is selected containing only the pure signal that will provide the reference standard. The software unmixing algorithm assigns the spatial distribution of each fluorophore in the image, which can then be false-coloured for extra clarity.

Space precludes giving a protocol for spectral unmixing here, but these can be found in [56, 57]. ThermoFisher sell a custom-made test sample for checking spectral systems, the Focal Check slide #2 (F36913) can be used. Each slide has bead mixtures of dye pairs as well as individually-stained control bead populations. The theory and practical implementation of spectral unmixing is described in [58]. For practical guidance on how to calibrate confocal spectrophotometers, see [59] and also [60].

5.4.3   Quantitative Confocal Microscopy and Quality Control

The confocal microscope is used primarily for optical sectioning, but can also be used to collect data quantitatively, for digital images are datasets of photon intensity [61]. In order to use a confocal microscope quantitatively, it must first be aligned and calibrated [62]. It is therefore worth carrying out routine quality control performance checks on the instrument. Those recommended are:

- Laser(s) power
- Pinhole(s) and collimator alignments
- Field of view intensity profile
- PMT gain sensitivity check with standard reflective mirror
- Objective alignment and chromatic correction—essential for colocalisation studies
- Check and record any error logs

In recent years some excellent guides to quality control protocols for assessing confocal microscope performance have been published [63, 64]. Confocal check [65] with the updated app Intensity check [66] are excellent starting points. The former paper contains a comprehensive bibliography of earlier work in this field, particularly that of Bob Zucker. For testing field illumination, see [67] and, since confocal microscopes are inherently noisy, also see [68] for checking the performance of the PMT detectors and SNR of the microscope with NoiSee. If using a CCD detector on a slit scanner or spinning disc microscope, Lambert and Waters [69] provide guidance on assessing camera performance for quantitative microscopy, whilst in the same volume [70] introduce the subject of quantitative microscopy with a good section on control samples. This early work has recently been expanded and updated by a group of very experienced facility managers in an excellent paper [52]. It is easy to make qualitative comparisons from confocal datasets. Quantitative microscopy is much more challenging, making it all the more imperative to read the references cited here.

5.4.4 Towards Super-Resolution: Image-Scanning and Pixel Re-assignment

With improved photomultiplier tube detectors, a great deal of effort has gone into improving the SNR of confocal microscopes. Improving SNR is usually achieved at the expense of image acquisition speed and resolving power—the so-called iron triangle [24, 71].

The resolving power of a confocal microscope is marginally improved over that of a widefield microscope because the overall image PSF is the product of both illumination and detection PSFs. The image in a confocal is blurred by addition of information from neighbouring points, therefore resolving power is improved by closing down the pinhole—ultimately to zero—but at the expense of rejecting 95% of signal intensity at a setting of 0.2 AU. With this smaller pinhole, besides resolving power the contrast and SNR are also increased.

Many years ago Colin Sheppard [72] realised collecting this rejected light could be used as a form of structural illumination to improve resolving power. This approach works because a displaced pinhole produces an image about equal to the resolving power of a pinhole aligned to the optical axis, but of lower intensity. Since the PSF is the product of illumination and detection, this is true also of the PSF arising from a displaced pinhole.

A high-sensitivity multi-channel GaAsP detector is able to detect these offset signals. The problem is that, with respect to the detector axis, each individual image is recorded from a slightly different viewing angle and suffers from parallax. Combining these without further processing would lead to a blurred final image, so each individual image must first be shifted onto a common axis. Software then sums the signal, but first shifted back on-axis to avoid parallax error. Zeiss market this equipment as the Airyscan (Fig. 5.12; [73]) a detector like an insect's compound eye consisting of 32 elements, each of them equivalent to a point detector sampled

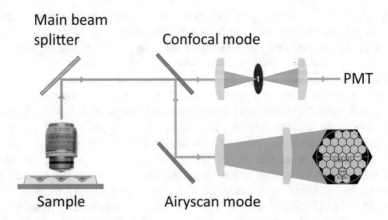

Fig. 5.12 Ray path of the Airyscan unit fitted into a Zeiss confocal microscope. See the text for details. Image copyright, author: J Sanderson

with a 0.2-AU pinhole. These elements are arranged in a circular geometry, such that the total detector area is equivalent to a 1.25-AU pinhole setting, collecting 50% more signal than at 1 AU. The spatial resolution of the final reconstructed image is defined by the sampling of the central pixel, comparable to images acquired with a conventional confocal at a 0.2 AU pinhole setting, while the total sensitivity of the system is equivalent to 1.25 AU. The Airyscan is a PMT detector, rather than a camera, to enable fast read-out. It scans the entire Airy disc, hence the name.

Simple pixel re-assignment is not the only method; slightly different image-scanning approaches may be taken, such as using digital micro-mirrors, re-scanning the image and shifting the phase of the Fourier image rather than shifting pixels directly in order to improve raster speed and axial resolution. Expanding the beam containing the emitted signal before descanning it [74] is the approach adopted by Olympus and Leica, since the Airyscan is patented by Zeiss. Deconvolving the images from each mini-detector ensures a 1.7× resolving power improvement is achieved, to give a lateral resolving power of 140 nm and an axial resolving power of 400 nm. Since the information still comes from a diffraction-limited pattern, it is ultimately limited, as with other structured illumination methods, to a two-fold increase in resolving power.

Take-Home Message
- Thick fluorescent samples require optical sectioning to remove blurring
- Single-beam laser-scanning confocal microscopes are the world's workhorses for acquiring 3-D image datasets, but they are not the only option.
- Confocal images are acquired by point illumination and point detection.
- Single beam-scanning confocals are noisy and relatively slow.
- Confocal images should be deconvolved.

References

1. Jacoby-Morris K, Patterson GH. Choosing fluorescent probes and labeling systems, Chapter 2. In: Brzostowski J, Sohn H, editors. Confocal microscopy: methods and protocols, methods in molecular biology, vol. 2304. Springer Nature US 2021; 2021.
2. Jensen EC. Use of fluorescent probes: their effect on cell biology and limitations. Anat Rec (Hoboken). 2012;295(12):2031–6.
3. Sahoo H. Fluorescent labelling techniques in biomolecules: a flashback. RSC Adv. 2012;2: 7017–29.
4. Toseland CP. Fluorescent labeling and modification of proteins. J Chem Biol. 2013;6(3):85–95.
5. Stelzer EHK. Contrast, resolution, pixelation, dynamic range, and signal-to-noise ratio. J Microsc. 1998;189(1):15–24.
6. Lichtman JW, Conchello J-A. Fluorescence microscopy. Nat Methods. 2005;2(12):910–9.
7. Maric K, Wiesner B, Lorenz D, Klussmann E, Betz T, Rosenthal W. Cell volume kinetics of adherent epithelial cells measured by laser scanning reflection microscopy: determination of water permeability changes of renal principal cells. Biophys J. 2001;80(4):1783–90.
8. Model MA. Cell volume measurements by optical transmission microscopy. Curr Protoc Cytom. 2015;72:12.39.1–9. https://doi.org/10.1002/0471142956.cy1239s72.
9. Sheppard CJR, Gan X, Gu M, Roy M. Signal-to-noise ratio in confocal microscopes, Chapter 22. In: Pawley JB, editor. Handbook of confocal microscopy. 3rd ed. New York: Springer; 2006.
10. Hell S, Reiner G, Cremer C, Stelzer EHK. Aberrations in confocal fluorescence microscopy induced by mismatches in refractive index. J Microsc. 1993;169(3):391–405.
11. Jacques SL. Optical properties of biological tissues: a review. Phys Med Biol. 2013;58(11): R37–61.
12. Ntziachristos V. Going deeper than microscopy: the optical imaging frontier in biology. Nat Methods. 2010;7(8):603–14. https://doi.org/10.1038/nmeth.1483.
13. Centonze VE, White JG. Multiphoton excitation provides optical sections from deeper within scattering specimens than confocal imaging. Biophys J. 1998;75(4):2015–24.
14. Diaspro A, et al. Fluorescence microscopy, Chapter 21. In: Hawkes PW, Spence JCH, editors. Springer handbook of microscopy. Cham: Springer; 2019. p. 1039–88. https://doi.org/10.1007/978-3-030-00069-1_21.
15. Minsky M. Memoir on inventing the confocal scanning microscope. Scanning. 1988;10:128–38.
16. Harvath L. Overview of fluorescence analysis with the confocal microscope, Chapter 20. In: Javois LC, editor. Methods in molecular biology, 115. Immunocytochemical methods and protocols. 2nd ed. Totowa, NJ: Humana Press; 1999. p. 149–58. https://doi.org/10.1385/1592592139.
17. Wilson T. Twenty-five years of confocal microscopy. Microsc Anal. 2012;2102:73–77. https://analyticalscience.wiley.com/do/10.1002/micro.265/full/iad081e22e52f50e3f3e52d62a56b2cd7.pdf
18. Amos WB, White JG. How the confocal laser scanning microscope entered biological research. Biol Cell. 2003;95(6):335–42. https://doi.org/10.1016/s0248-4900(03)00078-9.
19. White JG, Amos WB, Forham M. An evaluation of confocal versus conventional imaging of biological structures by fluorescence light microscopy. J Cell Biol. 1987;105(1):41–8.
20. White JG, Amos WB. Confocal microscopy comes of age. Nature. 1987;328(6126):183–4.
21. Rusk N. Nature milestones, light microscopy: milestone 4 the fluorescence microscope, 2009. https://doi.org/10.1038/ncb1941. Also see commentary Piston DW (2009) The impact of technology on light microscopy. Nat Cell Biol 11:S23–24. https://doi.org/10.1038/ncb1936
22. Brakenhoff GJ, van der Voort HTM, van Spronsen EA, Linnemans WAM, Nanninga N. Three-dimensional chromatin distribution in neuroblastoma nuclei shown by confocal scanning laser microscopy. Nature. 1985;317(6039):748–9. https://doi.org/10.1038/317748a0.

23. Ishikawa-Ankerhold HC, Ankerhold R, Drummen GP. Advanced fluorescence microscopy techniques-FRAP, FLIP, FLAP, FRET and FLIM. Molecules. 2012;17(4):4047–132. https://doi.org/10.3390/molecules17044047.
24. Sanderson J. Understanding light microscopy. Chichester: Wiley; 2019.
25. Demmerle J, Innocent C, North AJ, Ball G, Müller M, Miron E, Matsuda A, Dobbie IM, Markaki Y, Schermelleh L. Strategic and practical guidelines for successful structured illumination microscopy. Nat Protoc. 2017;12(5):988–1010.
26. Jost A, Heintzmann R. Superresolution multidimensional imaging with structured illumination microscopy. Annu Rev Mater. 2013;43:261–82. https://doi.org/10.1146/annurev-matsci-071312-121648.
27. Sheppard CJR. Structured illumination microscopy and image scanning microscopy: a review and comparison of imaging properties. Philos Trans A Math Phys Eng Sci. 2021;379 (2199):20200154. https://doi.org/10.1098/rsta.2020.0154.
28. Neil MA, Wilson T, Juškaitis R. A light efficient optically sectioning microscope. J Microsc. 1998;189(2):114–7.
29. Schaeffer LH, Schuster D, Schaffer J. Structured illumination microscopy: artefact analysis and reduction utilizing a parameter optimization approach. J Microsc. 2004;216(2):165–74.
30. Johnson K, Hagan GM. Artifact-free whole-slide imaging with structured illumination microscopy and Bayesian image reconstruction. GigaScience. 2020;9(4):giaa035. https://doi.org/10.1093/gigascience/giaa035.
31. Karras C, Smedh M, Förster R, Deschout H, Fernandez-Rodriguez J, Heintzmann R. Successful optimization of reconstruction parameters in structured illumination microscopy – a practical guide. Opt Commun. 2018;436(2019):69–75.
32. Shaw P. Deconvolution in 3-D optical microscopy. Histochem J. 1994;26(9):687–94. https://doi.org/10.1007/BF00158201.
33. Biggs DSC. 3D deconvolution microscopy. Curr Protoc Cytom. 2010;12:Unit 12.19.1-20. https://doi.org/10.1002/0471142956.cy1219s52.
34. Ellenberger SE, Young IT. Microscope image acquisition, Chapter 1. In: Baldock R, Graham J, editors. Image processing and analysis, the practical approach series. Oxford: Oxford University Press; 1999. p. 1–36.
35. Hiraoka Y, Sedat JW, Agard DA. Determination of three-dimensional imaging properties of a light microscope system: partial confocal behavior in epifluorescence microscopy. Biophys J. 1990;57(2):325–33.
36. Sage D, Donati L, Ferréol Soulez F, Fortun D, Schmit G, Seitz A, Romain Guiet R, Vonesch C, Unser M. DeconvolutionLab2: an open-source software for deconvolution microscopy. Methods. 2017;115:28–41.
37. Cannell MB, et al. Image enhancement by deconvolution. In: Pawley J, editor. Handbook of biological confocal microscopy. 3rd ed. New York: Springer; 2006. p. 488–500.
38. Cole RW, Jinadasa T, Brown CM. Measuring and interpreting point spread functions to determine confocal microscope resolution and ensure quality control. Nat Protocols. 2011;6 (12):1929–41. https://doi.org/10.1038/nprot.2011.407.
39. Weigert M, Schmidt U, Boothe T, Müller A, Dibrov A, Jain A, Wilhelm B, Schmidt D, Broaddus C, Culley S, Rocha-Martins M, Segovia-Miranda F, Norden C, Henriques R, Zerial M, Solimena M, Rink J, Tomancak P, Royer L, Jug F, Myers EW. Content-aware image restoration: pushing the limits of fluorescence microscopy. Nat Methods. 2018;15(12):1090–7. https://doi.org/10.1038/s41592-018-0216-7.
40. Pawley JB. Limitations on optical sectioning in live-cell confocal microscopy. Scanning. 2002;24(5):241–6. https://doi.org/10.1002/sca.4950240504.
41. Murray JM, Appleton PL, Swedlow JR, Waters JC. Evaluating performance in three-dimensional fluorescence microscopy. J Microsc. 2007;228(3):390–405.
42. Pawley JB. Pawley, JB (2006) Points, pixels and gray levels: digitizing image data, Chapter 4. In: Pawley J, editor. Handbook of biological confocal microscopy. 3rd ed. New York: Springer; 2006. p. 59–79.

43. Granzow N. Supercontinuum white light lasers: a review on technology and applications. Proc SPIE. 2019;2019:1114408. https://doi.org/10.1117/12.2533094.
44. Sandison DR, Williams RM, Wells KS, Strickler J, Webb WW. Quantitative fluorescence confocal laser scanning microscopy, Chapter 3. In: Pawley J, editor. Handbook of biological confocal microscopy. 2nd ed. New York: Plenum; 1995. p. 39–53.
45. Oppenheim AV, Willsky AS, Young IT. Sampling, Chapter 8. In: Signals and systems. Hoboken, NJ: Prentice Hall; 1983. p. 519.
46. Webb RH, Dorey CK. The Pixelated Image, Chapter 4. In: Pawley J, editor. Handbook of biological confocal microscopy. 2nd ed. New York: Plenum; 1995. p. 41–51.
47. Cox G, Sheppard CJR. Practical limits of resolution in confocal and non-linear microscopy. Microsc Res Tech. 2004;63(1):18–22.
48. Wilhelm S, Gröbler B, Gluch M, Heinz H. Confocal laser scanning microscopy: principles. Jena: Carl Zeiss Microscopy GmbH; 2011.
49. Gong W, Si K, Sheppard CJR. Improved spatial resolution in fluorescence focal modulation microscopy. Opt Lett. 2009;34(22):3508–10.
50. Gong W, Si K, Sheppard CJR. Divided-aperture technique for fluorescence confocal microscopy through scattering media. Appl Opt. 2010;49(4):752–7.
51. Sanderson J. Fundamentals of microscopy. Curr Protoc Mouse Biol. 2020;10:e76. https://doi.org/10.1002/cpmo.76.
52. Jonkman J, Brown CM, Wright GD, Anderson KI, North AJ. Tutorial: guidance for quantitative confocal microscopy. Nat Protoc. 2020;15(5):1585–611. https://doi.org/10.1038/s41596-020-0313-9.
53. Diel EE, Lichtman JW, Richardson DS. Tutorial: avoiding and correcting sample-induced spherical aberration artifacts in 3D fluorescence microscopy. Nat Protoc. 2020;15(9):2773–84. https://doi.org/10.1038/s41596-020-0360-2.
54. Hiraoka Y, Shimi T, Haraguchi T. Multispectral imaging fluorescence microscopy for living cells. Cell Struct Funct. 2002;27(5):367–74. https://doi.org/10.1247/csf.27.367.
55. Garini Y, Young IT, McNamara G. Spectral Imaging: principles and applications. Cytometry A. 2006;69(8):735–47. https://doi.org/10.1002/cyto.a.20311.
56. Kraus B, Ziegler M, Wolff H. Linear fluorescence unmixing in cell biological research. Mod Res Educ Top Microsc. 2007;2:863–72.
57. Lerner JM, Gat N, Wachman E. Approaches to spectral imaging hardware. Curr Protoc Cytom. 2010;12:Unit 12.20. https://doi.org/10.1002/0471142956.cy1220s53.
58. Zimmermann T, Marrison J, Hogg K, O'Toole P. Clearing up the signal: spectral imaging and linear unmixing in fluorescence microscopy. Methods Mol Biol. 2014;1075:129–48. https://doi.org/10.1007/978-1-60761-847-8_5.
59. Zucker RM, Rigby P, Clements I, Salmon W, Chua M. Reliability of confocal microscopy spectral imaging systems: use of multispectral beads. Cytometry A. 2007;71(3):174–89. https://doi.org/10.1002/cyto.a.20371.
60. Cole RW, Thibault M, Bayles CJ, Eason B, Girard A-M, Jinadasa T, Opansky C, Schulz K, Brown CM. International test results for objective lens quality, resolution, spectral accuracy and spectral separation for confocal laser scanning microscopes. Microsc Microanal. 2013;19(6):1653–68. https://doi.org/10.1017/S1431927613013470.
61. Cromey DW. Digital images are data: and should be treated as such. Methods Mol Biol. 2013;931:1–27. https://doi.org/10.1007/978-1-62,703-056-4_1.
62. Murray JM. Practical aspects of quantitative confocal microscopy, Chapter 18. In: Sluder G, Wolf DE, editors. Digital microscopy. Methods in cell biology, vol. 114. 4th ed. New York: Elsevier; 2013. p. 427–40.
63. Montero Llopis P, Senft RA, Ross-Elliott TJ, Stephansky R, Keeley DP, Koshar P, Marqués G, Gao YS, Carlson BR, Pengo T, Sanders MA, Cameron LA, Itano MS. Best practices and tools for reporting reproducible fluorescence microscopy methods. Nat Methods. 2021;18(12):1463–76. https://doi.org/10.1038/s41592-021-01156-w.

64. Sasaki A. Recent advances in the standardization of fluorescence microscopy for quantitative image analysis. Biophys Rev. 2022;14(1):33–9. https://doi.org/10.1007/s12551-021-00871-0.

65. Hng K, Dormann D. Confocal check: a software tool for the automated monitoring of confocal microscope performance. PLoS One. 2013;8(11):e79879. https://doi.org/10.1371/journal.pone.0079879.

66. Dormann D. Intensity check – the light measuring app for microscope performance checks and consistent fluorescence imaging. PLoS One. 2019;14(3):e0214659. https://doi.org/10.1371/journal.pone.0214659.

67. Brown CM, Reilly A, Cole RW. A quantitative measure of field illumination. J Biomol Tech. 2015;26(2):37–44. https://doi.org/10.7171/jbt.15-2602-001.

68. Ferrand A, Schleicher KD, Ehrenfeuchter N, Heusermann W, Biehlmaier. Using the NoiSee workflow to measure signal-to-noise ratios of confocal microscopes. Sci Rep. 2019;9:1165. https://doi.org/10.1038/s41598-018-37,781-3.

69. Lambert TJ, Waters JC. Assessing camera performance for quantitative microscopy. Methods Cell Biol. 2014;123:35–53. https://doi.org/10.1016/B978-0-12-420,138-5.00003-3.

70. Jonkman J, Brown CM, Cole RW. Quantitative confocal microscopy: beyond a pretty picture, Chapter 7. In: Waters JC, Wittmann, editors. Quantitative imaging in cell biology. Methods in cell biology, vol. 123. New York: Elsevier; 2014. p. 113–34. https://doi.org/10.1016/B978-0-12-420138-5.00007-0.

71. Weisshart K. The basic principle of Airyscanning. Zeiss Technology Note, 2014.

72. Sheppard CJR. Super-resolution in confocal imaging. Optik. 1988;80(2):53–4.

73. Huff J, Bergter A, Berkenbeil J, Kleppe I, Engelmann R, Krzic U. The new 2D Superresolution mode for Zeiss Airyscan. Nat Methods. 2017;14:1223. https://doi.org/10.1038/nmeth.f.404.

74. Lam F, Cladière D, Guillaume C, Wassmann K, Bolte S. Super-resolution for every-body: an image processing workflow to obtain high-resolution images with a standard confocal microscope. Methods. 2017;115:17–27.

Live-Cell Imaging: A Balancing Act Between Speed, Sensitivity, and Resolution

6

Jeroen Kole, Haysam Ahmed, Nabanita Chatterjee,
Gražvydas Lukinavičius, and René Musters

Contents

"In honour of Professor Anirban Banerjee, who taught us the wonders of observing living cells..."

The original version of the chapter has been revised. A correction to this chapter can be found at
https://doi.org/10.1007/978-3-031-04477-9_14

J. Kole
Confocal.nl, Amsterdam, Noord-Holland, The Netherlands
e-mail: jeroen.kole@scientifica.uk.com

H. Ahmed
UCB Pharma, Slough, Berkshire, UK
e-mail: haysam.ahmed@ucb.com

N. Chatterjee
Nikon India Private Limited, Bangalore, Karnataka, India
e-mail: nabanita.chatterjee@nikon.com

G. Lukinavičius
Chromatin Labeling and Imaging Group, Department of Nanobiophotonics, Max Planck Institute
for Biophysical Multidisciplinary Sciences, Göttingen, Germany
e-mail: grazvydas.lukinavicius@mpinat.mpg.de

R. Musters (✉)
M4N, Amsterdam University Medical Centers - VUmc, Amsterdam, The Netherlands

---

© The Author(s), under exclusive license to Springer Nature Switzerland AG 2022,
corrected publication 2023
V. Nechyporuk-Zloy (ed.), *Principles of Light Microscopy: From Basic
to Advanced*, https://doi.org/10.1007/978-3-031-04477-9_6

139

Abbreviations

AOTF	Acousto-optical tunable filter
BF	Brightfield
CCD	Charge-coupled device
CIN	Chromosome instability
CMOS	Complementary metal-oxide-semiconductor
CON	Confocal microscopy
CRC	Colorectal cancer
CRISPR	Clustered regularly interspaced short palindromic repeats
CRISPR-HOT	CRISPR-Cas9-mediated homology-independent organoid transgenesis
CSLM	Confocal scanning laser microscopy
DAPI	4′,6-diamidino-2-fenylindool
DIC	Differential interference contrast
EM-CCD	Electron multiplying charged-coupled device
EPI	Epifluorescence microscopy
FL	Fluorescence
FOV	Field of view
FRAP	Fluorescence recovery after photobleaching
FRET	Fluorescence resonance energy transfer
GFP	Green fluorescent protein
HCA	High-content analysis
HCS	High-content screening
HEPA	High-efficiency particulate air
HUVEC	Human umbilical vein endothelial cells
IR	Infrared
LED	Light-emitting diode
NA	Numerical aperture
NIR	Near-infrared
PFS	Perfect focus system
PMT	Photomultiplier tube
PSF	Point spread function
PUM-HD	Pumilio-homology domain
PZF	Polydactyl zinc finger
RCM	Re-scanning confocal microscopy
RFP	Red fluorescent protein
ROI	Region of interest
SIM	Structured illumination microscopy
SNR	Signal-to-noise ratio

STED	Stimulated emission depletion microscopy
STORM	Stochastic optical reconstruction microscopy
TALE	Transcription activator-like effector
TIRF	Total internal reflection fluorescence
TTL	Transistor-transistor logic
UV	Ultraviolet

What You Will Learn in This Chapter

Live-cell imaging is perhaps one of the most exciting and challenging activities in the field of microscopy. It is exciting as recent developments in microscope technology have enabled scientists to visualize cellular and subcellular processes in real time down to the molecular level. With this comes the prospect of studying the mechanisms of diseases in greater detail and finding possible therapeutic solutions. Nevertheless, live-cell imaging is equally challenging because cells themselves and in fact—all cellular processes—are extremely sensitive to the very impact of using light for their visualization. The aim of this chapter is to provide a practical overview for early PhD students as well as more experienced post-docs, who will spend considerable time mastering the most important challenges and prerequisites in the very rapidly evolving field of live-cell microscopy.

6.1 Essentials in Live-Cell Imaging

Modern live-cell microscopy encompasses methods ranging from transmitted light microscopy, wide-field epifluorescence microscopy, confocal scanning laser microscopy, multiphoton- and spinning disk confocal microscopy, to super-resolution microscopy, structured illumination microscopy as well as single-molecule localization microscopy and light-sheet microscopy (Chap. 3, 5, 6–9, and 11–13). These microscopy modalities can be performed on different dedicated microscope systems, but—more often today—can also be performed on the same, high-end microscope platform, and will thus strongly rely on a similar set of basic optical and mechanical components. As mentioned earlier, live-cell imaging provides scientists with the unique ability to study cellular dynamics and function in great detail and in real time. Therefore, it is of paramount importance to ensure cell viability and to confirm that physiological and biological processes—that are under investigation—are not altered in any (significant) way. Consequently, the three main experimental challenges in acquiring live-cell imaging data are: (i) to minimize photodamage, while retaining a useful signal-to-noise ratio; (ii) to provide a stable environment for cells (or multicellular preparations), in order to be able to replicate physiological cell dynamics; and last but not least (iii) to prevent focal drift.

6.1.1   The Problem of Phototoxicity

Most cells and tissues are never exposed to light during their normal life cycle, so live-cell microscopy is always a compromise between collecting enough photons while minimizing phototoxicity. While UV light is known to cause DNA damage [1], focused infrared (IR) light can cause localized heating and thereby excite fluorescent molecules to react with molecular oxygen, to produce reactive oxygen species and free radicals [2–4]. The latter in turn may interact with surrounding sub-cellular components resulting in cellular oxidative damage. Consequently, it is imperative to minimize light exposure by reducing light intensity, decreasing exposure time (which can be achieved using more sensitive detectors), and to shorten software overhead time by hardware triggering. In addition, reactive oxygen species scavengers, such as oxyrase or oxyfluor, have also been used to prevent photobleaching during live-cell recordings [5–7]. From a physiological perspective, however, it is probably best to minimize the deleterious effects of unnecessary light exposure in the first place. More about the effect of phototoxicity and a practical approach of how to avoid it is revealed in Experimental Boxes 6.1 and 6.2, respectively.

6.1.2   The Problem of Creating a Stable Environment

6.1.2.1 Laboratory Environment Conditions
While choosing a room for performing live-cell imaging experiments, it is advisable to allow enough space around the microscope for proper ventilation and access to cleaning as well as to reduce dust levels that can damage optical and electronic parts of the microscope (e.g., by installing HEPA filters). Additionally, to minimize cellular contamination by microorganisms it is important to wipe the microscope stage area and surrounding space with 70% ethanol periodically. To minimize mechanical vibrations due to environmental factors microscopes should not be exposed to air vents, air conditioners, or housed in the same room as a refrigerator. Furthermore, microscopes should be placed on gas-filled vibration isolation tables or low-cost vibration isolation pads.

6.1.2.2 Stable Live-Cell Incubation Conditions on the Microscope
In order to successfully image cellular dynamics in living cells, it is critical to maintain cells in a physiologically healthy and stable state. Some of the key factors of the cellular environment that definitely must be controlled are temperature, pH, osmolarity (by preventing evaporation of the cellular media).

6.1.2.3 Temperature
Incubators can be open systems, stage top, or cage type. Open environmental systems are useful for short-term live-cell experiments that require patch clamp or microinjections; however, for longer time-lapses, this system is impossible to keep sterile. For dedicated live-cell imaging, custom-made, fully climate-controlled boxes

Fig. 6.1 Oko-lab cage incubator enclosure on a Ti2-E for live-cell inverted widefield system in Nikon Imaging Center at Harvard Medical School. (Source: https://d33b8x22mym97j. cloudfront.net/production/ imager/productphotos/NIC-Systems/7182/Harvard-Live-Cell2_d5bf01b6690 ca8c589d10eb51a9ecfd9.jpg). (Reproduced with permission)

Fig. 6.2 Oko-lab stage-top incubator on a Nikon Ti microscope stage. Humidified 5% CO_2 enters from the insulated tube from the left. (Source: http://nic.ucsf.edu/ blog/2014/04/okolab-incubators). (Reproduced with permission)

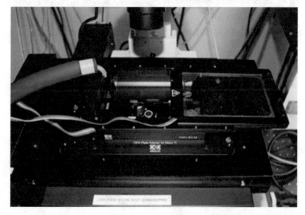

(made of Plexiglas) should enclose the entire microscope. These can also be linked to a tissue culture incubator on the microscope. Full enclosures provide superior thermal stability, but can also be difficult to work in. Fortunately, recent designs are becoming more compact (see Figs. 6.1 and 6.2).

Stage-top incubators are available in diverse designs from manufacturers such as Tokai Hit and Okolab among others. These chambers combine temperature and gas control as well as options for media perfusion and electrophysiology.

6.1.2.4 Osmolarity

Osmolarity is maintained by preventing evaporation of sample medium. Full enclosures and stage-top incubators often use pre-mixed CO_2 that is bubbled through water to humidify the environment over the sample. For cage-type incubators humidification is limited over the specimen area to prevent damage to other mechanical parts of the microscope.

6.1.3 The Problem of Focal Drift

6.1.3.1 Focus Drift

The term focus drift is often used to describe the inability of a microscope to maintain the selected focal plane over an extended period of time. This artifact occurs independently of the natural motion in living specimens and is primarily affected by changes in air and microscope temperature. Therefore, microscopes should be kept away from air conditioning and/or heating vents. In addition, heated objective collars are essential when immersion objectives are used, since the objective acts as a heat sink. In general, focus drift is more a problem when using high magnification and numerical oil immersion objectives (having a very shallow depth of focus) than it is for lower magnification objectives with wider focal depths ($10\times$ and $20\times$).

Experimental Box 6.1 Monitoring Cellular Events Under the Microscope: Timeline and Cell Health
Timeline of Cellular Events (Fig. 6.3)

Monitoring Cell Health Under the Microscope
Long time-lapses must be carefully observed for declining cell health due to photo toxicity to ensure data is collected from physiologically normal cells. Cell morphology can be easily monitored with transmitted illumination methods to assess cells that are stressed, dying, or dead. Common visual symptoms are blebbing, vacuole formation, detachment from substrate, enlarged mitochondria or broken mitochondrial network, and finally necrosis (Fig. 6.4).

6.2 Microscope Components and Key Requirements for Live-Cell Imaging

This paragraph provides a brief overview of the most essential parts and accessories of modern microscope platforms that are needed to successfully implement live-cell microscopy in laboratories.

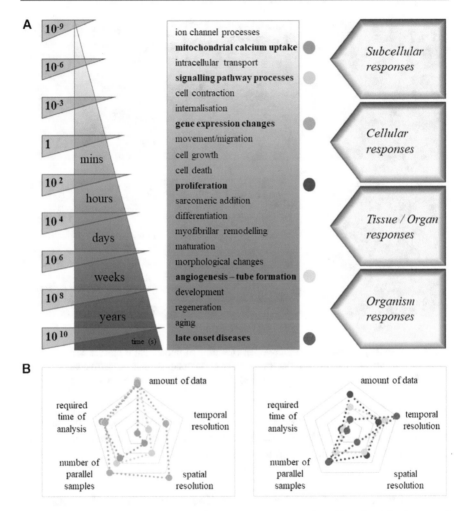

Fig. 6.3 (**a**) Estimated timeline of the main cellular processes monitored in cells, cell cultures, and tissues; (**b**) 2D spider charts of the labelled cellular processes (indicated by colored dots in panel **a**). The logarithmic scales demonstrate the values of different technical necessities, increasingly from the center to the edges in the graph. (Reproduced with permission)

6.2.1   Upright or Inverted Microscopes

The choice between upright and inverted microscopes depends on the type of specimen being imaged. When whole animals such as mice, drosophila are imaged upright stands are essential for accessing the regions of interest. However, for cellular imaging, inverted microscopes are particularly useful as it is easier to image specimens in culture medium except for patch clamping and other physiological experiments. The availability of hardware-based Z-drift control, such as Nikon's Perfect Focus System (PFS), is currently limited to inverted microscopes, which is a

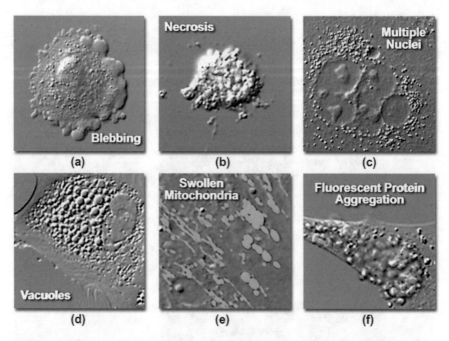

Fig. 6.4 Visual symptoms of unhealthy cells: clustering of fluorescent protein is also a common indicator of stress. (Source: https://www.microscopyu.com/applications/live-cell-imaging/maintaining-live-cells-on-the-microscope-stage). (Reproduced with permission)

strong reason why inverted systems are preferred over upright scopes. Both Inverted and upright microscope bodies are manufactured by all four main microscope manufacturers Leica, Nikon, Olympus, and Zeiss.

6.2.2 Microscope Components and Key Requirements for Live-Cell Imaging

6.2.2.1 Motorized Stage

Multipoint acquisition allows parallel data collection from multiple regions, which is especially important for long-term, live-cell imaging time-lapse experiments, monitoring multiple points of interest over time. The travel range of a typical motorized stage is 2 to 4 inches in both the X- and Y-direction. Linear encoder-equipped motorized stages use optical sensors to determine the position of the stage independent of stepper motors and demonstrate high accuracy of repeatability in the order of hundreds of nanometers. With microscopy evolving into nanoscopy, piezo nanopositioning stages are becoming essential for maintaining stability, nanometer precision, and speed. Applied Scientific Instruments, Physiks Instruments, Mad City Labs, Thorlabs and Prior among others manufacture piezo-controlled microscope stages.

6.2.2.2 Z-axis Control

In order to collect 3D image stacks in the Z-axis, motorized stepper focusing devices that drive either the entire nosepiece with objective(s) or the microscope stage itself, are available from all four major microscope manufacturers. Stepper motors have a long travelling distance, which allows for control over the entire size of the Z-image stacks. However, they are slower than piezo-electric Z-controllers. Therefore, it is recommended to use piezo-electric Z-drives for fast and precise 3D (volume) acquisition. On the other hand, the travelling distance for piezo-electric Z-drives is limited to 100–200 µm. For upright microscopes, it is also possible to use a piezo objective scanner for performing fast Z-stack acquisitions with nanometer precision.

6.2.2.3 Hardware-Based Autofocus

As mentioned before, focus drift is a major problem in live-cell imaging. Microscopes are subject to both thermal and mechanical drift. Thermal gradients and mechanical drift are primarily responsible for the progressive loss of focus in the Z-axial plane, which becomes a confounding problem, especially during long-term live-cell imaging. In order to specifically address this problem, a hardware-based

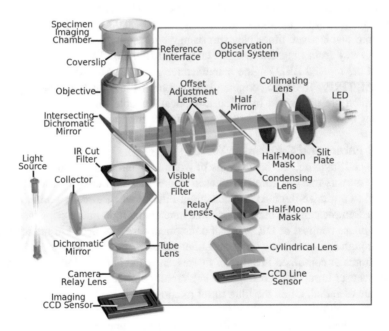

Fig. 6.5 Nikon imaging and Perfect Focus optical train side by side. Boxed region illustrates PFS optical path diagram. In brief, it uses a near IR laser or LED-based system in conjunction with a CCD image detection sensor and offset adjustment lens system to track the reflection from sample-coverslip interface and uses the drift from this reflection as feedback for the motorized Z-drive. The visible and IR filters prevent contaminating light from their respective imaging systems (imaging and PFS) interfere with each other. (Source: www.microscopyu.com) (Reproduced with permission)

autofocus system using a NIR-(LED)-light was first introduced by Nikon (Fig. 6.5). It is important to note that NIR-light does not interfere with most transmitted or fluorescent live-cell applications and does not contribute to photobleaching or phototoxicity.

6.2.2.4 Illumination Device

Traditionally, light sources for fluorescent microscopes were based on Mercury or Xenon-based arc lamps. Mercury arc lamps produce high levels of UV light, whereas Xenon lamps have peaks in the near IR region, and it is, therefore, advisable to use them with a near-IR filter. However, because of their limited lifespan (~200 h) and inconsistent brightness, they have been replaced by metal halides, where the light is delivered by using a liquid light guide. These metal halide bulbs are more endurable and produce more consistent illumination over time. Metal halides have similar spectral lines as mercury. To minimize phototoxicity, neutral density filters are used for lamp illumination sources, and shutters are used to restrict light exposure only during image capture. More recently, however, LED-based light engines for live-cell imaging have been introduced. A big advantage over other light sources is that these diode light sources (with multiple wavelengths) can be very rapidly switched on and off, which negates the use of additional mechanical shutters. Additionally, they have a much longer lifespan (10–20,000 h) and their intensity is also more stable over time. Not surprisingly, transmitted light sources are also moving away from tungsten halogen lamps to LED. CoolLED and Lumencor are some of the reputed LED engine manufacturers. For laser-based applications, such as CSLM, TIRF, FRAP, etc., commonly monochromatic solid-state laser sources are used, which are controlled by Acousto-Optical Tunable Filters (AOTF) for rapid switching between lasers, tuning their intensity and turning them on or off (Chap. 5).

6.2.2.5 Filters and Condenser Turret

For live-cell imaging, the bandwidths of excitation and emission filters as well as dichroic mirrors must be carefully chosen in order to limit unnecessary light exposure and to optimize fluorescence detection. Filters can be directly purchased from Chroma, Semrock among others. For some transmitted light microscopy techniques, such as phase contrast or DIC, special components are introduced in the light path, both through the condenser turret and filter turret(s). For fast imaging of multiple fluorophores or combined imaging of fluorescence and DIC, motorized filter turrets and condenser turrets that enable to switch rapidly between optical components and filters, are essential. Often the filter turret rotation speed is a rate-limiting factor in a typical live-cell experiment. To address this issue, Nikon's Ti2-E uses camera-based triggering, thereby removing software overhead time and accelerating filter turret switching speed significantly. Alternately, multi-pass band filters are also used for collecting fast cellular processes (<10 s), although their light collection efficiency is reduced. Fast switching can also be achieved by using lasers coupled through AOTF or rapid LED-switching and an emission image splitter, so that no filter changes are required. It is also important to realize that phase rings, DIC-prisms, and analyzers in the fluorescence imaging light path will also reduce light efficiency.

6.2.2.6 Shutter

Fast electronic shutters should be used for both transmitted and fluorescence micros-copy in order to limit light exposure to cells between imaging acquisition. Shutters may be built in into the microscope body or added between light source and microscope. It is worth noting that software overhead time can significantly delay the shutter speed, so shutters can be triggered via TTL with a detector (e.g., CCD).

6.2.2.7 Objective Lens

Brightness increases with the fourth power of NA and decreases with the square of magnification of an objective. The primary goal in live-cell imaging is to collect the maximum number of photons without inflicting photo damage, which is why it is recommended to use the highest NA objectives with lower magnifications whenever possible. Low magnification objectives capture a wider field of view and improve temporal resolution. High NA objectives are required for capturing weak fluorescent signals. It is worth noting that high NA objectives are also generally corrected for different optical aberrations, such as spherical aberration and chromatic aberration for up to four wavelengths, and which reduces light transmission (Chaps. 1–3). For live-cell imaging, it is usually better to choose objectives with lower magnification and highest fluorescence transmission (NA), which in turn can significantly reduce exposure time and thus reduce phototoxicity. All four major microscope companies manufacture high NA objectives but, unfortunately, they are not interchangeable. For example, Nikon infinity-corrected objectives are not interchangeable with Olympus infinity-corrected objectives; not only because of differences in tube length but also because of the fact that the mounting threads (fittings) are not the same (different pitch or diameter). Therefore, objectives need to be matched to a particular microscope from a single manufacturer. Likewise, objective immersion oils are also not interchangeable between companies, as they lead to optical mismatches and axial chromatic aberrations. Finally, heated objective collars are essential when immer-sion objectives are used, since the objective acts as a heat sink.

6.2.2.8 Detector

Cooled EM-CCD monochrome cameras have long been used for biological imaging, due to their low read noise and low signal recording ability. Over the last decade, however, the camera field has evolved greatly, and scientific CMOS cameras are becoming very popular due to their high speed and sensitivity as well as their large detector areas. Sensitive detectors are useful for collecting high signal-to-noise images from even weakly illuminated samples. Furthermore, large format cameras with small pixel size are useful for capturing large field of view at high spatial resolution, which allows for higher data throughput and temporal resolution with less phototoxicity. In Experimental Box 6.2, a practical example of the beneficial use of a CMOS camera in combination with long-term, high-resolution re-scanning confocal microscopy (RCM) is revealed. Andor, Photometrics and Hamamatsu—amongst many others—are reputed names in the camera manufacturing field. The most important properties to look for in cameras used for live-cell microscopy include quantum efficiency, frame rate, read noise, camera chip size, pixel size, dynamic range, bit depth, and binning capacity (Chap. 4). Confocal scanning laser microscopes use PMT detectors that are less sensitive and have lower quantum efficiencies (Chap. 5).

6.2.2.9 Image Acquisition Software

Automated control of the microscope and all its components is an essential part of live-cell imaging. Not surprisingly, all four major microscope manufacturers have developed their own software packages that provide a turnkey solution for image acquisition and analysis: Nikon (NIS Elements), Leica (LAS X), Olympus (CellSense), and Zeiss (Zen-Blue) software. Other commercial software packages include Metamorph (Molecular Devices), SlideBook (3i), Image-Pro (Media Cybernetics), Velocity (Perkin Elmer). Two open-source software packages, μManager [http://www.micro-manager.org] and ScanImage [http://www.scanimage.org] offer more flexibility to scientific researchers than the commercially available turnkey solutions. μManager is mainly used for camera image acquisition, although it can control scanning systems as well. It has full control over hardware components of most microscopes, including the platforms of all four major microscope manufacturers. ScanImage provides a software framework for controlling confocal scanning laser microscopes [8]. Laboratories that develop novel imaging technologies—or those with frequent changing needs that cannot be satisfied by the existing commercial packages—often take advantage of LabView (National Instruments), MATLAB (Mathworks) and Python, in order to write their own code.

Experimental Box 6.2 Re-scanning Confocal Microscopy (RCM) long-Term, High-Resolution Live-Cell Imaging: Laser Excitation Energy Load and Phototoxicity Compared to Other Commonly Used Microscopy Techniques

Introduction

The excitation light used in fluorescence microscopy can have a devastating impact on the health and viability of living cells and organisms (see also Experimental Box 6.1). The damaging effect of light on cellular macromolecules can impair physiology and even lead to cell death. While phototoxicity is known to occur in many live-cell experiments its effects are often underestimated [9]. One key strategy to reduce phototoxicity is lowering the intensity of the light the cells are exposed to. The required light intensity is dependent on several factors. Two key factors to consider are the sensitivity of the detector and the light efficiency of the microscope in general. Here we performed an experiment on several commonly used imaging modalities (EPI—widefield epifluorescence microscopy, CON—confocal scanning laser

---

(continued)

Table 6.1 Several imaging modalities and the light dose received by the sample. Light dose was calculated as (*Measured power/FOV*) * *Exposure time* giving the light dose in μJ/μm². From this, it can clearly be observed that RCM uses the least excitation energy while confocal and STED require the highest

	Exposure time (s)	FOV (μm²)	Measured power (P) @ 561 (μW)	Light dose @ 561(μJ/μm²)
EPI	0.1	17,689	910	0.0051
CON	6	3364	16	0.0285
STED	12	900	40	0.5333
RCM	6	8100	1	0.0007
SIM	0.3	4225	32	0.0023

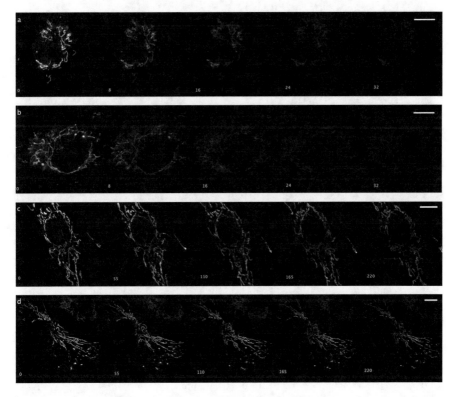

Fig. 6.6 Time-lapse of HUVEC cells stained with PK Mito Red. Panel (**a**) Single xy-slice using the confocal microscope (CON); Panel (**b**) single xy-slice using the STED modality; Panel (**c**) RCM z-stack; and Panel (**d**) SIM z-stack. Note the numbers displaying the respective number of frames acquired. Scale bar = 10 μm

Experimental Box 6.2 (continued)

microscopy, STED—stimulated emission depletion microscopy, RCM—rescanning confocal microscopy and SIM—structured illumination microscopy) and measured the light dose the cells were exposed to. The results clearly show that there are major differences between the different imaging modalities, underlining that one has to choose carefully the most suitable method for a live-cell imaging experiment.

Materials & Methods

HUVEC were cultured in round glass bottom dishes (Willco Dish, 35 mm) suitable for high-resolution microscopy. Cells were stained with PK Mitored diluted 1:100.000 in medium [10]. After 2 h the cells were imaged on an Olympus IX83 microscope, equipped with a 100X 1.5NA objective and a Hamamatsu Flash 4 V3 camera (EPI), a Leica TCS SP8 STED 3X microscope platform, equipped with a 100X 1.4NA objective and a HyD detector (CON & STED), a Nikon N-SIM microscope system, equipped with a 100X 1.49NA objective and a Hamamatsu Flash 4 V3 camera (SIM) or on an Olympus IX83 microscope, equipped with a 100X 1.5NA objective, RCM2-module (Confocal.nl) and a Hamamatsu Flash 4 V3 camera (RCM). Images were taken at a laser power to obtain sufficient SNR. Subsequently, z-stacks were imaged over time (11 planes per z-stack). A 561 nm laser was used to excite the mitochondrial dye. Post-acquisition processing was performed in Fiji for visualization optimization. All datasets were shown with a similar background and signal value, so they have approximately the same SNR. The exact laser power was measured with a Thorlabs S170C sensor at the sample plane after the objective (Table 6.1 and Fig 6.6).

Results

Data interpretation

From Fig. 6.6 we can observe that the sample bleaches very quickly (within a few tenths of frames) using Confocal and STED imaging (panels a and b), while bleaching is reduced to a minimum in the RCM (panel c) and SIM image (panel d). This supports the notion that there is a certain phototoxicity threshold below which there are no observed phototoxicity effects [9]. In addition, the STED laser seems to have less effect on bleaching compared to excitation laser (bleaching in Fig. 6.6 panels (a) and (b) are comparable). These results demonstrate that RCM utilizes the least amount of excitation energy. Confocal, SIM and STED require roughly the same amount of energy, but since SIM has a much shorter acquisition time, the received light dose is much less than confocal and STED. Epifluorescence illumination (EPI) is also relatively mild on the sample, as long as exposure time is kept at a minimum (see Table 1).

Live-cell imaging video

In the final part of this experimental box, we illustrate that RCM can be used to acquire long-term live-cell imaging videos with high spatial resolution

(continued)

Experimental Box 6.2 (continued)

(please follow this link to see our supplemental movie: https://vimeo.com/44 7800606). The key factor in making such movies is to use an extremely low amount of excitation energy in combination with photostable and bright fluorescent molecules. Furthermore, a sensitive photodetector (in this case a sCMOS camera) has to be used, which maximizes the detected signal—thereby offering a high SNR—while using only a very low amount of laser power.

Experimental parameters which were used to acquire the 61-h time-lapse of living HO1N1 cells

HO1N1 cells were acquired from Dandan Ma (AUMC, location VUMC, Amsterdam) and placed in glass-bottom 8-well μ-Slides (Ibidi, Cat. no. 80827). The cells were seeded at 50% confluency and allowed to adhere for 24 h before being transiently transfected overnight with CellLight Bacmam 2.0 Mitochondria-RFP (Invitrogen, Cat. no. C10505). The next day the cells were imaged using a Nikon Ti2 microscope equipped with a 40X PLAN APO 1.3 NA objective. The right port of the microscope was coupled to a re-scanning confocal microscope module (RCM, Confocal.nl) equipped with a Hamamatsu Flash 4 V3 sCMOS camera. For excitation, a Toptica CLE 561 nm laser was used and the laser power was measured to be 1 μW at the sample plane (measured using a Thorlabs PM100D with S170C sensor). The chosen ROI was 100 μm^2 and pixel size was 107 nm. One image was taken every 10 s, for a period of 61 h, creating over 22,000 frames. The images were further processed using SVI Huygens deconvolution. The results in the video show an improvement in both the resolution and SNR of the raw RCM data.

Acknowledgments

The authors would like to thank Marko Popovic (Nikon Center of Excellence, Amsterdam UMC) for providing access to the confocal, STED, RCM, and SIM microscope set-ups. We are also grateful to Philippa Phelp (Boon Lab, Amsterdam UMC) for providing us with the HUVEC cells.

6.3   Fluorescent Proteins, Probes, and Labelling Techniques for Live-Cell Imaging

Fluorescence microscopy can be used to observe dynamic processes in living cells and organisms. At the dawn of fluorescence microscopy, brave experiments were limited to observing autofluorescent specimens. In 1914, Stanislaus von Prowazek introduced the first fluorescent stain to label non-fluorescent organisms (protozoa). Later—in 1941—Albert Coons in collaboration with Louis Fieser developed the immunofluorescence method by coupling fluorescent dyes to antibodies and using them to detect antigens in tissues. The enormous leap forward in fluorescence

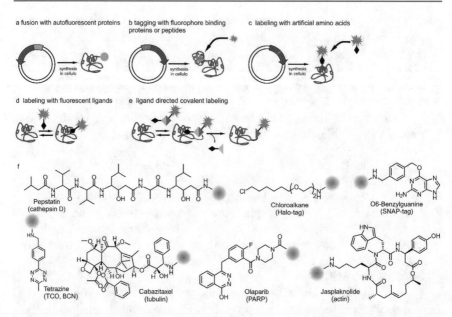

Fig. 6.7 The most common methods used for labelling proteins inside living cells. The protein labelling methods: (**a**) relying on fusions with autofluorescent proteins; (**b**) using specific peptides which can be labelled afterward or "self-labelling" tags; (**c**) based on the introduction of unnatural amino acids, carrying a reactive group, followed by a labelling reaction; (**d**) exploiting small molecule ligands; or (**e**) ligand directed covalent labelling; and (**f**) chemical structures of small molecule ligands used for protein targeting

microscopy has been made by Osamu Shimomura who discovered the Green Fluorescent Protein (GFP) in the jellyfish Aequorea victoria in 1962 [11]. It took another 30 years until the GFP gene was sequenced and cloned by Douglas C. Prasher [12]. Two years after that, the first transgenic organism expressing GFP fusion protein (Fig. 6.7, panel a) was created by Martin Chalfie [13]. In addition, Roger Tsien laid the fundament for multicolour fluorescence microscopy by generating the blue-shifted variant of GFP [14].

The discovery of super-resolution microscopy techniques (Chap. 13) has imposed new requirements on the labelling methods and fluorophore properties. The resolution limit has leaped beyond the Abbe diffraction limit and is approaching 1 nm. Thus special attention has to be paid to the size of the fluorescent probe and labelling density, which are starting to be the limiting factors determining attainable resolution. For example, multiple studies report measured tubulin diameter larger than 30 nm, which contrasts with the ground truth data obtained using cryo-electron microscopy ~25 nm. High-resolution imaging requires high labelling density to be able to resolve all features of the structure of interest and this can be achieved only using small fluorescent molecules. If selected carefully, small tags and ligands perturb the target biomolecules to less extent compared the large tags. The vast majority of the labelling methods can be classified into four categories:

1. Inserting or attaching specific tags which can be used for visualization of the target biomolecule.
2. Using fluorescently tagged proteins which have affinity toward specific parts of the target biomolecule.
3. Exploiting fluorescent dyes and probes which specifically bind to the target biomolecule.
4. Introducing reactive moieties or fluorescent components during synthesis of biomolecule.

6.3.1 Protein Labelling Methods

Although fluorescently labelled proteins are well-known tools for highlighting the protein of interest, their application is limited by their photophysical properties (Fig. 6.7, panel a). Self-labelling proteins are more attractive because the fluorescent spectrum can be easily tuned by the exchange of the organic dye used for labelling (Fig. 6.7, panel b). The most frequently used protein labelling tags are SNAP-tag and Halo-tag [15]. They offer high labelling specificity and reaction rate reaching 10^8 M^{-1} s^{-1} [16]. SNAP-tag relies on O6-benzyl guanine fluorescent derivatives and Halo-tag accepts halo-alkane substrates (Fig. 6.7, panel b and f). Both compounds are relatively easy to derivatize with fluorescent dyes. An alternative to self-labelling proteins is peptides such as Flash-tag, His-tag, or Flag-tag [17]. Their small size offers little interference with the target protein function and contributes to the resolution increase in super-resolution imaging. The most recent example is His-tag labelling which demonstrated excellent quality images in living cells [18]. However, the wide use of this method is hampered by the membrane impermeability of the fluorescent tris-NTA-group, which calls for cell squeezing to deliver these dyes inside the cell.

Protein tagging with self-labelling proteins or short peptides usually is performed at the N- or C-terminus. This reduces flexibility and becomes an issue when distance measurements within the same protein must be performed. Insertions in the loop regions of large self-labelling proteins, or even short peptides, might strongly interfere with the protein of interest's function. An alternative elegant approach to accomplish this challenging task is the introduction of artificial amino acids (Fig. 6.7, panel c). Genetic code engineering allows the introduction of an amino acid carrying a "click" compatible functional group into any site of the protein. Subsequently, this protein can be labelled via "click" chemistry with a cell-permeable fluorophore (Fig. 6.7, panel c and f). The downside of this approach is that living cells must be engineered by the introduction of special aminoacyl-tRNA synthetase - tRNA pairs in order to support the artificial amino acid incorporation.

Protein labelling based on ligand binding is the most promising approach in the field (Fig. 6.7, panel d). It eliminates protein overexpression phenotype and does not require loop engineering, nor the use of aminoacyl-tRNA synthetase - tRNA pairs. The main pitfall of this method, however, is finding the ligands which are selective and could be modified by attaching a fluorophore. Based on this principle several research groups introduced a series of highly biocompatible probes targeting actin, tubulin, DNA, poly(adenosine diphosphate-ribose) polymerase and lysosomes (Fig. 6.7, panel f). Although this advanced chemical synthesis has been employed to generate multiple cell-permeable fluorescent dyes, only a few have been shown to not interfere with the processes in living cells. Nevertheless, short-term exposure of cells to this class of fluorophores does not seem to produce significant changes in living cells.

Ideally, the ligand should not interfere with the function of the protein. However, often the targeting ligand can be identified among the well-characterized drugs, which show high potency in living cells and organisms. The possible solution for this problem, proposed by Itaru Hamachi [19], is the introduction of a cleavable linker, which can separate the ligand from the fluorophore after the binding event. The cleaved part containing fluorophore then stays attached to the protein of interest, while the ligand can be washed-off. Initially, this approach suffered from a slow linker cleavage rate, but this problem has been successfully addressed and now reaches $10^4 \, \mathrm{M}^{-1} \, \mathrm{s}^{-1}$ (Fig. 6.7, panel e).

6.3.2 DNA Labelling Methods

DNA staining with fluorescent dyes has a long history. The first reports of fluorescent DNA labelling and microscopy imaging can be attributed to Caspersson T., who used quinacrine mustard, an intercalating covalent DNA binder, for the fluorescent

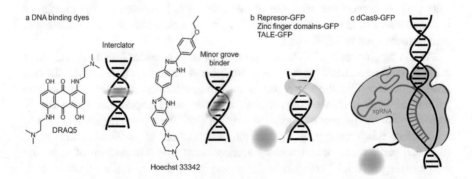

Fig. 6.8 The most common methods used for labelling DNA inside living cells. DNA labelling can be achieved via these methods: (**a**) exploiting fluorescent dyes which bind to the DNA structure; (**b**) relying on the fluorescently tagged proteins, which are interacting with specific DNA sequences; and (**c**) catalytically inactivated Cas9 protein fusion with autofluorescent proteins, programmable via sgRNA

chromosome banding in 1968 [20]. Lammler and Schutze used Hoechst 33258, a minor grove binder, to stain nuclei in animal tissues, which was patented by German company Hoechst AG in 1967 (Fig. 6.8, panel a). Later, trypanosomiasis treatment drug search led to the synthesis of DAPI stain by the Otto Dann's laboratory in 1971. Interestingly, DAPI molecule can intercalate or bind to the minor groove of DNA depending on the sequence context. The need for a far-red DNA stain was satisfied by the introduction of intercalating DRAQ5 fluorophore (Fig. 6.8, panel a). Even though these dyes were discovered a long time ago, they remain popular until today.

The first sequence-specific live-cell imaging approach involved the insertion of the large arrays of the lac operator (lacO) sequences, which can be detected after binding with the fluorescently tagged Lac repressor (LacR) protein (Fig. 6.8, panel b). The interaction is highly specific and shows low Kd, which results in a good contrast and allows long-term observations of the dynamics of the tagged loci. An alternative system is based on tet operator/Tet repressor (tetO/TetR). Development of genome editing tools has opened new horizons in imaging specific sequences in living cells. The first experiments have utilized polydactyl zinc finger DNA-binding domains fused to GFP, which were targeted to a 9-bp sequence in the major satellite repeats, localized in heterochromatin domains containing the centromere [21]. Later, this approach was greatly expanded by the introduction of the transcription activator-like effectors (TALEs) systems and the clustered regularly interspaced short palindromic repeat–CRISPR-associated CRISPR-Cas9 (Fig. 6.8, panel b and c) [22]. However, imaging of the non-repetitive sequences on the genomic DNA remained a great challenge, until the combination of dCas9 with the bacteriophage-derived RNA stem-loop motifs MS2 and PP7 made it possible (Fig. 6.8, panel c). Multiple genomic loci in living cells can be imaged using multicolor versions of CRISPR, specifically dCas9 from three bacterial orthologues or the CRISPRrainbow system, which utilizes engineered sgRNAs binding to combinations of different fluorescent proteins. The final optimizations, which decreased the background and enhanced the signal, were introduction of multiple fluorescent reporter binding sites on the sgRNA and introduction of molecular beacon approach. It must be noted that all of these modifications resulted in the great expansion of the molecular size of the probe, which is beneficial for single-site imaging but detrimental to its localization precision.

6.3.3   RNA Labelling Methods

Specific RNA labelling is a challenging task because of the numerous structures which can be formed by this biomolecule. However, this very property was exploited for the first tagging and imaging experiment of a specific RNA species. A tandem array of bacteriophage MS2 derived stem-loops, that was specifically recognized by MS2 coat protein fusion with autofluorescent protein, was introduced into the target mRNA (Fig. 6.9, panel a) [23]. Multiple binding sites allow observation down to the single-molecule resolution. Later this principle was employed by creating multiple mRNA tagging systems PP7, U1, and λN-boxB. The next logical step in simplifying

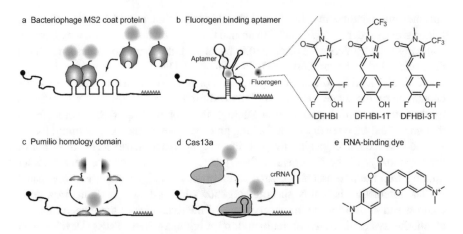

Fig. 6.9 The most common methods used for labelling RNA inside living cells. Labelling of RNA can be accomplished using the following methods: (**a**) using bacteriophage MS2 RNA binding coat protein fusion with the autofluorescent protein; (**b**) relying on the introduced specific RNA sequence, which folds into aptamer and is able to bind fluorescent dye; (**c**) exploiting RNA interacting programmable Pumilio homology domains, fused to split autofluorescent proteins; (**d**) using catalytically inactive dCas13a protein fusion with autofluorescent proteins, programmable via sgRNA; and (**e**) using RNA interacting fluorescent dyes

the RNA labelling methods was the introduction of an RNA aptamer that could fluoresce on its own or after binding the fluorogenic dye. The latter strategy was successfully employed in creation of Spinach and Broccoli aptamers, which can bind the isolated GFP chromophore and its versions [24, 25]. Further improvement resulted in a selection of more photostable chromophores: Mango III aptamer binds thiazole orange one and Riboglow aptamer recognizes cobalamin-fluorescein/rhodamine/cyanine conjugates (Fig. 6.9, panel b).

The high selectivity of RNA labelling could be achieved by employing programmable proteins recognizing specific sequences. This eliminates the need of inserting artificial tag into the native RNA molecule. Indeed, two approaches were demonstrated. The first exploits Pumilio homology domain (PUM-HD), which can be programmed to target any eight-base RNA sequence (Fig. 6.9, panel c). The specificity and SNR of this method is increased by employing two domains that are fused to split GFP and become fluorescent only after binding in close proximity. The alternative system exploits the catalytically inactive Cas13a, a class 2 type VI-A CRISPR-Cas RNA-guided RNA ribonuclease, which can be programmed using crRNA (Fig. 6.9, panel d). The dCas13a is expressed as a fusion with autofluorescent protein and introduction of crRNA leads to the assembly of the complex on the specific site of the target RNA molecule [26].

The difficulty in designing RNA selective dyes is reflected by the fact that SYTO RNAselect is the only commercial probe available for RNA imaging in live cells and its structure remains unpublished [27]. However, attempts have been made to improve this situation by chemically synthesizing several types of RNA stains. For example, Wang and co-workers designed a series of crescent-shaped probes for the imaging of nucleolar RNA in live cells (Fig. 6.9, panel e) [28]. In addition, Li and Chang screened a library of styryl derivatives and identified a potential probe F22 fluorescing in the red region, which showed the characteristic nucleolar and cytosol staining in living Hela cells [29]. An interesting design was also reported by Turro and co-workers, who covalently linked the intercalating dye ethidium bromide and the fluorescein dye [30]. The resulting phenanthridine derivative covalently linked to a fluorescein moiety (FLEth) probe showed selectivity toward RNA and characteristic staining of nucleolus and cytoplasm in mammalian breast cancer cell lines.

6.3.4   Fluorescence Resonance Energy Transfer (FRET)

Although visualizing structures in living cells often relies on using fluorescence—which provides high selectivity and contrast—the optical resolution of the light microscope limits determinations of protein proximities to approximately 200–300 nm (Chap. 1). However, in addition to braking the diffraction limit by using super-resolution microscopy (Chap. 13), this degree of diffraction-unlimited spatial resolution can also be achieved in light microscopy, by using appropriate sets of fluorescently labelled proteins, a technique called fluorescence resonance energy transfer (FRET) [31]. FRET is a physical process by which radiation-less transfer of energy occurs from a fluorophore in the excited state to an acceptor molecule (i.e., matching fluorophore) in close proximity. The range over which resonance energy transfer can occur is limited to approximately 0.01 μm and the efficiency of energy transfer is extraordinarily sensitive to the distance between fluorophores. FRET between fluorescent dyes in aqueous media is decreased to 50% at a 5–6 nm distance [32]. Nevertheless, measurement of intra- and intermolecular FRET under the microscope provides a particularly powerful, non-invasive approach to visualize the spatiotemporal dynamics within and between proteins in a living cell. One of the many elegant examples of FRET microscopy in live-cell imaging has been reported by Banerjee and co-workers, who used multiple FRET pairs to visualize dynamic co-localization of signalling molecules in the inhibitor kappa B kinase signalosome [33]. For further reading on the various applications of FRET imaging in biomedical research, the reader is referred to the vast amount of existing literature.

Experimental Box 6.3 Label-Free Holographic Imaging
In order to be visualized by using a standard or fluorescent light microscope, cells must be stained—or genetically modified—to absorb, emit or scatter light. Unfortunately, the invasive preparations necessary to make cells visible are most likely to affect cellular behavior, compromising the in vivo relevance of in vitro live-cell observations (see also Experimental Boxes 6.1 and 6.2).
   Gentle cell imaging

(continued)

Fig. 6.10 An example of a quantitative phase image of living cells in 3D created by HoloMonitor®. The height of the cell and its color tone correspond to the optical thickness of the cell. (Source: https://phiab.com) (Reproduced with permission)

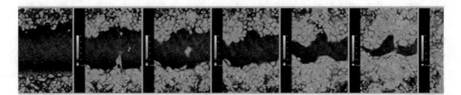

Fig. 6.11 Time-lapse image sequence created by using HoloMonitor® in a wound-healing assay. (Source: https://phiab.com/) (Reproduced with permission)

Experimental Box 6.3 (continued)

Unstained cells do, however, slow down and distort the light passing through them, just like beach waves are distorted by shallower water. By using a phase-contrast microscope these *phase-shift distortions*—created by living cells—can be observed, making unstained cells clearly visible. Just like water waves, light waves of a specific wavelength have two basic characteristics: i.e., amplitude and phase. Amplitude corresponds to light intensity and is the height of the wave, measured from crest to trough. Phase describes whether a wave is currently at its crest, in its trough, or somewhere in between. When light passes through a cell submerged in cell media, the light

(continued)

Experimental Box 6.3 (continued)

amplitude is relatively unaffected. However, the more optically dense cell slows down and delays the light, slightly relative to the surrounding ambient light, creating a *phase-shift* that makes cells visible in a phase-contrast microscope. Importantly, conventional phase-contrast microscopy cannot quantify phase shifts, only visualize them.

Quantitative phase imaging

Using a digital image sensor, low power diode illumination and sophisticated computer algorithms, the HoloMonitor® live-cell imager from Phase Holographic Imaging PHI AB, Lund, Sweden has the ability to both quantify and visualize *phase-shifts* [https:/phiab.com/]. HoloMonitor® employs a technique called quantitative phase imaging (QPI) or *quantitative phase contrast microscopy,* to distinguish it from its soon 100-year-old non-quantitative predecessor—the phase-contrast microscope [34–36] (Fig. 6.10).

Gentle time-lapse imaging

As the cell does not absorb any light energy, the cells are completely unaffected when observed using HoloMonitor®—no energy exchange, no change. This allows HoloMonitor® to gently acquire time-lapse image sequences over extended periods of time without compromising cellular behavior. HoloMonitor® provides both quantitative and beautiful time-lapse images of living cells, transforming phase microscopy and label-free live-cell imaging into a quantitative tool for detailed analysis of living cells on a population and single-cell level (Fig. 6.11).

6.4   Live-Cell Imaging in 3D Cell Cultures: Spheroids and Organoids

Cell-based assays have traditionally relied on 2D cell cultures, which represent a simple and easy-to-image model to study the cellular response to chemical stimulation or stress. However, due to their inability to capture the complex nature of organs, new assays based on 3D cell cultures have evolved rapidly over the past years. Based on the type of cells and their cellular organization there are currently two types of 3D self-organized cell culture models, i.e., spheroids and organoids (Fig. 6.12). Spheroids and organoids are more realistic models of both healthy and pathological tissues [37]. In addition, they are suitable for high-content screening (HCS). Therefore, the effects of potential therapeutical molecules can be investigated in an environment similar to the target tissue and in high-throughput settings, thereby improving the physiological significance of these assays. Microscopy-based HCS is further highlighted in Experimental Box 6.4. These 3D

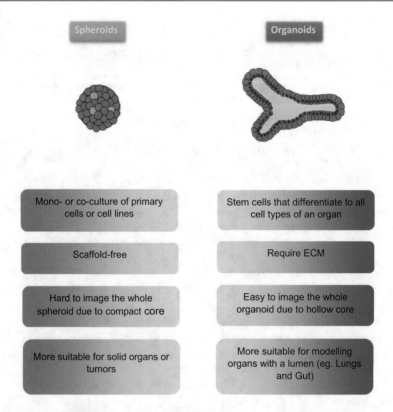

Fig. 6.12 Schematic comparison of 3D spheroid (left) and organoid (right) models

cell-based assays will ultimately reduce the effort and costs of drug development as well as the failure rate of clinical studies.

As research models, spheroids and organoids offer a compromise between 2D cell lines and living animals. Live-cell imaging has been essential to probe their complex morphology and demonstrate that they faithfully reflect their in vivo counterparts. Essentially, spheroids are clumps of poorly organized cells that have become a popular model in oncology research (Fig. 6.12). Due to their solid spherical morphology, both oxygen and nutrients decrease toward the center, decreasing cell viability from the outer cell layers to their hypoxic and necrotic cores [38]. Spheroids do not need a supportive matrix to grow and are more irregularly arranged cell aggregates with a rather poor organization of relevant tissue. Organoids, on the other hand, originate from stem cells, which give rise to different organ-specific cell types and ensure the cell culture's high self-renewal capabilities. Organoids do require a matrix to grow and possess a much more ordered assembly that typically recapitulates the 3D complex tissue structures (Fig. 6.12). When embedded in a suitable matrix and cultured with specific biochemical factors that mimic the in vivo stem cell niche, stem cells possess an intrinsic ability to differentiate and self-organize into 3D structures that resemble the in vivo organ [39]. Not surprisingly, 3D organoid models have recently gained a lot of popularity as new in vitro tools for drug testing, disease modeling and tissue engineering [40].

Thin section preparation, combined with classic immunohistochemistry, has been extensively used to roughly screen tissue architecture in 2D and to observe the distribution of single and multiple markers. 2D imaging though informative, however, does not allow a full appreciation of the complexity of 3D structures. Providentially, the past decade has emerged as a new era of volume or 3D imaging with novel microscopic approaches that can cross scales from cell to tissue. Noninvasive optical sectioning methods such as confocal imaging or multiphoton microscopy, and more recently light-sheet technology, now make it possible to visualize fine cellular details as well as overall tissue architecture within a single biological sample.

6.4.1 Challenges

Live-cell imaging is, of course, not trivial to do well, in particular when studying primary cells. Spatiotemporal resolution, signal-to-noise ratio, and acquisition time must be balanced to obtain the best images over time while avoiding light toxicity. A major challenge, however, is obtaining high-resolution images. Due to the size of the 3D cultures, live-cell microscopy to subcellular or even cellular resolution is extremely challenging when using conventional imaging techniques such as widefield fluorescence microscopy, laser-scanning confocal microscopy, or spinning-disk confocal microscopy (Chaps. 3, 5 and 11).

There are three major limitations that make high-quality images hard to acquire. The first limitation is phototoxicity following the repeated exposure of fluorescently labelled cells to illumination (from lasers) as mentioned earlier in this chapter and also illustrated in Experimental Boxes 6.1 and 6.2. When exposed to high amounts of laser light, fluorescent molecules react with molecular oxygen to produce free radicals. These can cause severe damage throughout the cell, in particular to the cellular DNA and the mitochondria. Recording intervals at least 20–30 min apart should be set when using the automated microscope to avoid high phototoxicity. To evaluate artifacts due to phototoxicity effects of a negative control, which is not imaged, should therefore be kept during live-cell imaging. The second limitation is the resolution required to separate subcellular biological structures, which is often near the resolving power of the microscopy setup being used. The lateral and axial resolution of a setup is determined by its PSF, which in turn depends on the NA of the objective. In 2D imaging, good lateral resolution is the key parameter, which can be achieved at relatively low NA. In 3D, however, refractive index mismatch can cause spherical aberrations, which lead to widening of the axial PSF and thus considerably impacting the quality of the imaging data obtained. High-resolution live-cell imaging, therefore, requires a high NA objective lens. Thirdly, when using ultra-low-attachment plates, the spheroids can drift out of the field of view, thereby causing distorted projection images upon acquiring z-stacks. Leary and co-workers recently showed that this limitation can be overcome by using 3D-printed molds containing 4 micro-posts per well (microwells) in agarose at defined spatial locations [41]. In addition, they addressed the issue of non-uniform fluorescence loss by performing ratio imaging.

6.4.1.1 Multiphoton Imaging of Spheroids

Multiphoton microscopy is regarded as the method of choice for imaging of living, intact biological tissues from the molecular level through to the whole organism (Chap. 9). The technique is uniquely suited to perform experimental measurements with minimal invasion over prolonged periods of time. It offers the researcher the ability to observe dynamic biological processes in substantial detail on time scales ranging from microseconds to weeks. In comparison to similar optical imaging techniques, multiphoton microscopy holds inherent advantages for imaging living 3D tissues such as improved penetration depth and reduced photodamage. In contrast to single-photon microscopy, two-photon microscopy is associated with a smaller volume of excitation as the two-photon effect only occurs in the focal spot of the objective, where the photon flux is high. In single-photon microscopy, the excitation also occurs above and below the focal point, resulting in significant phototoxicity and out-of-focus light. In a very elegant study by Grist and co-workers, long-term live-cell imaging of tumor spheroids in a microfluidic system was carried out using a two-photon confocal scanning laser microscope [42]. The spheroids were observed for 72 h at 20 min intervals under different oxygen concentration conditions to determine the effect of oxygen concentration on tumors and their susceptibility to treatment (Fig. 6.13).

6.4.1.2 Confocal and Spinning-Disk Live-Cell Imaging of Organoids

A considerable advantage of organoids is their accessibility for live observation to study dynamic processes even at high resolution. Confocal live-cell microscopy, for instance, enables the study of complex cellular processes in space and in time in organoids, where specific cell types or subcellular organelles and activities have been marked with fluorescent reporters. In an interesting work recently published, Artegiani and co-workers coupled a novel genome editing method—dubbed "CRISPR-HOT"—that allows efficient generation of knock-in human organoids representing different tissues without extensive cloning [43]. This method allowed for fluorescently tagging of non-constitutively expressed differentiation markers and visualizing the differentiation process via a confocal scanning laser or a spinning-disk microscope over a period of up to 72 h.

In another example from Clevers and co-workers, CRISPR–Cas9-mediated genome editing was used to introduce sequential mutations into human colon organoids to model the adenoma-carcinoma sequence [44]. Chromosome instability (CIN) and aneuploidy, both hallmarks of colorectal cancer (CRC), were then analyzed using a fluorescently tagged histone 2B (H2B)-encoding lentivirus. The use of 4D live-cell imaging over several days allowed a precise determination of chromosome segregation defects, including aberrant chromosome number and erroneous mitotic events, and pinpointed the mutations sufficient to acquire CIN. This

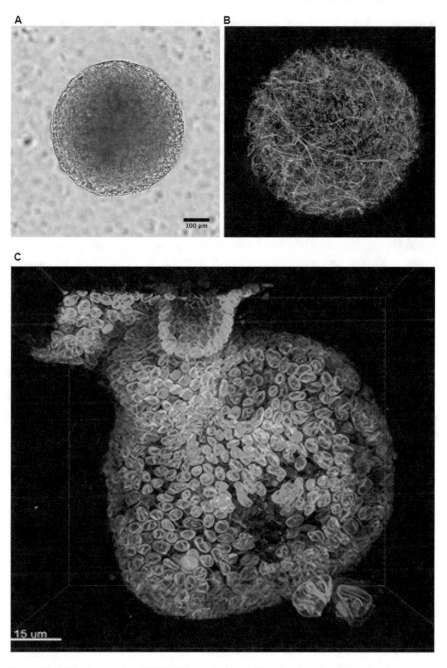

Fig. 6.13 MicroBrain spheroid (BF, left panel **a**) and 3D volume-rendered image of a spheroid generated by NIS Elements.AI (FL, right panel **b**). Liver epithelial organoid (FL, panel **c**)

powerful combination of engineered organoids and dynamic imaging will no doubt continue to aid our understanding of disease progression as well as cellular mechanisms orchestrating development and homeostasis.

6.4.1.3 Light-Sheet Imaging of Spheroids and Organoids

In recent years, light-sheet microscopy has emerged as a particularly fast and gentle technology for live-cell imaging (Chap. 12). Because the excitation is restricted to a thin sheet of light that can rapidly scan sizable biological specimens, light exposure and phototoxicity are minimized. With only a section of the sample being illuminated at any time and rapid frame-wise data capture, light-sheet fluorescence microscopy creates a photonic load several orders of magnitude lower than standard confocal fluorescence imaging. This sophisticated imaging technique has therefore been used to investigate dynamic processes on varying scales, including tracking microtubules plus tips of the mitotic apparatus as well as lineage tracing of cells in spheroids. In addition, Held and co-workers employed light-sheet microscopy to image organoids embedded in a freely rotatable hydrogel cylinder to facilitate imaging from various angles [45]. The authors succeeded in imaging labelled cells within re-aggregated kidney organoids over 15 h and tracking their fate while simultaneously monitoring the development of organotypic morphological structures.

Experimental Box 6.4 Microscopy-Based High-Content Screening (HCS)

With the advent of microscope automation, fluorescent probes, and image analyses methodologies, it is now possible to perform high-content analysis of visual phenotypes, to extract quantitative, multi-parametric information from images with minimal user bias [46]. High-throughput imaging can be classified into screening and profiling [47]. Screening—per definition—uses a priori knowledge to interrogate a phenomenon, measure multiple, visually discernible phenotypes and choose a subset of hits to pursue a biological question. Image-based screening has been applied to study alterations in protein localization in various cancer cells and complex organismal phenotypes. Profiling is a much more exploratory systems-level technique that uses an unguided approach to capture a broad spectrum of measurements from samples and maps. The latter is based on similarity and offers a greater chance to discover unknown mechanisms. Both screening and profiling applications are used in drug discovery, functional genomics, and disease phenotyping. Microscopy-based analyses also facilitate longitudinal single-cell analysis among populations, to identify cell-to-cell variability, otherwise obfuscated in population averaging experiments, leading to insights into complex biological processes at single-cell resolution [48–50].

There are many commercial solutions available for high-content screening. High-content imagers are typically box-type systems available from, e.g.,

(continued)

Experimental Box 6.4 (continued)

Molecular Devices (ImageXpress), Thermo-Fischer (CellInsight, ArrayScan), Perkin-Elmer (Opera Phenix), and Zeiss (Celldiscoverer 7). High-content analysis platforms based on microscopes are offered by, e.g., Nikon (HCA system) and Olympus (ScanR). In contrast to the box systems mentioned above, Nikon's HCA system builds on the completely motorized Ti2-E body and provides a very flexible platform for researchers to customize hardware components for their evolving needs. Users can add components including— but not limited to—light sources, optics, detection systems, filters, photo stimulation devices, confocal, and super-resolution modules. An automated plate loader controlled by NIS-Elements software can be added to this platform. Furthermore, in addition to Nikon's silicon objective lenses, an automated water immersion dispenser for Nikon's water immersion objectives for long-term experiments is also available (Fig. 6.14).

High-content screening is limited by the number of images that can be acquired in a short span of time. Nikon's Ti2-E microscope has the largest field of view camera port in the market. The Ti2's large FOV coupled with hardware triggering of native devices significantly reduces the number of images and acquisition time by minimizing software latencies. NIS-Element's JOBS and General analyses modules allow conditional and customizable workflow determined by real-time data analyses. In addition, Nikon's Perfect Focus system allows for fast accurate focusing across multiple points. Large format sensor cameras—like Nikon's DSQi2—with small pixel sizes provide improved spatial and temporal resolution.

Summary and Take-Home Messages

The three main experimental challenges in live-cell imaging are to minimize photodamage, to provide a stable environment for long-term experiments in cells and multicellular preparations and—last but not least—to prevent focal drift. Generally, it is best to use as little as possible excitation light in combination with fast and sensitive detectors. LED light sources as well as CMOS cameras have therefore become crucial components of live-cell imaging microscopes. As modern high-end microscopy platforms for live-cell imaging are evolving to allow for automated imaging of multiple fluorescent reporters, under low excitation conditions at multiple positions, in 3D, and over longer time-periods, the possibilities of these sophisticated platforms will only expand. We envision that high-throughput platforms will enable simultaneous studies of cellular morphology and live-cell activity alterations within 2D and 3D cell cultures, with both high speed and resolution, while at the same time also performing sensitive molecular drug screens in vitro. Undoubtedly, the next great challenge will therefore be the extraction and analysis of the immense amount of imaging data generated. Image-based cell profiling techniques are now being developed to find, segment, and count densely packed cells in large tissues, or to quantify phenotypic differences in 2D as well as 3D cell populations assayed in

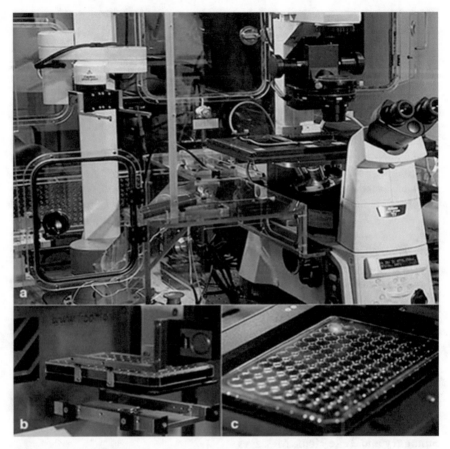

Fig. 6.14 High-content microscope platform based on Nikon Ti-E inverted microscope integrated with a robotic plate loader. Ti2-E can double the throughput of this design (**a**) for moving samples from incubator (**b**) to microscope stage (**c**). The system displayed in this figure belongs to the Finkbeiner laboratory. (Source: https://www.nature.com/articles/nmeth.f.400/figures/1) (Reproduced with permission)

high-throughput settings. It is obvious that—while pushing cellular and subcellular resolution—an unbiased quantitative understanding of dynamic intra- and intercellular relationships within cell cultures, spheroids, and organoids, will be equally essential to fully comprehend the complexity that can be monitored using live-cell microscopy.

Take-Home Messages
- Choose a stable and expandable microscope platform, with solid and well-integrated live-cell imaging components (e.g., stage-incubator, temperature controller, humidifier), that best matches your experimental model(s) as well as your specific scientific questions.

- In general, in order to minimize photodamage during live-cell imaging experiments, use as little as possible excitation light (e.g., low laser light, spinning disk, or LED light-engines) in combination with a fast and sensitive detector, that also supports a large FOV (e.g., CMOS camera).
- Choose your fluorescent dyes and/or probes wisely: there are numerous dedicated probes available for live-cell imaging applications (some have been designed specifically for certain live-cell applications).
- Use software-controlled microscopy automation and—if possible—hardware-triggering to support fast and reproducible acquisition with minimal phototoxicity.

Appendix: Microscope Company and Resources List with Internet-Links

- Andor—www.andor.oxinst.com
- Applied Scientific Instrumentation (ASI)—www.asiimaging.com
- Chroma—www.chroma.com
- Confocal.nl—www.confocal.nl
- CoolLED—www.coolled.com
- Hamamatsu—www.hamamatsu.com
- Intelligent Imaging Innovations (3i)—www.intelligent-imaging.com
- Leica Microsystems—www.leica-microsystems.com (and www.leica-microsystems.com/science-lab/science-lab-home)
- Lumencor—www.lumencor.com
- Mad City Labs (MCL) GmbH—www.madcitylabs.eu
- μManager—www.micro-manager.org
- Mathworks—www.mathworks.com
- Media Cybernetics—www.mediacy.com
- Molecular Devices—www.moleculardevices.com
- National Instruments—www.ni.com
- Nikon Instruments—www.microsope.healthcare.nikon.com (and www.microscopyu.com)
- Okolab—www.oko-lab.com
- Olympus Microscopy—www.olympus-lifescience.com (and www.olympus-lifescience.com/en/learn)
- Perkin-Elmer—www.perkinelmer.com
- Phase Holographic Imaging (PHI)—https:/phiab.com/
- Photometrics—www.photometrics.com
- Physics Instruments (PI)—www.physikinstrumente.com
- Prior Scientific—www.prior.com
- ScanImage—www.scanimage.org
- Semrock—www.semrock.com
- Thermo-Fischer—www.thermofisher.com
- Thorlabs—www.thorlabs.com

- Tokai Hit—www.tokaihit.com
- Zeiss—www.zeiss.com/microscopy

References

1. Pattison DI, Davies MJ. Actions of ultraviolet light on cellular structures. EXS. 2006;96:131–57.
2. Dixit R, Cyr R. Cell damage and reactive oxygen species production induced by fluorescence microscopy: effect on mitosis and guidelines for non-invasive fluorescence microscopy. Plant J. 2003;36(2):280–90.
3. Grzelak A, Rychlik B, Bartosz G. Light-dependent generation of reactive oxygen species in cell culture media. Free Radic Biol Med. 2001;30(12):1418–25.
4. Godley BF, Shamsi FA, Liang F-Q, Jarrett SG, Davies S, Boulton M. Blue light induces mitochondrial DNA damage and free radical production in epithelial cells. J Biol Chem. 2005;280(22):21061–6.
5. Wittmann T, Bokoch GM, Waterman-Storer CM. Regulation of leading edge microtubule and actin dynamics downstream of Rac1. J Cell Biol. 2003;161(5):845–51.
6. Zhai Y, Kronebusch PJ, Borisy GG. Kinetochore microtubule dynamics and the metaphase-anaphase transition. J Cell Biol. 1995;131(3):721–34.
7. Rodionov VI, Borisy GG. Microtubule treadmilling in vivo. Science (New York, NY). 1997;275(5297):215–8.
8. Pologruto TA, et al. ScanImage: flexible software for operating laser scanning microscopes. Biomed Eng Online. 2003;2:13.
9. Icha J, Weber M, Waters JC, Norden C. Phototoxicity in live fluorescence microscopy, and how to avoid it. BioEssays. 2017;39(8):170003.
10. Shen F-F, Chen Y, Dai X, Zhang H-Y, Zhang B, Liu Y, Liu Y. Purely organic light-harvesting phosphorescence energy transfer by beta-cyclodextrin pseudorotaxane for mitochondria targeted imaging. Chem Sci. 2020;12(5):1851–7.
11. Shimomura O, Johnson FH, Saiga Y. Extraction, purification and properties of aequorin, a bioluminescent proten from theluminous hydromedusan, Aequorea. J Cell Comp Physiol. 1962;59:223–39.
12. Prasher DC, Eckenrode VK, Ward WW, Prendergast FG, Cormier MJ. Primary structure of the Aequorea victoria green-fluorescent protein. Gene. 1992;111:229–33.
13. Chalfie M, Tu Y, Euskirchen G, Ward WW, Prasher DC. Green fluorescent protein as a marker for gene expression. Science. 1994;263:802–5.
14. Heim R, Prasher DC, Tsien RY. Wavelength mutations and posttranslational autoxidation of green fluorescent protein. Proc Natl Acad Sci U S A. 1994;91:12501–4.
15. Keppler A, Gendreizig S, Gronemeyer T, Pick H, Vogel H, Johnsson K. A general method for the covalent labelling of fusion proteins with small molecules in vivo. Nat Biotechnol. 2003;21:86–9.
16. Wilhelm J, Kühn S, Tarnawski M, Gotthard G, Tünnermann J, Tänzer T, Karpenko J, Mertes N, Xue L, Uhrig U, Reinstein J, Hiblot J, Johnsson K. Kinetic and structural characterization of the self-labeling protein tags HaloTag7, SNAP-tag, and CLIP-tag. Biochemistry. 2021;60(33):2560–75.
17. Griffin BA, Adams SR, Tsien RY. Specific covalent labelling of recombinant protein molecules inside live cells. Science. 1998;281:269–72.
18. Uchinomiya SH, Nonaka H, Fujishima SH, Tsukiji S, Ojida A, Hamachi I. Site-specific covalent labeling of His-tag fused proteins with a reactive Ni(II)-NTA probe. Chem Commun (Camb). 2009;39:5880–2.
19. Tamura T, Hamachi I. Chemistry for covalent modification of endogenous/native proteins: from test tubes to complex biological systems. J Am Chem Soc. 2019;141(7):2782–99.

20. Caspersson T, Farber S, Foley GE, Kudynowski J, Modest EJ, Simonsson E, Wagh U, Zech L. Chemical differentiation along metaphase chromosomes. Exp Cell Res. 1968;49:219–22.
21. Lindhout BI, Fransz P, Tessadori F, Meckel T, Hooykaas PJ, van der Zaal BJ. Live cell imaging of repetitive DNA sequences via GFP-tagged polydactyl zinc finger proteins. Nucleic Acids Res. 2007;35(16):e107.
22. Chen B, Gilbert LA, Cimini BA, Schnitzbauer J, Zhang W, Li G-W, Park J, Blackburn EH, Weissman JS, Qi L-S, Huang B. Dynamic imaging of genomic loci in living human cells by an optimized CRISPR/Cas system. Cell. 2013;155(7):1479–91.
23. Bertrand E, Chartrand P, Schaefer M, Shenoy SM, Singer RH, Long RM. Localization of ASH1 mRNA particles in living yeast. Mol Cell. 1998;2(4):437–45.
24. Paige JS, Nguyen-Duc T, Song W, Jaffrey SR. Fluorescence imaging of cellular metabolites with RNA. Science. 2012;335(6073):1194.
25. Filonov GS, Moon JD, Svensen N, Jaffrey SR. Broccoli: rapid selection of an RNA mimic of green fluorescent protein by fluorescence-based selection and directed evolution. J Am Chem Soc. 2014;136(46):16299–308.
26. Abudayyeh OO, Gootenberg JS, Essletzbichler P, Joung HS, Belanto JJ, Verdine V, Cox DBT, Kellner MJ, Regev A, Lander ES, Voytas DF, Ting AY, Zhang F. RNA targeting with CRISPR-Cas13. Nature. 2017;550(7675):280–4.
27. Suseela YV, Narayanaswamy N, Pratihar S, Govindaraju T. Far-red fluorescent probes for canonical and non-canonical nucleic acid structures: current progress and future implications. Chem Soc Rev. 2018;47(3):1098–131.
28. Liu W, Zhou B, Niu G, Ge J, Wu J, Zhang H, Xu H, Wang P. Deep-red emissive crescent-shaped fluorescent dyes: substituent effect on live cell imaging. ACS Appl Mater Interfaces. 2015;7(13):7421–7.
29. Li Q, Chang YT. A protocol for preparing, characterizing and using three RNA-specific, live cell imaging probes: E36, E144 and F22. Nat Protoc. 2006;1(6):2922–32.
30. Stevens N, O'Connor N, Vishwasrau H, Samaroo D, Kandel ER, Akins DL, Drain CM, Turro NJ. Two color RNA intercalating probe for cell imaging applications. J Am Chem Soc. 2008;130(23):7182–3.
31. Day RN, Davidson MW. Fluorescent proteins for FRET microscopy: monitoring protein interactions in living cells. BioEssays. 2012;34(5):341–50.
32. Berney C, Danuser G. FRET or no FRET: a quantitative comparison. Biophys J. 2003;84:3992–4010.
33. Gamboni F, Escobar GA, Moore EE, Dzieciatkowska M, Hansen KC, Mitra S, Mydam TA, Silliman CC, Banerjee A. Clathrin complexes with the inhibitor kappa B kinase signalosome: imaging the interactome. Phys Rep. 2014;2(7):12035.
34. Zhang Y, Judson RL. Evaluation of holographic imaging cytometer HoloMonitor M4® motility applications. Cytometry A. 2018;93(11):1125–31.
35. Janicke B, Kårsnäs A, Egelberg P, Alm K. Label-free high temporal resolution assessment of cell proliferation using digital holographic microscopy. Cytometry A. 2017;91(5):460–9.
36. Sebesta M, Egelberg PJ, Langberg A, Lindskov J-H, Alm K, Janicke B. HoloMonitor M4: holographic imaging cytometer for real-time kinetic label-free live-cell analysis of adherent cells. Boston: Phase Holographic Imaging PHI Inc; 2016.
37. Torras N, Garcia-Diaz M, Fernandez-Majada F, Martinez E. Mimicking epithelial tissues in three-dimensional cell culture models. Front Bioeng Biotechnol. 2018;6:197.
38. Lin RZ, Chang HY. Recent advances in three-dimensional multicellular spheroid culture for biomedical research. Biotechnol J. 2008;3(9–10):1172–84.
39. Rossi G, Manfrin A, Lutolf MP. Progress and potential in organoid research. Nat Rev Genet. 2018;19(11):671–87.
40. Fang Y, Eglen RM. Three-dimensional cell cultures in drug discovery and development. SLAS Discov. 2017;22(5):456–72.
41. Leary E, Rhee C, Wilks BT, Morgan JR. Quantitative live-cell confocal imaging of 3D spheroids in a high-throughput format. SLAS Technol. 2018;23(3):231–42.

42. Grist SM, Nasseri SS, Laplatine L, Schmok JC, Yao D, Hua J, Chrostowski L, Cheung KC. Long-term monitoring in a microfluidic system to study tumour spheroid response to chronic and cycling hypoxia. Sci Rep. 2019;9(1):17782.
43. Artegiani B, Hendriks D, Beumer J, Kok R, Zheng X, Joore J, de Sousa C, Lopes S, van Zon J, Tans S, Clevers H. Fast and efficient generation of knock-in human organoids using homology-independent CRISPR-Cas9 precision genome editing. Nat Cell Biol. 2020;22(3):321–31.
44. Bolhaqueiro ACF, Ponsioen B, Bakker B, Klaasen SJ, Kucukkose E, van Jaarsveld RH, Vivié J, Verlaan-Klink I, Hami N, Spierings DCJ, Sasaki N, Dutta D, Boj SF, Vries RGJ, Lansdorp PM, van de Wetering M, van Oudenaarden A, Clevers H, Kranenburg O, Foijer F, Snippert HJG, Kops GJPL. Ongoing chromosomal instability and karyotype evolution in human colorectal cancer organoids. Nat Genet. 2019;51(5):824–35.
45. Held M, Santeramo I, Wilm B, Murray P, Lévy R. Ex vivo live cell tracking in kidney organoids using light sheet fluorescence microscopy. PLoS One. 2018;13(7):e0199918.
46. Boutros M, Heigwer F, Laufer C. Microscopy-based high-content screening. Cell. 2015;163(6):1314–25.
47. Caicedo JC, Singh S, Carpenter AE. Applications in image-based profiling of perturbations. Curr Opin Biotechnol. 2016;39:134–42.
48. Arrasate M, Finkbeiner S. Automated microscope system for determining factors that predict neuronal fate. Proc Natl Acad Sci U S A. 2005;102(10):3840–5.
49. Snijder B, Sacher R, Rämö P, Liberali P, Mench K, Wolfrum N, Burleigh L, Scott CC, Verheije MH, Mercer J, Moese S, Heger T, Theusner K, Jurgeit A, Lamparter D, Balistreri G, Schelhaas M, De Haan CAM, Marjomäki V, Hyypiä T, Rottier PJM, Sodeik B, Marsh M, Gruenberg J, Amara A, Greber U, Helenius A, Pelkmans L. Single-cell analysis of population context advances RNAi screening at multiple levels. Mol Syst Biol. 2012;8:579.
50. Snijder B, Pelkmans L. Origins of regulated cell-to-cell variability. Nat Rev Mol Cell Biol. 2011;12(2):119–25.

Structured Illumination Microscopy

7

Nicholas Hall and Ian Dobbie

Contents

What You Will Learn in This Chapter

The advent of super-resolution techniques has fundamentally changed how biology is done and what biological questions it is possible to ask and answer now that we can image beyond the resolution limit. There are a number of super-resolution techniques, each with their own strengths and preferred use cases. Structured illumination microscopy (SIM) relies on spatially structured illumination light to encode super-resolution information in a resolution limited image. A number of these resolution limited images are used together to extract the super-resolution information, producing a reconstructed image with twice the resolution of the original images. The number of images required to obtain a super-resolution image is considerably less than most other super-resolution techniques, making SIM ideally suited to imaging dynamic biological processes. Herein we will cover how to think of images as information in Fourier space (including an introduction to Fourier

N. Hall (✉)
TCAD, Synopsys Inc, Glasgow, UK
e-mail: nickhall@synopsys.com

I. Dobbie
Integrated Imaging Center, Department of Biology, Johns Hopkins University, Baltimore, MD, USA
e-mail: ian.dobbie@jhu.edu

© The Author(s), under exclusive license to Springer Nature Switzerland AG 2022
V. Nechyporuk-Zloy (ed.), *Principles of Light Microscopy: From Basic to Advanced*, https://doi.org/10.1007/978-3-031-04477-9_7

mathematics sufficient to understand this), the mathematics behind SIM and how images are reconstructed, and the strengths and limitations of SIM.

7.1   What You Should Already Know

There are a few of concepts that the reader is assumed to be familiar with and/or have been covered in previous chapters. These are:

- Geometric optics leading to the rise of the diffraction limit
- Fluorescence microscopy
- An understanding that functions can be represented by Fourier decomposition

These will be explained briefly when and where they are relevant. However, the full scope and complexity will not be covered. Readers who are unfamiliar with any of these are advised to pause here and familiarise themselves. Some suggested sources are provided in the *Further Reading* section at the end of the chapter.

7.2   Fourier Decomposition of Images

In order to understand SIM, we must first take a step back and ensure we understand a more fundamental concept, Fourier analysis. What follows will not be a series of extensive mathematical derivations. These can be found in previous publications. Rather specific formulae will be stated when appropriate to demonstrate particular important details.

Readers should already be familiar with the concept of Fourier decomposition; describing a function as an infinite series of frequency components, also referred to as reciprocal components, called a Fourier series. Typically, this explanation starts with time-varying functions since most people have an intuitive understanding of a single time-varying function being composed of a number of frequency components. For example, it is common knowledge that a single musical chord is composed of a number of different notes (i.e. frequency components) played simultaneously. However, there is nothing unique about time-varying functions. If the function is spatially varying rather than time-varying, all of the mathematics remain unchanged. In this case, the Fourier components would be described by reciprocal quantities that we call *spatial frequencies*.

More generally, we can describe a reciprocal variable for the Fourier series to correspond to the original variable in the function. Likewise, in the same way, we describe a vector space for the original variable or variables—such as the 3-dimensional vector space described by the *xyz* Cartesian vectors—we can describe a vector space for the reciprocal variables which we call reciprocal space. Therefore, we can say:

Any function can be described by an infinite Fourier series of reciprocal variables.

There is a mathematical operation to convert from the original vector space to the corresponding reciprocal space known as the Fourier transform which is written as [1]

$$\widetilde{\mathbf{F}}(\mathbf{k}_x) = \int_{-\infty}^{\infty} f(\mathbf{x}) e^{-i\mathbf{k}_x \mathbf{x}} d\mathbf{x}, \tag{7.1}$$

where $\mathbf{x}$ and $\mathbf{k}_x$ are the original and reciprocal variables, respectively. This yields a continuous function, $\widetilde{\mathbf{F}}(\mathbf{k}_x)$ described in terms of the reciprocal variable, $\mathbf{k}_x$. In general $\widetilde{\mathbf{F}}$ is used to denote the Fourier transform of f. Likewise, the operation to move from reciprocal space back to the original vector space, the inverse Fourier transform, is written as

$$f(\mathbf{x}) = \frac{1}{2\pi} \int_{-\infty}^{\infty} \widetilde{\mathbf{F}}(\mathbf{k}_x) e^{i\mathbf{k}_x \mathbf{x}} d\mathbf{k}_x, \tag{7.2}$$

Figure 7.1 shows the top hat function and its Fourier transform, a sinc function. Notice that the Fourier transform has positive and negative frequency components and it is symmetrical around $\omega = 0$. This symmetry is the result of a more general property of Fourier transforms of real functions which is that:

$$\widetilde{\mathbf{F}}(\mathbf{k}_x) = \overline{\widetilde{\mathbf{F}}(\mathbf{k}_x)}, \tag{7.3}$$

where $\overline{\widetilde{\mathbf{F}}(\mathbf{k}_x)}$ is the so-called complex conjugate of $\widetilde{\mathbf{F}}(\mathbf{k}_x)$. A complex conjugate of a complex number $a + ib$ is $a - ib$. Essentially the value of the Fourier transform at some value of $\mathbf{k}_x$ is the complex conjugate of the value of the Fourier transform at $-\mathbf{k}_x$. Now, the Fourier transform of a top hat function symmetric around 0 has no imaginary component (i.e. $b = 0$ in $a + ib$) and so this property becomes $\widetilde{\mathbf{F}}(\mathbf{k}_x) = \widetilde{\mathbf{F}}(-\mathbf{k}_x)$, and is symmetric around $\omega = 0$.

Although up to this point, we have dealt with one-dimensional functions, there is nothing in the mathematical construction of the Fourier transforms which requires this. Take, for example, the Mona Lisa shown in Fig. 7.2a. Although it might not be immediately obvious, there exists a mathematical function which describes the intensity variations in this image. We can, therefore, consider the function which describes the Mona Lisa as the visualisation of a 2D function in (x, y). As such, we can also perform a Fourier transform on this function to acquire Fig. 7.2b, which is a function in $(\mathbf{k}_x, \mathbf{k}_y)$ where $\mathbf{k}_x$ and $\mathbf{k}_y$ are the spatial frequencies which are reciprocal variables to the x and y variables, respectively. Once again, these are equivalent representations of the same information. We only use different variables to describe said information.

It is worth noting that the Fourier space representation is equivalent to the real space representation (i.e. the actual function), just described in terms of reciprocal variables. A mathematical theorem, the derivation of which is not presented here,

Fig. 7.1 (a) A continuous top hat function centred at the origin, $f(x)$. (b) The continuous Fourier transform of $f(x)$, $\widetilde{\mathbf{F}}(\mathbf{k}_x)$

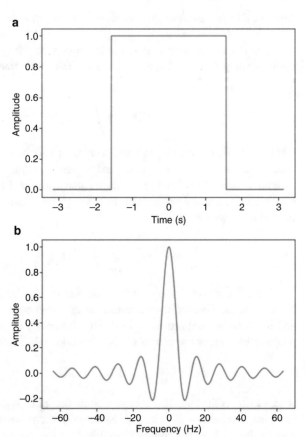

called Parseval's theorem states that the integral of the square of a function is equal to the integral of the square of its Fourier transform. Now, when the function in question describes amplitude—in the case of light, the amplitude of the electromagnetic field—then the square of the function is its intensity. Therefore, the integral of the square of the function f is the total energy of the signal or image the function describes. Consequently, the integral of the square of both the real space function and its Fourier transform is the total energy of the signal or image the function describes. This information allows us to conclude that the Fourier space representation is equivalent to the real space representation, just described in terms of reciprocal variables.

We should also note that the previous discussion refers to continuous functions, that have a known value at all possible positions in space, and that these need infinite Fourier components to be faithfully reproduced. However, a modern microscope image is represented by pixels in a camera image, which is only defined at the centre positions of the pixels. This type of function can be completely defined by a finite Fourier series with the same number of components as the positions in the original discrete function.

Fig. 7.2 (**a**) An greyscale image of the Mona Lisa. (**b**) The 2D Fourier transform of the image

a)

b)

Before we proceed, we also need one last mathematical concept; *convolution*. The convolution operation is defined as

$$f(x) \circledast g(x) = \int_{-\infty}^{\infty} f(\tau)g(x-\tau)d\tau = \int_{-\infty}^{\infty} f(x-\tau)g(\tau)d\tau, \tag{7.4}$$

where $\circledast$ is the convolution operator. The resulting function expresses how the shape of one function modifies the other. It may not be intuitively apparent what this means, which is fine as a complete understanding of convolution is not required at this stage. There are two relevant properties of convolution which are required going forward.

The first is what happens when some function, $f(x)$, is convolved with a Dirac delta function, $\delta(x-a)$. The Dirac delta function is a real-valued function which is 0 everywhere except at the position a, where it is 1. In other words, it is an infinitely

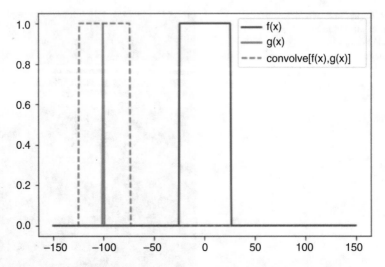

Fig. 7.3 The convolution of a top hat function centred at the origin, $f(x)$, with a Dirac delta function centred on -100, $g(x) = \delta(x + 100)$. The resulting convolution is a top hat function centred on -100, i.e. $f(x + 100)$

narrow peak at a particular position, a, and 0 everywhere else. When some $f(x)$ is convolved with the Dirac delta function, we obtain

$$f(x) \circledast g(x) = \int_{-\infty}^{\infty} f(\tau)\delta(\tau - a)d\tau = f(x - \tau), \qquad (7.5)$$

In other words:

The convolution of a function $f(x)$ with a Dirac delta function results in $f(x)$ being shifted to be centred at the position of the Dirac delta function.

Figure 7.3 show the convolution of a top hat function centred at the origin, $f(x)$ with a Dirac delta function centred on -100, $g(x) = \delta(x + 100)$. As we can see, function obtained as a result of this convolution is identical to the top hat function, except it is centred on $x = -100$ instead of $x = 0$.

The other property of convolutions is the *convolution theorem*. Say we once again have the convolution of two functions, $f(x)$ and $g(x)$. The convolution theorem states that:

$$\mathcal{F}\{ f(x) \circledast g(x) \} = \widetilde{\mathbf{F}}(k_x) \cdot \widetilde{\mathbf{G}}(k_x), \qquad (7.6)$$

where $\mathcal{F}$ denotes performing a Fourier transform (otherwise known as the Fourier operator), $\cdot$ is the dot product (i.e. point-wise multiplication), and $\widetilde{\mathbf{F}}(k_x)$ and $\widetilde{\mathbf{G}}(k_x)$

are the Fourier transforms of the functions $f(x)$ and $g(x)$, respectively, as before. What the convolution is essentially stating is:

The Fourier transform of a convolution of two functions is equivalent to the product of their Fourier transforms

From Eq. (7.6), we can obtain the corollary:

$$\mathcal{F}\left\{\widetilde{\mathbf{F}}(k_x)\circledast\widetilde{\mathbf{G}}(k_x)\right\} = f(x)\cdot g(x). \tag{7.7}$$

7.3   Optics in Reciprocal Space

Most readers should be familiar with the concept of the resolution limit in microscopy. The typical formulation of the resolution limit is the minimum separation at which two point objects can be meaningfully separated as independent objects. The often-quoted value for this resolution limit from the Rayleigh Criterion is

$$r_l \approx \frac{1.22\lambda}{2NA}, \quad r_a \approx \frac{2\lambda n}{NA^2}, \tag{7.8}$$

where r_l and r_a are the lateral and axial resolution limits, respectively, λ is the wavelength of light, NA is the numerical aperture of the imaging lens, and n is the refractive index of the media [2–5]. These formulations require several approximations and simplifications, but they are nonetheless useful quantities as the theoretical, ideal resolution limit of a microscopy system.

Rather than exploring the derivation of these diffraction limits from a geometric perspective, it is useful for us to instead discuss their origins using the notions of reciprocal space we have established previously. This will allow us to more intuitively understand the principles of SIM. First, consider a single slit with monochromatic light shining through it as shown in Fig. 7.4 which should be familiar to readers. Consider a point which is vertically displaced y from the centre of the slit with width d and a distance L from the slit. There is a path difference, shown in red in Fig. 7.4, between the top and bottom of the slit determined by $d\sin\theta'$.

In the extreme case of a maxima in the diffraction pattern, this path difference must be an integer multiple of the wavelength of light, λ, i.e. $d\sin\theta' = m\lambda$, where m is some integer. Using the simple geometry $\tan\theta = \frac{y}{L}$, the approximation that $\theta \approx \theta'$ and the paraxial approximation that $\sin\theta \approx \tan\theta \approx \theta$ then the maxima, or anti-nodal, intensity positions are given by

$$y = \frac{m\lambda L}{d}, \tag{7.9}$$

Similarly, the minima, or nodal, intensity positions are given by

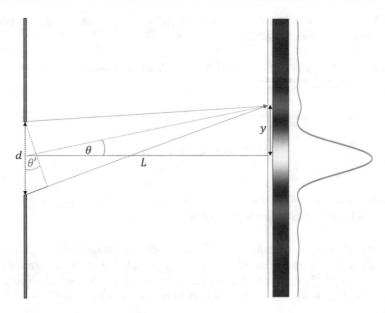

Fig. 7.4 Diagram of a single slit diffraction pattern

$$y = \frac{n\lambda L}{d}, \tag{7.10}$$

composed of binary minima and maxima, as the intensity line profile to the left of Fig. 7.4 shows. In fact, the intensity of the diffraction pattern, λ, is directly proportional to sinc$(y)^2$.

This follows if we consider a situation we have already discussed in Sect. 7.2; the top hat function. Previously we considered the amplitude of a time-varying top hat signal pulse. As shown in Fig. 7.1, the Fourier transform of this time-dependent function is a sinc function. Now, consider the transmissivity profile of a single slit. It is 0 everywhere except the slit opening, essentially a top hat centred around the centre of the slit. The amplitude profile of the electromagnetic field of the light is therefore also a top hat. As previously established, the Fourier transform of a top hat function is a sinc function, but we observe the intensity of the diffraction pattern not its amplitude. Since intensity is directly proportional to amplitude squared, it should not be surprising that the diffraction pattern is a sinc^2 function.

Now, consider the setup shown in Fig.7.5a where we place a lens at L with a focal length, $f = \frac{L}{2}$. At L we would expect to recover an image of the Fourier transform of the slit. If we then place an identical lens (i.e. a $\times 1$ magnification setup) at $\frac{3}{2}L$, then at $2L$ we might expect to recover a perfect image of the single slit, as shown in Fig. 7.5b. However, as the inclusion of the word "perfect" in the previous sentence might imply, this is not what happens. Instead at $2L$ a blurred image is observed similar to Fig. 7.5c.

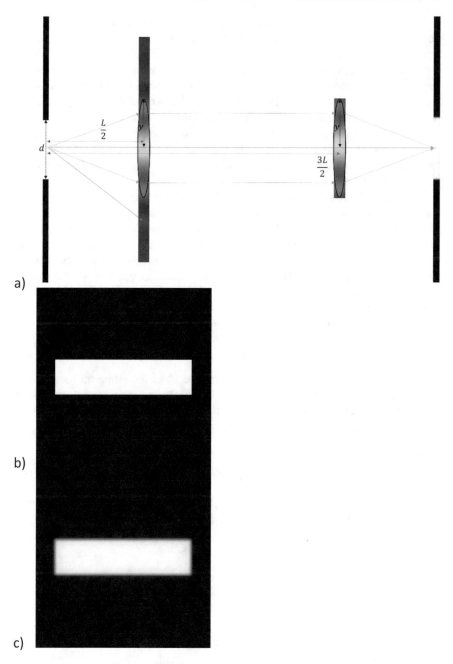

a)

b)

c)

Fig. 7.5 A simple imaging system observing a single slit. (**a**) Diagram of a single slit imaging setup with a ×1 magnification. Diffraction orders captured by the imaging setup are shown in green. Diffraction orders not captured by the imaging setup are shown in red. (**b**) Ideal image of the single slit. (**c**) Diffraction limited image of the ideal single slit

This occurs because a lens at L has a finite aperture. Therefore, only a finite portion of the diffraction pattern is collected by the imaging lens. In other words, only a finite range of spatial frequencies are collected by the imaging lens. This is shown in Fig.7.5a where certain diffraction orders are within the lens apertures and therefore propagate through the system, whereas others are not collected and therefore are missing when the final image is formed. When these spatial frequencies are used to construct an image at $2L$ by the second lens, the absence of the higher order spatial frequencies leads to an incomplete reconstruction of the original object, in this case a single slit. In effect, the imaging lens acts as a low-pass filter on the spatial frequencies. The spatial frequencies which are captured by the imaging system as said to be within the *observable region*.

In the lateral reciprocal plane, $k_x k_y$,—that is, the spatial frequency plane which is reciprocal to the lateral optical plane xy—the observable region has a radius ω_l as shown in Fig. 7.6a. In the simple example considered previously, this would be the radius of the imaging lens at L. Modern microscopy systems are considerably more complex than this and contain a number of lenses which collect the diffracted spatial frequencies. As such, we define a quantity, numerical aperture or *NA*, which describes the effective aperture of the entire imaging setup. The radius of the observable region in the $k_x k_y$ plane is defined as

$$\omega_l = \frac{2NA}{\lambda}, \tag{7.11}$$

which observant readers will note is the reciprocal of the Abbe diffraction limit [6]. The bounds of the observable region in k_z differs since the intensity spectrum in k_z is independent of the object nature and the point spread function (PSF) [5]. The bounds of the observable region in k_z can be shown to be

$$k_z = \pm \frac{\|k_{xy}\|}{2\lambda_k} \left(\omega_l - \|k_{xy}\| \right), \tag{7.12}$$

for $\|k_{xy}\| \leq \omega_l$, where $\|k_{xy}\|$ is the length of the vector denoting the lateral spatial frequencies, (k_x, k_y) and $\lambda_k = \frac{2\pi}{\lambda}$ [5]. Figure 7.6b shows the projection of the observable region onto the axial spatial frequency plane, $k_x k_z$. The maximum extension of this observable region, ω_a, is given by

$$\omega_a = \frac{\omega_l^2}{8\lambda_k}, \tag{7.13}$$

From Eq. (7.12), we can see that unlike the observable region for the lateral spatial frequencies, the observable region in k_z is bandpass limited at both high and low spatial frequencies. This results in the "missing cone" phenomenon which limits the axial spatial frequencies which are able to be collected, in turn limiting the axial resolution [7, 8]. This, coupled with the fact that for all practical cases $\omega_a \ll \omega_l$,

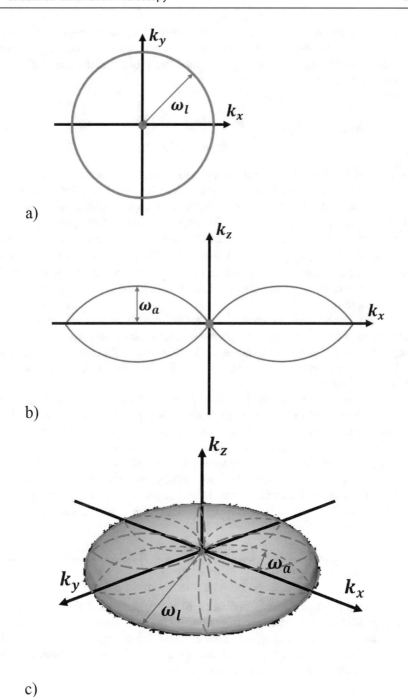

Fig. 7.6 Visualisation of the observable region of a conventional imaging system. (**a**) The projection of the observable region onto the $k_x k_y$ plane. (**b**) The projection of the observable region onto the $k_x k_y$ plane. (**c**) The full observable region for a conventional widefield microscope in reciprocal space as a 3D render

explains why the axial resolution is always worse than the lateral resolution of a system. Figure 7.6c shows the complete 3D observable region, obtained by rotating Fig. 7.6c around the k_z axis resulting in a torus-like 3D observable region which they can be described by

$$O_0(\boldsymbol{k}) = \begin{cases} 1, & \text{if } \|\boldsymbol{k}_{xy}\| < \omega_l \text{ and } \|\boldsymbol{k}_z\| < \dfrac{\|\boldsymbol{k}_{xy}\|}{2\lambda_k} \left(\omega_l - \|\boldsymbol{k}_{xy}\|\right). \\ 0, & \text{otherwise} \end{cases} \quad (7.14)$$

Only spatial frequencies within this observable region contribute to the observed image. If we consider an illumination wavelength of 561 nm and a numerical aperture of 0.7—typical values for a widefield air objective imaging system—we obtain $\omega_l = 2.50 \times 10^6 \,\text{m}^{-1}$. Since spatial frequencies have a corresponding peak-to-peak separation, for this imaging system we can see that objects which are laterally separated by less than $\frac{1}{\omega_l} = 400$ nm cannot be separated since no lateral spatial frequency with a peak-to-peak separation of less than 400 nm is captured by the imaging system. So produced images cannot contain objects with less than that lateral separation.

Individual objects which are below the resolution limit will still be observed, since a portion of their intensity is captured within the observable region and form an image. However, their apparent size will still be defined by the resolution limit, since the spatial frequencies required to construct an image with dimensions smaller than this are not available.

For a practical demonstration of the effects this observable region has on the image of a sample, take the ideal image presented in Fig. 7.7a. As shown in Fig. 7.7c, this image has a complete set of the spatial frequencies present in the underlying structure. Once a diffraction limit is imposed on the ideal image, Fig. 7.7b shows that although much of the large structures remain clearly visible, many of the fine structures such as spacing between individual leaves, small ripples of water, or patches of lichen on the rocks are no longer observable. In the power spectrum shown in Fig. 7.7d, the only spatial frequencies with non-zero values are those within the observable region.

7.4 Principles of SIM

What follows is not a rigorous derivation of the mathematical framework which underpins SIM imaging and reconstruction. Rather it is meant to provide readers with an intuitive understanding of the principles which are employed in SIM. For a more rigorous explanation of the mathematics, consult the papers references in the Further Reading section at the end of the chapter.

Fig. 7.7 The impact of a finite observable region on the resolution of an image. (**a**) an ideal image, (**b**) a resolution limited image, (**c**) the power spectrum of the ideal image. (**d**) The power spectrum of the resolution limited image. The only spatial frequencies which have any power in them are those within the observable region

a)

b)

c)

d)

This implementation of SIM is best understood through the moiré effect. Two patterns, such as those shown in Fig. 7.8a, b, which are simply two-dimensional positive sinusoids of the form:

$$f(x,y) = 1 + \cos\left(\tau_x x + \phi_x, \tau_y y + \phi_y\right), \qquad (7.15)$$

where x and y are the lateral coordinates, and τ_x and τ_y are the frequencies of the sinusoid in the x and y axis, respectively, and are multiplicatively superimposed on one another will produce a beat pattern—moiré fringes—such as those shown in Fig. 7.8c. Using the convolution theorem, the Fourier transform of this superposition is equivalent to the convolution of the Fourier transforms of the two original patterns [9], as shown in Fig. 7.8d–f.

For SIM, one of the patterns is the underlying biological structure—or more specifically the spatial distribution of fluorophores—and the other is the spatially structured excitation illumination. The moiré fringes arising from the superposition of these two structures can be coarser than either of the original patterns, meaning that information arising from biological structures beyond the resolution limit of the microscope and a known illumination pattern can be observed. These moiré fringes contain information about these super-resolution structures and the super-resolution information can be extracted from the moiré fringes, effectively extending the observable region of a microscope beyond the diffraction limit [10].

In order to provide an intuitive understanding of SIM, let us look at a simulated example of 2D SIM. Here we will return to the ideal image from Fig. 7.7a. As we have already seen, only a portion of the spatial frequencies present are within the observable region, and therefore we obtain a diffraction limited image as shown in Fig. 7.7b. We are imaging our sample with fluorescence, where the sample intensity distribution is a multiplication of the illumination distribution with the sample fluorescence distribution. If we multiplicatively superimpose on the ideal image from Fig. 7.7a, a sinusoidal pattern similar to that presented in Fig. 7.8a with a known frequency, τ, and phase, ϕ, we obtain $D(r)I(r)$ which is shown in Fig. 7.9a, where $D(r)$ is the underlying sample structure and $I(r)$ is the illumination pattern. From the convolution theorem we know that the Fourier transform of this image is a convolution of the image Fourier transform and the sinusoidal pattern Fourier transform, which results in an image with three identical copies of the same Fourier spectrum albeit it centred on different frequencies, as shown in Fig. 7.9b. We therefore obtain a resultant Fourier spectrum, $\widetilde{\mathbf{F}}'$ of the form:

$$\widetilde{\mathbf{F}}'(k) = e^{-i\phi}\widetilde{\boldsymbol{D}}(k + \tau) + \widetilde{\boldsymbol{D}}(k) + e^{i\phi}\widetilde{\boldsymbol{D}}(k - \tau), \qquad (7.16)$$

where $\widetilde{\boldsymbol{D}}$ is the Fourier transform of the sample structure. However, the observable region still limits the spatial frequencies which can be captured by the imaging system, as shown in Fig. 7.9c. This results in a Fourier spectrum described by

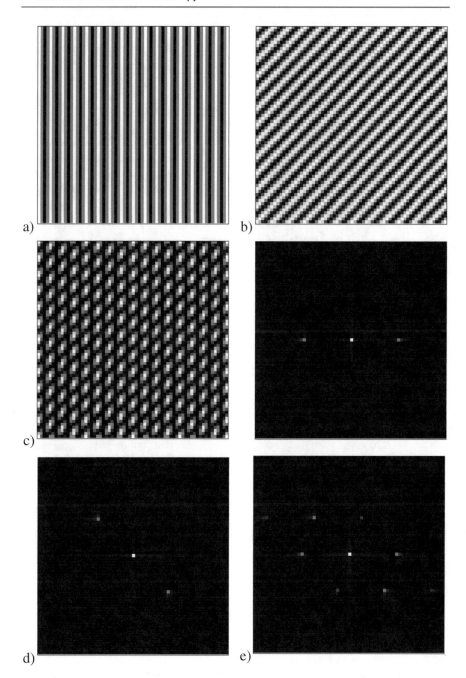

Fig. 7.8 Visualisation of moiré fringes. (**a–b**) Shows two spatially structured images. (**c**) Shows the resultant moiré fringes arising from the interference of (**a, b**). (**d–f**) Shows the respective Fourier transforms of (**a–c**). Note that (**f**) is a convolution of (**d**) and (**e**)

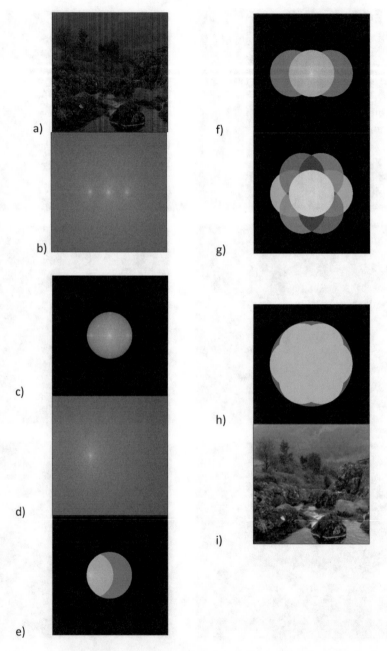

Fig. 7.9 Workflow of simulated 2D-SIM resolution enhancement. (**a**) The ideal image from Fig. 7.7a with a sinusoids pattern multiplicatively imposed on it. (**b**) Complete Fourier transform of (**a**). (**c**) The frequency components of the Fourier transform within the observable region. (**d**) Contribution to the Fourier transform in (**b**) arising from the convolution of the underlying image structure, $\widetilde{D}(k)$, and the Fourier illumination component centred on $[k_x = \tau, k_y = 0]$. (**e**) The frequency components of the convolution within the observable region. Spatial frequencies which are normally outside of the observable region, the super-resolution spatial frequencies, are shaded in

$$\widetilde{\mathbf{F}}'(\mathbf{k}) = \widetilde{\mathbf{O}}_{-1}(\mathbf{k})e^{-i\phi}\widetilde{\mathbf{D}}(\mathbf{k}+\tau) + \widetilde{\mathbf{O}}_0(\mathbf{k})\widetilde{\mathbf{D}}(\mathbf{k}) + \widetilde{\mathbf{O}}_1(\mathbf{k})e^{i\phi}\widetilde{\mathbf{D}}(\mathbf{k}-\tau), \qquad (7.17)$$

where $\widetilde{\mathbf{O}}_m$ is the optical transfer function of the imaging system, the Fourier transform of the point spread function (PSF). Let us now consider the contribution from the only one of the components, $e^{-i\phi}\widetilde{\mathbf{D}}(\mathbf{k}+\tau)$, which is shown in Fig. 7.9d. When we consider the region of this component which is within the observable region, Fig. 7.9e, clearly some spatial frequencies are ones which are already captured in conventional, widefield imaging. However, due to the lateral shift of the Fourier spectrum there are many spatial frequencies which were not previously within the observable region which are now captured by the imaging system, shown shaded in green in Fig. 7.9e. These spatial frequencies, beyond the resolution limit of the imaging system, contain the super-resolution information.

By isolating them and laterally shifting them to the correct position relative to the central contribution, the observable region is laterally extended. This enhanced resolution information is on and near to the line subtended by the Fourier components of the structured illumination pattern, shown shaded in green in Fig. 7.9f. To cover the whole range of angles the illumination pattern is rotated and the observable region can be extended in multiple directions, shown shaded in Fig. 7.9g. Rotating the illumination pattern such that the various Fourier components are spaced $\frac{\pi}{3}$ from one another gives a near isotropic improvement in lateral resolution. The comparison between a true isotropic doubling of the resolution limit and the extended observable region created by 2D-SIM is shown in Fig. 7.9h, with the difference in observable region extent shaded in magenta. Figure 7.9i shows a 2D-SIM reconstruction of the diffraction limited image shown in Fig. 7.7b. Clearly Fig. 7.9i recovers many of the fine details which are unobservable in Fig. 7.7b and appears far closer to the ideal image shown in Fig. 7.7a.

Readers may be wondering how the contributions from the 3 Fourier components can be separated. This problem is actually relatively simple to understand. If we consider any single Fourier spatial frequency within the observable region $\mathbf{k}_i$, then according to Eq. (7.17), the amplitude at the frequency $\mathbf{k}_i$ is a sum of three contributions. In other words, it is a linear equation with 3 unknowns. As is well documented, if one has N linear equations with N unknowns, then one can solve this system of linear equations and obtain the value for each of the unknowns. Since each of the contributions has not only a phase component, $e^{im\phi}$, but this phase component is different for each contribution, by varying the phase of the structured illumination

---

Fig. 7.9 (continued) green. (**f**) The extended observable region when the super-resolution spatial frequencies are laterally shifted to their correct locations. (**g**) The extended observable region when sinusoidal patterns of multiple direction are imposed on the ideal image, imaged through the observable region and the super-resolution frequencies are relocated to their correct position. (**h**) Comparison between a true isotropic doubling of the resolution limit and the extended observable region shown in (**g**). The regions where the extended observable region does not match the doubled resolution limit are shown in purple. (**i**) A 2D-SIM reconstruction of the diffraction limited image shown in Fig. 7.7b, imaged through the extended observable region

pattern for different images it is possible to solve this system of linear equations for each spatial frequency and thereby determine the contribution of each Fourier component. Since the contributions are laterally translated by a known amount, $\pm \tau$, it is possible to translate them to their correct position, leading to an extended observable region.

From this explanation it should be intuitive why, for each angle we wish to extend the observable region in 2D-SIM, we require 3 images. However, for 3D-SIM we have 7 Fourier components and so it might seem like there has been an error when we say we only need 5 images per angle. Here, we utilise the property of the Fourier transform of real functions described in Eq. (7.3) and can conclude that the $+k_z$ contributions are identical to the $-k_z$ contributions. We therefore only have 5 unknowns rather than 7, and therefore only need 5 images to solve this system of linear equations.

7.4.1 Reconstructing a SIM Image

The general process of reconstructing a SIM image from multiple raw widefield images takes several steps, described below but there are a few general points to be made first. The exact mathematical description of the reconstruction will not be covered here, but it is readily available in the literature and there a number of open-source implementations [10–14].

The SIM reconstruction relies upon information shifted in frequency by the stripes (or other patterns) in the illumination, then propagated into the recorded image. The most critical factor in gaining good SIM reconstructions is to have high contrast in the original illuminating stripes and in the produced fluorescent images. This is influenced by both the optical setup and the sample. Imaging at depth with SIM is challenging as sample-induced aberration, usually spherical aberration, reduces and finally eliminates stripe contrast as you image deeper into the sample. Two common techniques to reduce, or eliminate spherical aberration, are to vary the refractive index of the immersion oil or to use an objective with a correction collar. Either of these techniques can dramatically improve stripe contrast and hence reconstruction quality. However, it should be noted that mismatch between the immersion and sample refractive indices often mean this correction only applies over a limited depth range. Dispersion, variation of the refractive index with wavelength, in samples can also mean the required correction differs at different wavelengths. Additionally imaging samples with high out of focus background is much more challenging than samples with very low background.

The steps in the SIM reconstruction process are:

1. The images are converted into reciprocal space using a Fourier transform.
2. The multiple images at a single orientation are used as a set of linear equations to extract the components due to the different Fourier elements in the excitation light.

3. The components are then divided by the optical transfer function to normalise the amplitude of information at different spatial frequencies as the microscope detected the different spatial frequencies with a varying sensitivity.
4. The normalised components are then shifted to their correct positions and combined to generate an extended resolution image in reciprocal space. The exact shift used is usually optimised on an image by image basis by finding the highest correlation between the expected overlapping regions in the images from the multiple components. This allows correction for errors introduced by small changes in the optical setup due to issues such as temperature variations.
5. Steps 2-4 are repeated for each orientation. As previously mentioned, often there are three orientations as this minimises the number of images required while producing an almost isotropic result.
6. All orientations are combined to produce a single reciprocal space image.
7. The combined reciprocal image then has a frequency based filter applied, usually a Weiner filter is used.
8. Finally the reciprocal space image is inverse Fourier transformed to generate the final reconstruction.

The spatial filter has two effects, it produces an image which is comparable to other microscope images, it effectively reverses the division by the OTF which was applied earlier, but with a synthetic OTF which has twice the frequency limit, as the image has double the resolution. Secondly, the filter strongly suppresses high frequency noise that is amplified in the initial processing as the original noisy data is divided by the OTF. The OTF has low amplitude near the edge of the observable region meaning noise here contributes more to the final image.

In 2D SIM the above processes are performed in 2D, whereas in 3D SIM the processing uses 3D images throughout. In 3D SIM this means that the processing can account for the out of focus contributions and properly double the resolution in Z as well as in the XY directions. This does depend on the original 3D image stack including in focus information for all significant contributions. Meaning thicker samples may require large stacks to ensure that all significant fluorescence has both in focus and out of focus information included.

The initial stages of the processing are entirely linear and well constrained. The final filtering step does introduce a non-linear element but is extremely important to suppress high frequency noise components. The effective cut-off frequency of the final filtering can be tuned to play off resolution against noise, with stronger filtering eliminating more high frequency noise at the cost of lower resolution in the final reconstruction.

The output reconstruction quality depends on many factors and data quality should be carefully checked to ensure that experimental conclusions are adequately supported by the imaging data. There are resources for sample preparation, imaging and data quality assessment available [15–17].

7.5   Strengths and Limitations of SIM

Each super-resolution technique has benefits and drawbacks which need to be considered when employing them including; the resolution required (both laterally and axially), the acquisition speed, photodamage incurred, depth of imaging, and multi-colour capability [16, 18].

Although it has a modest resolution improvement—only ×2 compared to conventional microscopy much less than stimulated emission depletion (STED) microscopy and single-molecule localisation microscopy (SMLM) which offer many times that, SIM a widely applied super-resolution technique for biology for a number of reasons [16, 19]. The number of images required in order to reconstruct a super-resolution SIM image is typically 9 for 2D SIM and 15 for 3D SIM, determined by the number of lateral Fourier components in the illumination pattern—3 for 2D SIM and 5 for 3D SIM—and the number of stripe angles used to rotate the illumination pattern through, typically 3. This relatively low number of images per reconstruction has a number of benefits. Firstly, the temporal resolution is considerably higher for SIM than for point-scanning super-resolution techniques, such as STED, or SMLM techniques which require a great deal more images per reconstruction [16, 20]. Secondly, each of these images is fundamentally still a widefield-style image and therefore has a lower light-dosage per image than STED or SMLM techniques. The lower light-dosage per image combined with fewer total images required per reconstruction contributes to a low photodamage impact. Finally, SIM is easily expandable to multi-colour imaging using a wide range of standard fluorophores not requiring specific dyes, or imaging buffers [21, 22]. Overall, SIM is a super-resolution technique well suited to biological imaging, particularly imaging dynamic biological processes.

Take Home Message
- Structured illumination microscopy is one of the most widely used super-resolution techniques in modern microscopy.
- The low-light dosage, low number of images required to achieve super-resolution and multi-colour extensibility make it well suited to biological imaging, particularly in live samples.
- The quality of your final super-resolution images is critically dependent on having a high-contrast structured illumination pattern in the images used for the reconstruction.
- The resolution improvement relative to other super-resolution techniques is modest.

References

1. Körner TW. Fourier transforms. In: Fourier analysis. Cambridge: Cambridge University Press; 1988. p. 219–20.

2. Rayleigh L. XII. On the manufacture and theory of diffraction gratings. Lond Edinb Dublin Philos Mag J Sci. 1874;47(310):81–93.
3. Rayleigh L. Investigations in optics, with special reference to the spectroscope. Mon Notices R Astron Soc. 1880;40:254.
4. Pawley J. Handbook of biological confocal microscopy, vol. 236. New York: Springer; 2006.
5. Roy Frieden B. Optical transfer of the three-dimensional object. JOSA. 1967;57(1):56–66.
6. Abbe E. Beiträge zur Theorie des Mikroskops und der mikroskopischen Wahrnehmung. Archiv für mikroskopische Anatomie. 1873;9(1):413–8.
7. Behan G, et al. Three-dimensional imaging by optical sectioning in the aberration-corrected scanning transmission electron microscope. Philos Trans R Soc A. 2009;367(1903):3825–44.
8. Arnison MR, Sheppard CJR. A 3D vectorial optical transfer function suitable for arbitrary pupil functions. Opt Commun. 2002;211(1–6):53–63.
9. McGillem CD, Cooper GR. Continuous and discrete signal and system analysis. Oxford: Oxford University Press; 1991.
10. Gustafsson MGL. Surpassing the lateral resolution limit by a factor of two using structured illumination microscopy. J Microsc. 2000;198(2):82–7.
11. Gustafsson MGL, et al. Three-dimensional resolution doubling in wide-field fluorescence microscopy by structured illumination. Biophys J. 2008;94(12):4957–70.
12. Müller M, et al. Open-source image reconstruction of super-resolution structured illumination microscopy data in ImageJ. Nat Commun. 2016;7(1):1–6.
13. Lal A, Shan C, Xi P. Structured illumination microscopy image reconstruction algorithm. IEEE J Sel Top Quantum Electron. 2016;22(4):50–63.
14. O'Holleran K, Shaw M. Optimized approaches for optical sectioning and resolution enhancement in 2D structured illumination microscopy. Biomed Opt Express. 2014;5:2580–90.
15. Ball G, et al. SIMcheck: a toolbox for successful super-resolution structured illumination microscopy. Sci Rep. 2015;5:15915.
16. Schermelleh L, et al. Super-resolution microscopy demystified. Nat Cell Biol. 2019;21(1):72.
17. Spilger R, et al. Deep probabilistic tracking of particles in fluorescence microscopy images. Med Image Anal. 2021;72:102128.
18. Hell SW, et al. The 2015 super-resolution microscopy roadmap. J Phys D Appl Phys. 2015;48 (44):443001.
19. Schermelleh L, Heintzmann R, Leonhardt H. A guide to super-resolution fluorescence microscopy. J Cell Biol. 2010;190(2):165–75.
20. Leung BO, Chou KC. Review of super-resolution fluorescence microscopy for biology. Appl Spectros. 2011;65(9):967–80.
21. Wu Y, Shroff H. Faster, sharper, and deeper: structured illumination microscopy for biological imaging. Nat Methods. 2018;15(12):1011–9.
22. Allen JR, Ross ST, Davidson MW. Structured illumination microscopy for superresolution. ChemPhysChem. 2014;15(4):566–76.

Further Reading

Prior Knowledge

Hecht E. Optics. 4th ed. Cambridge: Addison Wesley Longman; 1998.
Illing L. Fourier analysis. J Pure Appl Math Ser. 2008;7:1–21.
Morin D. Fourier analysis. Cambridge: Harvard University; 2009.

Mathematical Derivations of 2D and 3D SIM

Gustafsson MGL. Extended resolution fluorescence microscopy. Curr Opin Struct Biol. 1999;9(5): 627–8.

Gustafsson MGL. Surpassing the lateral resolution limit by a factor of two using structured illumination microscopy. J Microsc. 2000;198(2):82–7.

Gustafsson MGL, et al. Three-dimensional resolution doubling in wide-field fluorescence microscopy by structured illumination. Biophys J. 2008;94(12):4957–70.

Heintzmann R, Cremer CG. Laterally modulated excitation microscopy: improvement of resolution by using a diffraction grating. Opt Biopsies Microsc Tech III. 1999;3568:185–96.

Stimulated Emission Depletion Microscopy

8

Silvia Galiani, Jana Koth, Ulrike Schulze,
and B. Christoffer Lagerholm

Contents

What You Will Learn in This Chapter

Stimulated emission depletion (STED) microscopy is in its simplest form an extension of confocal fluorescence microscopy that offers much enhanced spatial resolution in both 2D and 3D. This chapter provides a basic overview of the theory behind STED microscopy and the technology developments and modern design of the STED microscope. Like with any advanced imaging technology, it is important to implement simple testing procedures of the overall performance. This chapter

S. Galiani · U. Schulze
Wolfson Imaging Centre Oxford, MRC Weatherall Institute of Molecular Medicine, University of Oxford, Headley Way, Oxford, UK

MRC Human Immunology Unit, MRC Weatherall Institute of Molecular Medicine, University of Oxford, Headley Way, Oxford, UK

J. Koth · B. Christoffer Lagerholm (✉)
Wolfson Imaging Centre Oxford, MRC Weatherall Institute of Molecular Medicine, University of Oxford, Headley Way, Oxford, UK
e-mail: christoffer.lagerholm@imm.ox.ac.uk

© The Author(s), under exclusive license to Springer Nature Switzerland AG 2022
V. Nechyporuk-Zloy (ed.), *Principles of Light Microscopy: From Basic to Advanced*, https://doi.org/10.1007/978-3-031-04477-9_8

provides detailed examples of the testing procedures that have proven useful to ensure optimal performance of a range of STED microscopes. Finally, this chapter includes a few application image examples.

8.1   Introduction

STED (stimulated emission depletion) microscopy is a far-field super-resolution fluorescence microscopy technique that allows fluorescence imaging with, in principle, unlimited spatial resolution. The concept of STED microscopy was first introduced in 1994 [1], and the first successful experimental implementation of STED microscopy applied to a biological specimen was published in 2000 [2]. STED microscopy has subsequently become a well-established super-resolution microscopy technique suitable for imaging a broad range of biological samples with a resolution that is much smaller than the conventional diffraction-limited resolution that can be obtained by a conventional confocal microscope. STED microscopy can further be efficiently combined with fluorescence correlation spectroscopy (FCS), thus enabling super-resolution studies of, for example, diffusion in cell membranes [3, 4]. The implementation of STED-FCS, which follows the same general approach as for conventional FCS, has the unique ability to enable studies at observation volumes much smaller than the diffraction limit. For further details on implementation and practical guide to STED-FCS, we refer the reader to Sezgin et al. [5].

8.2   Basic Principle of the STED Microscope

STED microscopy enables imaging beyond the diffraction limit by exploiting the property that there are two distinct relaxation routes: (1) spontaneous emission, and (2) stimulated emission, whereby an excited fluorescent molecule can return to the ground state. This is illustrated in terms of a Jablonski diagram in Fig. 8.1a. In STED microscopy, as is the case for conventional confocal microscopy, photons from the excitation laser are absorbed by fluorophores in the specimen resulting in a molecular transition to a higher excited state. In the case of conventional confocal microscopy, subsequent to internal conversion, the excited molecules would ideally return, with a high yield, to the ground state via "spontaneous emission" whereby photons of light, red-shifted relative to the excitation wavelength, are emitted and detected as conventional fluorescence. In the case of STED microscopy, however, a significant fraction of excited molecules, as defined by the shape of the STED point spread function (PSF), is instead stimulated to return to the ground state by "stimulated emission" upon exposure by the STED laser. In this context, it is important to remember that relaxation by stimulated emission is not a dark process but rather is a process where relaxation to the ground state, as stimulated by the STED laser, results in the emission of a photon of identical nature (i.e., wavelength, polarization, phase) to the photon from the STED laser. Furthermore, as the photon emitted from the STED laser is not absorbed by the excited dye, this process results in an exact

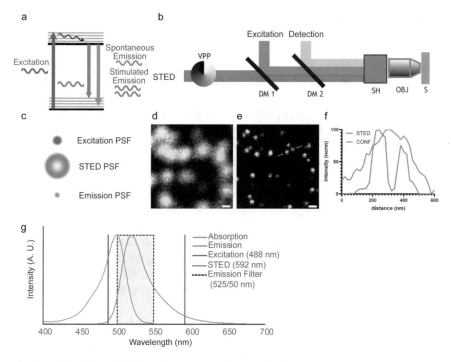

Fig. 8.1 Principles of STED microscopy. (**a**) Schematic of STED microscopy in terms of a Jablonksi diagram illustrating the transitions in electronic state during conventional fluorescence (i.e., spontaneous emission) and during stimulated emission. (**b**) Schematic of optical path of a 2D STED microscope. The excitation laser path (excitation; blue) and the STED beam (STED; pink), donut shaped via a vortex phase plate (VPP), are super-positioned in the sample plane (S) via scanning head (SH) and objective (OBJ). (**c**) Schematic of 2D cross section of excitation point-spread function (PSF)(blue), 2D-STED PSF with zero intensity at the center (red), and effective emission PSF (green) that would result from the super-position of the excitation and STED PSFs. (**d**) Example confocal and (**e**) STED image of the same field of view of fluorescent beads. Also shown in (**d, e**) are white arrows indicating the location of the fluorescent intensity profiles shown in f) for the confocal image (blue line) and STED image (pink line). Scale bars = 200 nm. (**g**) Schematic of STED microscopy in terms of asorption and emission spectra for sample stained with Alexa Fluor 488 showing the typical selection of the excitation laser (488 nm), the STED laser (592 nm), and the emission bandpass filter settings (525/50 nm) for sample stained with Alexa Fluor 488

doubling of the photon emitted by the STED laser. A key aspect in the design of the detection window in STED microscopy is thus to ensure that the emission bandpass filter does not overlap with the STED laser such that the photons that result from the stimulated emission process are, as far as the detector is concerned, effectively a dark state. A second key aspect in STED microscopy, and the reason for the requirement of exquisite time control, is that the STED laser is able to stimulate a vast majority of specific molecules to return to the ground state prior to the occurrence of relaxation by conventional spontaneous emission. Because most fluorophores have a mean excited state lifetime of a few nanoseconds, this introduces a requirement that the

STED depletion process must ideally occur within a fraction of a ns after the excitation process.

8.3   Basic Design of the STED Microscope

The basic layout of a STED microscope is the same as for traditional confocal microscopy with an additional requirement of the incorporation of a STED laser, optical elements that enable exquisite shaping of the STED laser beam, and electronics for exquisite control of the time synchronization of the STED laser relative to the excitation laser. A 2D STED implementation via a vortex phase mask is shown in Fig. 8.1b. The excitation path is based on a confocal optical scheme. The laser light is focused onto the sample via the objective lens, and the resulting fluorescence is collected back pixel by pixel via the same objective lens and detected via a high sensitivity point detector such as avalanche photo diodes (APDs). A pinhole aperture positioned before the detector guarantees the optical section of the sample. The shape of the STED beam is adjusted by the vortex plate, and the resulting shaped beam is spatially and temporally adjusted to closely overlap the excitation beam in the sample plan.

In the case of 2D STED, the function of the beam shaping optical elements is to engineer the point-spread function (PSF) of the STED laser to a donut pattern with a zero intensity foci in the middle (Fig. 8.1c), while any number of alternate beam shapes have also been demonstrated, including for 3D STED [6]. A prerequisite for STED microscopy is that the centroid of the conventional diffraction-limited PSF of the excitation laser (Fig. 8.1c) is perfectly co-aligned in space with the centroid of the PSF of the STED laser, and furthermore that the arrival time of the excitation laser and STED laser at the sample plane is precisely controlled. With such exquisite spatial and temporal control, it is possible to create an effective emission PSF (or observation volume; Fig. 8.1c) that is much smaller than the conventional diffraction limit. With such a STED microscope, it is possible to acquire both a conventional confocal image of e.g. fluorescent beads with an inactive STED laser beam at conventional diffraction-limited resolution (Fig. 8.1d) and a super-resolution STED image of the same specimen with an active STED laser beam with a much improved resolution (Fig. 8.1e) as is clearly demonstrated by the representative example of fluorescence intensity line profile in Fig. 8.1f.

An important characteristic of STED microscopy is also that it is possible to tune the resolution of the STED microscope by adjusting the STED laser power. As a result, the diffraction barrier no longer limits the resolution of a STED microscope that increases with increasing intensity of the STED laser:

$$d_{\text{STED}} = \frac{d}{\sqrt{1 + A \frac{I_{\text{STED}}}{I_S}}} \tag{8.1}$$

where d is the diameter of the diffraction-limited excitation laser spot, $A > 0$ is a geometrical parameter that takes into account the shape of the STED laser beam, I_{STED} is the STED laser intensity, and I_S is the so-called saturation intensity, characteristic for a specific fluorescence label representing how efficiently it can be depleted. To increase the lateral resolution of a standard confocal microscope a donut-shaped STED beam is generated via a vortex phase mask; to increase the axial resolution of a standard confocal microscope a bottleneck STED beam is created via a pi phase mask. A combination of both masks will allow tuning the resolution in both lateral and axial direction. While theoretically, the spatial resolution can be pushed to unlimited scales, the signal-to-noise ratio is a limiting factor. In practice, this means that a typical resolution up to about 50 nm can be routinely achieved for cell studies.

In order to answer to the multiple demands coming from biological and biomedical applications, the most common implementation is a multicolor STED microscope. Such a microscope allows combining different fluorescent markers in order to disclose morphology, proximity, and co-localization of molecules at a nanoscale level. A schematic of single-color STED microscopy in terms of the absorption and emission spectra of the representative green fluorophore Alexa Fluor 488 is shown in Fig. 8.1g. In STED microscopy, just as in the case of conventional confocal microscopy, the excitation laser is selected to match closely to the peak of the absorption spectra, while the emission bandpass filter is typically selected to collect emitted light broadly around the peak of the emission spectra. Meanwhile, the selection of the STED laser, typically 592 nm for green dyes, is dictated by a requirement that the laser wavelength is red-shifted relative to the peak fluorescence emission (and the emission bandpass filter) of the specific probe. These same characteristics are representative for STED microscopy of e.g. intermediate red dyes such as Alexa Fluor 555 or TMR, which can be used for STED microscopy with a 660 nm STED laser, or more red-shifted dyes such as Alexa Fluor 594, Atto 590, Abberior Star Red, or Abberior Star 635P, which can be used for STED microscopy with a 775 nm STED laser. While different excitation and STED beams could be combined to obtain a multicolor STED image, the most reliable implementation for co-localization studies employs a single STED beam to deplete multiple excitations, so that the alignment of the distinct color channels is determined by the center of the STED PSF.

8.4   Microscope Performance Tests

To ensure the best performance of a STED microscope, it is absolutely essential to check for laser stability, point spread function (PSF) distribution of excitation and STED beams, and the co-alignment of the excitation and STED laser beams. Here, we suggest a maintenance routine to guarantee reliable and comparable STED measurements. In order to follow the presented routine it is useful to be equipped with a power meter to check the stability of the STED laser, a gold bead sample to check the alignment of the excitation and STED PSF of the beams, and a fluorescent bead sample to check alignment and resolution with fluorescence detection.

8.4.1 Laser Power Measurements

Since the STED resolution depends on the STED laser intensity power, the STED laser source needs to remain stable over time. A daily or even hourly variation of the STED power will provide different resolution effects on same dyes and so incomparable images that will nullify any data analysis. It is good practice to check weekly the laser power provided by the STED beam at different intensity levels. This measurement can be done via a power meter detector positioned on top or at the back aperture of the objective lens used to run the experiment. For these measurements, we regularly use a power sensor with a spectral range of 400–1100 nm and a maximum power rating of 500 mW (Thorlabs; S130C) and a power meter (Thorlab; PM100D). We typically measure laser powers at least every week in order to verify laser coupling stability. Different laser intensity levels can be set via software and the corresponding power recorded. A variation of more than 10% has to be considered as a possible cause for inconsistent data.

8.4.2 Gold Beads

In order to obtain a reliable STED effect, the alignment of both the excitation and STED PSF should be symmetric and straight along all directions. Such PSFs can be visualized via imaging of gold beads in reflection mode. The excitation beam should show a classical diffraction-limited point spread function (PSF) characterized by Gaussian distribution along lateral and axial directions. The STED beam in the absence of masks should have the very same profile as the excitation beam. Once a mask is applied, the beam must remain straight and show a characteristic zero intensity point in the center as shown in Fig. 8.2a. Uneven intensity distribution and tilt of the beam suggest that the optical path must be corrected (e.g., align phase plate with the beam, align mirrors, damaged lenses, or objective lens). Gold beads measurement allows also to check on the spatial alignment of the excitation and the STED beam; the beam should be coaxial along the xy-axis and positioned at the very same z-plane.

8.4.3 Fluorescent Beads, Nanorulers, and Immunostained Cell Samples

The fine alignment between excitation and STED beams should always subsequently be validated on fluorescent calibration samples such as fluorescent bead samples (Fig. 8.2b). In acquiring such STED images, just like for conventional confocal images, it is of course essential to strive to acquire images at high signal-to-noise

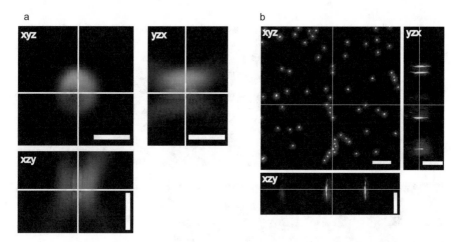

Fig. 8.2 Validation of super-position of excitation and STED beams for a well-aligned STED microscope. (**a**) Validation of spatial alignment of STED microscope with 80 nm diameter gold beads showing respectively the expected pattern of light scattered from the excitation beam (red) and the STED beam (cyan) along each axis as shown. Scale bar = 500 nm. (**b**) Validation of spatial alignment of STED microscope with fluorescent GATTA beads (GattaQuant) labeled with Atto 647 N (Bead R) showing the super-position of the confocal point-spread function (PSF; blue) and the STED PSF (orange glow) along each axis as shown. Scale bar = 1 μm

ratio (SNR) by use of either line (or frame) averaging or frame (or line) accumulation or both, and to use proper sampling according to the Nyquist theorem. In terms of sampling, this typically entails acquisition of images with a projected pixel size of at least 3× smaller than either the anticipated or determined lateral resolution. In accordance with an expected lateral resolution of around 50 nm, proper sampling would require a projected pixel size of around 15 nm.

From images of sub-diffraction-limited beads, preferably of a size of 20–40 nm in diameter, it is possible to directly visualize the effective observation volume of the STED microscope. When the highest STED laser power is applied, the maximum resolution of the microscope is measured. Tuning the STED power and combining different mask will allow an indicative calibration of the resulting observation volume. If the system is equipped with a variable pinhole aperture repeating this measurement with both a closed pinhole and an open pinhole setting allows for an evaluation of the alignment of the detection path. As a prerequisite, when the pinhole is fully opened, the STED PSF should be aligned with the excitation PSF in all directions. Upon closing the pinhole, a well-aligned STED system is characterized by a symmetric intensity distribution along the lateral direction and a straight beam along the axial direction as shown in the example in Fig. 8.2b. Furthermore, when the confocal and STED images of fluorescent beads are super-imposed, the super-resolution signal should appear in the center of the confocal image in all directions as shown in Fig. 8.2b.

Fluorescent bead samples of dimensions much smaller than the diffraction limit, usually of 20–40 nm in diameter, are also commonly used to evaluate the resolution of a STED microscope as a function of the STED laser power (Fig. 8.3). One version

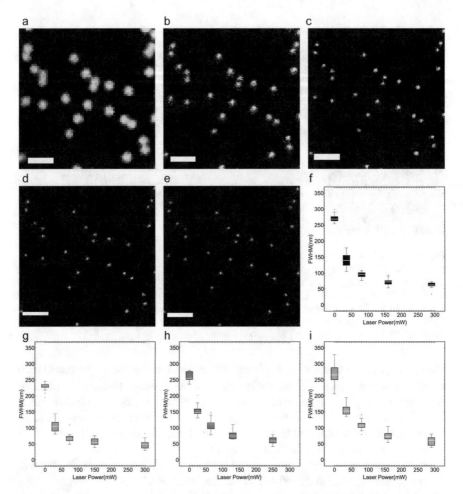

Fig. 8.3 Performance validation of STED microscope in 2D STED imaging mode with 23 nm diameter DNA origami GATTA beads (GattaQuant). (**a–e**) STED image examples of GATTA beads labeled with Atto647N (Bead R) that were imaged on Leica STED 3X microscope with a STED white 100X 1.40 NA Plan APO oil immersion objective, 633 nm laser excitation, a pulsed 775 nm STED laser, and time-gated detection ($0.5 \leq t_d \leq 7$ ns) with a spectral emission window of 649–701 nm. Scale bars = 1 μm. The resolution of the resulting STED images evaluated quantitatively by using the line profile tool in ImageJ (version 1.53j) to create line intensity profiles across select apparent single GATTA beads ($N = 15$ for each image condition) and by fitting the resulting line intensity profiles to one-dimensional spatial Gaussians in Wolfram Mathematica (version 12.0.0.0). (**f**) Box-and-whisker plot of FWHM values from images of GATTA beads labeled with Atto647N (Bead R). (**g**) Box-and-whisker plot of FWHM values from images of GATTA beads labeled with Oregon Green 488 (Bead B). The corresponding images were acquired on the same microscope system as for images in (**a–e**) except with 488 nm excitation, a CW 592 nm STED laser, and time-gated detection ($1.5 \leq t_d \leq 6.5$ ns) with a spectral emission window of 494–584 nm. (**h**) Box-and-whisker plot of FWHM values from images of GATTA beads labeled with Atto 542 (Bead G). The corresponding images were acquired on the same microscope system as for images in (**a–e**) except with 561 nm excitation, a CW 660 nm STED laser, and time-gated detection ($1.5 \leq t_d \leq 6.5$ ns) with a spectral emission window of 570–663 nm. i) Box-and-whisker plot of FWHM values from images of GATTA beads labeled with Atto 594 (Bead O). The corresponding images were acquired on the same microscope system as for images in (**a–e**) except with 561 nm excitation, a pulsed

of this uses DNA origami GATTA beads that can be ordered with the same fluorophores that are to be used in a given biological experiment. This approach has the great advantage that the ultimate resolution calibration can then be run on the very same fluorescent molecules that are to be employed in the actual experiment. Another version of the resolution calibration of a STED microscope as a function of the STED laser power is to use dot-like features, such as nuclear pores, stained with the same dye that is to be used in the actual experiment (Fig. 8.4), or other structures of dimensions much smaller than the diffraction limit such as single microtubule filaments [7]. Both of these sample types thus preclude any fluorophore-dependent effects that may originate from the consequence that the STED resolution depends on the saturation intensity, I_S, of the fluorescent dye used to label the sample (as shown in Eq. 8.1). The same STED power can thus provide a quite different STED effect depending on the dye, to the extreme situation in which the STED laser quickly photobleaches certain dyes such that no super-resolution effect can be obtained. This is because besides the absolute resolution provided by the instrument, a key parameter in STED microscopy (just like in conventional confocal micros-copy) is also the signal-to-noise ratio (SNR) as a high-quality image requires sufficient contrast to resolve the signal above the noise of the background.

From the imaging of single sub-diffraction-limited beads or sub-diffraction-limited sized cellular structures, as shown in Figs. 8.3 and 8.4, it is possible to extract quantitatively the resolution of the STED microscope from measurements of the full width at half maximum (FWHM) of line intensity profiles across the objects of interest. This is most commonly done by fitting such intensity line profiles, $I(x)$, to a one-dimensional spatial Gaussian

$$I(x) = I_{bkgd} + I \exp\left[\frac{-(x - x_0)^2}{2\sigma^2}\right] \tag{8.2}$$

where I_{bkgd} is the intensity of the background fluorescence, I is the peak intensity, x_0 is the position of the peak, σ is the Gaussian width of the peak, and the FWHM is given by

$$\text{FWHM} = 2\sqrt{2 \ln 2}\sigma. \tag{8.3}$$

An interesting alternative to these calibration measurements of sub-diffraction-limited beads and structures are also DNA origami-based GATTA-STED nanorulers consisting of two fluorescent marks of dense arrangements of multiple dye fluorophores at a specified distance of separation. Examples of confocal and STED images of such nanorulers specified to consist of two fluorescent marks separated by

---

Fig. 8.3 (continued) 775 nm STED laser, and time-gated detection ($0.5 \leq t_d \leq 7$ ns) with a spectral emission window of 603–666 nm. The STED laser power, projected pixel size, and FWHM results of the analysis of intensity line profiles are shown in Table 8.1. $N = 15$ for all analysis in Figures (**f–i**) for each image condition

Table 8.1 Image conditions and FWHM analysis results for all GATTA Beads in Fig. 8.3

Sample	STED laser power (mW; % of max available power)	Projected pixel size ($p_x = p_y$) (nm)	FWHM (mean ± standard deviation) (nm)
Bead R (Atto647N)	0	70	270 ± 12
	33 (10%)	39	140 ± 20
	82 (25%)	28	95 ± 9.4
	158 (50%)	21	73 ± 12
	291 (100%)	15	62 ± 9.7
Bead B (Oregon Green 488)	0	70	230 ± 13
	30 (10%)	44	110 ± 19
	77 (25%)	30	69 ± 14
	151 (50%)	22	56 ± 11
	300 (100%)	16	48 ± 14
Bead G (Atto 542)	0	70	260 ± 13
	27 (10%)	50	150 ± 18
	67 (25%)	35	110 ± 18
	131 (50%)	25	77 ± 13
	248 (100%)	18	61 ± 10
Bead O (Atto 594)	0	70	270 ± 31
	33 (10%)	36	160 ± 17
	82 (25%)	27	110 ± 15
	158 (50%)	20	75 ± 14
	291(100%)	15	59 ± 14

90 and 50 nm are shown in Fig. 8.5. By using these samples, it is in principle possible to directly determine the minimum STED laser power of the STED microscope that is required to resolve the two fluorescent marks as two separate structures by visual examination of a series of images acquired at different STED laser powers. Alternatively, for a full quantitative analysis, intensity line profiles across the two separate structures can be fit to the sum of two spatial Gaussians

$$I(x) = I_{bkgd} + I_1 \exp\left[\frac{-(x - x_{01})^2}{2\sigma^2}\right] + I_2 \exp\left[\frac{-(x - x_{02})^2}{2\sigma^2}\right] \quad (8.4)$$

where I_{bkgd} is the intensity of the background fluorescence, I_1 and I_2 are the respective peak intensities of each respective fluorescent mark, x_{01} and x_{02} are the respective positions of each peak, and σ is the Gaussian width of the peaks. From such fits, the FWHM is again given by Eq. (8.3), while the separation distance between the two fluorescent marks is given by

$$\text{Peak Separation} = |x_{01} - x_{02}| \pm \sqrt{2\sigma^2} \quad (8.5)$$

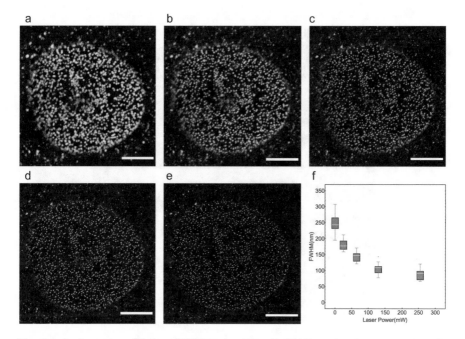

Fig. 8.4 Performance validation of STED microscope in 2D STED imaging mode with cells immunostained for nuclear pore complex protein Nup-153. (**a–e**) HeLa cells were fixed with paraformaldehyde, permeabilized with Triton X-100 (0.1%), blocked with bovine serum albumin (10%), and immunostained with primary antibody to Nup-153 (Abcam; ab24700; 1:500 dilution) and secondary goat anti-mouse IgG labeled with Alexa Fluor 532 (Thermo Fisher Scientific; A11002; 1:500 dilution), mounted in Prolong Gold mounting media, and imaged on a Leica STED 3X microscope with a STED white 100X 1.40 NA Plan APO oil immersion objective, 514 nm laser excitation, a continuous wave (CW) 660 nm STED laser, and time-gated detection ($1.5 \leq t_d \leq 6.5$ ns) with a spectral emission window of 550–624 nm. The STED laser power and projected pixel size ($p_x = p_y$) for each image was respectively (**a**) 0 mW and 74 nm, (**b**) 26 mW (10% of maximum available laser power) and 46 nm, (**c**) 65 mW (25%) and 32 nm, (**d**) 129 mW (50%) and 23 nm, and (**e**) 255 mW (100%) and 17 nm. The resolution of the resulting STED images was evaluated quantitatively as described in Fig. 8.3 by using the line profile tool in ImageJ (version 1.53j) to create line intensity profiles across select apparent single NUP complexes ($N = 15$ for each image condition). The full width-half maximum (FWHM) results of the intensity line profiles from this analysis are shown in traditional box-and-whisker plots in (**f**). The mean FWHM values ($\pm$ standard deviation) were respectively (**a**) 250 $\pm$ 32 nm, (**b**) 180 $\pm$ 16, (**c**) 140 $\pm$ 15 nm, (**d**) 110 $\pm$ 17 nm, and (**e**) 88 $\pm$ 17 nm. Scale bars = 5 μm

8.4.4 Time Alignment and Gating

STED microscopy can be implemented via pulsed or continuous (CW) laser sources. The classical implementation employs pulsed lasers both for the excitation and STED beam. A temporal misalignment of the excitation and STED beams will result in a confocal blurring around the STED image. Shifting the time synchronization of the STED laser with respect to the excitation laser will allow a perfect time alignment between the beams resulting in a well-contrasted super-resolution image. Over the

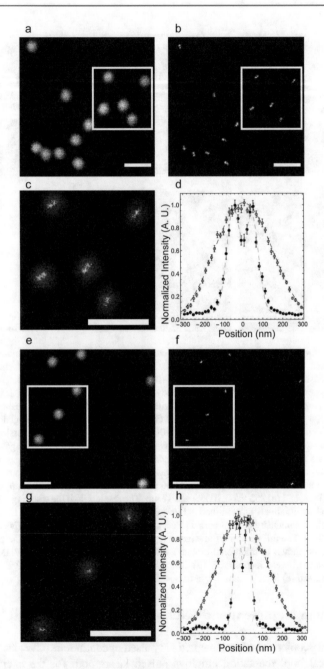

Fig. 8.5 Performance validation of STED microscope with GATTA-STED nanorulers (GattaQuant). Confocal, STED images, and image analysis thereof of GATTA-STED nanorulers consisting of two fluorescent marks of dense arrangements of the dye ATTO 647N that are separated by, respectively, (**a–d**) 90 nm (STED 90R) and (**e–h**) 50 nm (STED 50R). Images were acquired on a Leica STED 3X microscope with a STED white 100X 1.40 NA Plan APO oil immersion objective, 633 nm laser excitation, a pulsed 775 nm STED laser, and time-gated detection ($0.3 \leq t_d \leq 8$ ns) with a spectral emission window of 640–752 nm. The images were

years, a CW implementation has been favored over the pulsed one due to a reduced intensity of the STED beam delivered to the sample. In this case, a pulsed excitation laser is combined with a CW STED laser. In such an implementation, the confocal blurring will always appear and a time gating of the detected photons is needed as shown for the image examples in e.g. Figs. 8.3 and 8.4.

8.5 Application Examples

8.5.1 STED Imaging in Cells

STED microscopy is nowadays a well-known technique to disclose features and conduct co-localization studies at a nanoscale level. Here, we discuss the detailed procedure to acquire reliable multicolor STED images on a well-aligned and calibrated system.

The simplest multicolor STED imaging is done with dyes that have distinct excitation and preferably also distinct emission spectra but which can be depleted with a single STED laser. One of the best examples of such dye combinations are intermediate red dyes such as Alexa Fluor 594 or Atto 590 in combination with far red dyes such as Abberior Star Red or Abberior Star 635P (see example images in Fig. 8.6). Just as was the case for the STED images of the calibration samples shown in Figs. 8.3 and 8.4, it is also possible in these instances to tune the resolution by adjusting the STED laser power. One caveat, however, is that the SNR of the resulting images typically decreases with increasing STED laser power although one possibility in these instances is to use image deconvolution in order to enhance the contrast of such images as shown in Fig. 8.6b, h. A major advantage of this

---

Fig. 8.5 (continued) acquired in line scanning mode with a projected pixel size of $(15)^2$ nm^2 and with a STED laser power of respectively (**a, e**) 0 mW, (**b**) 82 mW (25%), and (**f**) 240 mW (80% of maximum available power). The resolution of the resulting STED images was evaluated quantitatively by using the line profile tool in ImageJ (version 1.53j) to create line intensity profiles across select apparent single nanorulers in the orientation of the two fluorescent marks as detected in the STED images ($N = 15$ for each image condition) and by fitting the resulting line intensity profiles for the confocal data to a one-dimensional spatial Gaussian, and the STED data to the linear sum of two one-dimensional spatial Gaussians (Wolfram Mathematica; version 12.0.0.0). (**a, e**) Confocal image data. (**b, f**) STED image data. (**c, g**) Magnified superimposed confocal (blue) and STED data (orange glow) of nanorulers from ROIs outlined by white squares in (**a, b, e,** and **f**). Scale bars = 1 μm. (**d, h**) Average (± standard error of the mean) of the intensity line profiles for (**d**) STED 90R and (**h**) STED 50R nanoruler image data from the confocal image data (open circles) and the STED image data (closed circles). Also shown in (**d, h**) are the best fits to the average confocal data (blue dashed line) and the average STED data (dashed orange line). The average FWHM (± standard deviation) of the confocal data for the STED 90R nanoruler was 326 ± 20 nm, while the average peak separation between the two distinct fluorescent marks in the STED data was 86 ± 20 nm. The average FWHM (± standard deviation) of the confocal data for the STED 50R nanoruler was 286 ± 22 nm, while the average peak separation between the two distinct fluorescent marks in the STED data was 54 ± 6 nm

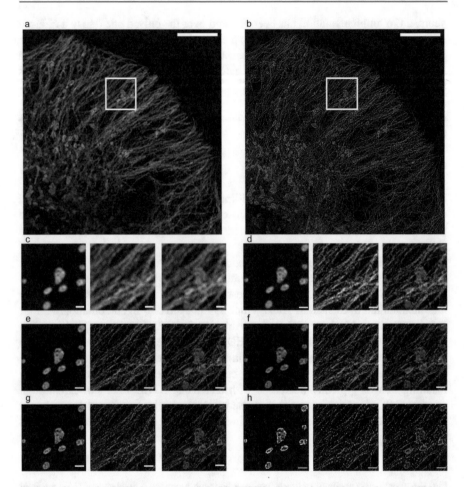

Fig. 8.6 Two-color STED image example in 2D STED imaging mode with a single pulsed 775 nm STED laser. (**a–h**) HeLa cells were fixed and stained as for image data in Fig. 8.3 except that immunostaining was done with primary antibodies to tyrosine tubulin (yellow; Sigma-Aldrich; T9028 (Clone TUB-1A2); monoclonal mouse IgG3; 1:500 dilution) and TOM20 (pink; Santa Cruz Biotechnology; sc-11,415; polyclonal rabbit IgG; 1:100 dilution) and secondary antibodies of respectively goat anti-mouse IgG labeled with Abberior Star Red (Abberior GmbH; STRED-1001; 1:500 dilution) and goat anti-rabbit IgG labeled with Atto 590 (Sigma-Aldrich; 68919; 1:500 dilution). Stained cells were imaged on a Leica STED 3X microscope with a STED white 100X 1.40 NA Plan APO oil immersion objective, in frame sequential mode with time-gated detection ($0.5 \leq t_d \leq 7$ ns), and respectively 561 nm (Atto 590) and 660 nm (Abberior Star Red) excitation and spectral emission windows of 589–643 nm (Atto 590) and 673–748 nm (Abberior Star Red). (**a**) Confocal image of tyrosine tubulin (yellow) and TOM20 (pink). The outlined ROI (white square) in (**a**) is shown in magnified view for (**c**) confocal image settings (left: TOM20; center: tyrosine-tubulin; right: overlay of TOM20 (pink) and tyrosine tubulin (yellow)); (**d**) 2D STED image with 775 nm STED laser power of 33 mW (10% of maximum STED laser power); (**e**) 2D STED image with 775 nm STED laser power of 82 mW (25%); (**f**) 2D STED image with 775 nm STED laser power of 158 mW (50%); and (**g**) 2D STED image with 775 nm STED laser power of 291 mW (100%). Also shown in (**b, h**) is the 2D STED image with 775 nm STED laser power of 291 mW (100%) after deconvolution using Huygens Professional at default settings for the Classical Maximum Likelihood Estimator algorithm (Scientific Volume Imaging). Scale bars in (**a, b**) are 10 μm, and (**c–h**) are 1 μm

approach to multicolor STED imaging is that in this instance there is no chromatic shift between the color channels. This is because in STED microscopy, it is the center of the STED PSF that determines the absolute alignment of the multicolor images [8]. This is in stark contrast to conventional confocal imaging where chromatic aberrations cannot be avoided. Thus, multicolor STED imaging with a single STED laser is an optimal arrangement for super-resolution co-localization studies as has been demonstrated previously [8]. Good practice, however, nevertheless always requires that single color control samples to optimize the detection channels for minimum cross talk, while it is also essential to acquire images at high SNR.

Multicolor STED microscopy is further possible with multiple STED and excitation beams (see example images in Fig. 8.7). But in cases of detailed co-localization studies, the chromatic aberration shift of the system in this instance must be evaluated by use of a calibration sample, usually a 100 nm diameter TetraSpeck bead sample. When the system is aligned the fluorescence emitted in the different spectral range should overlap both in confocal and STED modality. Furthermore, to avoid bleaching of the red-like dyes before their visualization, it is essential that the STED images are acquired starting with the most red-shifted fluorophore and ending with the most blue-shifted fluorophore. Additionally, if a 3D reconstruction of the sample is required, then the entire stack of images needs to be acquired channel by channel starting with the most red-shifted dye and ending with the most blue-shifted dye. Noteworthy here is that, due to the high intensity associated with the STED laser beam, photobleaching can also be a limiting factor in STED microscopy. Consequentially, it might be easy to acquire a single image of the sample at maximum STED power and resolution but much more challenging to acquire a z-stack of the same sample that will guarantee a 3D reconstruction of the region of interest. Depending on the needed resolution to disclose the required features of a study, the employed STED power might therefore be reduced following the resolution calibration. When a multicolor STED image is required for a certain experiment, the scientist would further need to decide for example if the images should provide the same resolution per each fluorescent dye or if it is acceptable to acquire the images at the maximum achievable resolution per each dye. The system needs to be calibrated and the STED power must be tuned accordingly. Depending on the analysis that is run on the acquired images, one solution might be favorable to the other.

8.5.2 STED Imaging in Deep Tissue

The very principle that STED imaging is based upon makes it unfortunately susceptible to light aberrations and scattering that occur with increasing depth in thicker samples. Yet a precise spatial and temporal overlap of the excitation and STED depletion laser beams is required to obtain the narrow excitation focus as explained in this chapter. Thus, light scattering as induced by thick or optically dense samples reduces the ability to optimize the beam shape used for the STED depletion beam. The deeper within or the denser a sample is, the lower the STED efficiency and

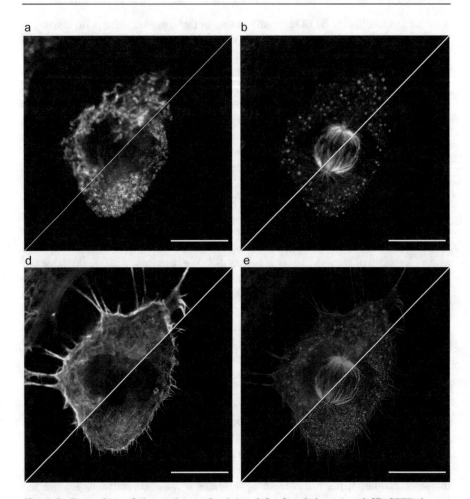

Fig. 8.7 Comparison of three-color confocal (top left of each image) and 2D STED images (bottom right of each image). (**a**) Mitochondria stained as described for data in Fig. 8.3 but with primary antibody to TOM20 (Santa Cruz Biotechnology;sc-11415; polyclonal rabbit IgG; 1:100 dilution) and secondary antibody labeled with Abberior Star 635P (Abberior GmbH; ST635P-1002; 1:500 dilution). (**b**) Tyrosine-Tubulin stained with primary antibody (Sigma-Aldrich; T9028 (Clone TUB-1A2); monoclonal mouse IgG3; 1:500 dilution) and secondary antibody labeled with Alexa 555 (Thermo Fisher Scientific; A32727; 1:500 dilution). (**c**) F-actin stained with phalloidin labeled with Oregon Green 488 (Thermo Fisher Scientific; O7466; 1 unit per coverslip). (**d**) Superimposed three-color images of mitochondria (red), tyrosine tubulin (green) and F-actin (blue). Stained cells were imaged on a Leica STED 3X microscope with a STED white 100X 1.40 NA Plan APO oil immersion objective with a projected pixel size of 30 nm, in frame sequential mode in the following order: (1) Abberior Star Red emission channel (**a** excitation at 633 nm; STED depletion at 775 nm at STED laser power of 82 mW (25%); spectral emission window of 656–737 nm; time-gated detection of $0.5 \leq t_d \leq 7$ ns), (2) Alexa Fluor 555 emission channel (**b** excitation at 561 nm; STED depletion at 660 nm at STED laser power of 67 mW (25%); spectral emission window of 566–613 nm; time-gated detection of $1 \leq t_d \leq 8.5$ ns), and (3) Oregon Green emission channel (**c** excitation at 488 nm; STED depletion at 592 nm at STED laser power of 77 mW (25%); spectral emission window of 494–549 nm; time-gated detection of $1.5 \leq t_d \leq 6.5$ ns). Scale bars = 10 μm

therefore the higher the STED power requirements to obtain any super-resolution images. In addition, the often weaker signal requires longer acquisition times and leads to increased photobleaching and a higher signal-to-noise level. These limitations mean that albeit it is possible, STED imaging is much more challenging in thicker or optically denser samples when compared to thin samples close to the cover glass. The achievable light penetration depth and resolution depends on the optical properties of the sample, but for each, there is ultimately a limit at which STED imaging may not yield higher resolution images, then airy-scan imaging, conventional confocal imaging, and ultimately multi-photon imaging.

At present most super-resolution studies have been imaging samples of up to 10 μm thickness (see review: [9]), which is a standard tissue section thickness, but in a biological context is not considered deep tissue imaging. Although STED imaging of bright and very photostable fluorescent beads has been demonstrated as deep as 155 μm [10], most biological samples have more light scattering properties throughout the tissue and have fluorescence less bright and less photostable than such optimal bead standard samples. Nevertheless, various studies have demonstrated the usability and superior imaging resolution of STED in common thick biological specimen such as in several hundred micrometer thick brain slices (for protocol: [11]) or actin filament dynamics in live mouse brains up to 40 μm away from the implanted cranial cover glass window [12, 13]. Willig and Nägerl [11] have been able to visualize via STED time-lapse imaging the dynamics of dendritic spines of pyramidal neurons in living hippocampal mouse brain slices with a resolution of ~100 nm. Since the STED microscope setup is very similar to conventional laser scanning confocal and multi-photon microscopy, similar sample approaches (e.g., upright microscopes, sample mounting, objective types) and limitations (e.g., sample movement due to growth or drift, sample optical density, and viability issues) apply.

To increase the axial resolution (z dimension) in 3D stacks of samples, adaptive optics (AO) are being developed for STED [14, 15]. To achieve a higher axial resolution, multiple laser beams need to be shaped independently, so that the point-spread function is narrowed along the axial dimension. The additional beam shaping elements make maintaining the beam alignments in 3D STED even more difficult than in 2D STED with increasing aberration at increased sample depth [16]. Another approach to improve STED imaging in thicker samples is to use two-photon (2P) STED [17–19] to compensate for the changing and increasing light scattering along the z-axis of the sample. 2P STED can improve the spatial resolution achieved by regular two-photon microscopy (two to sixfold) at moderate imaging depths [17].

Taken together, STED imaging in deeper areas of biological samples in vivo and ex vivo is achievable, but requires considerably more effort to test and establish compared to conventional STED imaging with samples very close to the cover glass. It is therefore advisable for anyone interested in deep tissue STED imaging to seek out established users of the technique to learn and troubleshoot the methodology and potentially test out any envisaged sample.

Take-Home Message

The super-resolution effect in STED microscopy is obtained by the combination of stimulated emission depletion (STED) laser beam, which overlaps in space and time with a standard confocal excitation beam, and which has been engineered to specifically deplete all excited fluorophores at the periphery of a local zero (by the process of stimulated emission) while all fluorophores at the center of the local zero of the STED beam are permitted to relax by conventional spontaneous emission, thus resulting in a much improved effective emission PSF (schematic in Fig. 8.1c and image examples in Figs. 8.3, 8.4, 8.5, 8.6, and 8.7).

STED microscopy has become a mature super-resolution microscopy technique that is routinely applied for single- and multicolor imaging in thinner specimen (<10 μm) such as fixed cells. As with any advanced technique however it is essential to fully validate the performance of the STED microscope prior to imaging of the specimen of interest. This chapter introduces a range of such calibration samples and shows representative data that should be obtainable on a well-aligned system. Much work is also currently being done on improving the performance and usability of STED microscopy in thicker tissue samples.

References

1. Hell SW, Wichmann J. Breaking the diffraction resolution limit by stimulated emission: stimulated-emission-depletion fluorescence microscopy. Opt Lett. 1994;19(11):780–2. https://doi.org/10.1364/ol.19.000780.
2. Klar TA, Jakobs S, Dyba M, Egner A, Hell SW. Fluorescence microscopy with diffraction resolution barrier broken by stimulated emission. Proc Natl Acad Sci U S A. 2000;97(15): 8206–10. https://doi.org/10.1073/pnas.97.15.8206.
3. Eggeling C, Ringemann C, Medda R, Schwarzmann G, Sandhoff K, Polyakova S, et al. Direct observation of the nanoscale dynamics of membrane lipids in a living cell. Nature. 2009;457 (7233):1159–62. https://doi.org/10.1038/nature07596.
4. Kastrup L, Blom H, Eggeling C, Hell SW. Fluorescence fluctuation spectroscopy in subdiffraction focal volumes. Phys Rev Lett. 2005;94(17):178104. https://doi.org/10.1103/PhysRevLett.94.178104.
5. Sezgin E, Schneider F, Galiani S, Urbančič I, Waithe D, Lagerholm BC, Eggeling C. Measuring nanoscale diffusion dynamics in cellular membranes with super-resolution STED–FCS. Nat Protoc. 2019;14(4):1054–83. https://doi.org/10.1038/s41596-019-0127-9.
6. Tang J, Ren J, Han KY. Fluorescence imaging with tailored light. Nanophotonics. 2019;8(12): 2111–28. https://doi.org/10.1515/nanoph-2019-0227.
7. Wegel E, Gohler A, Lagerholm BC, Wainman A, Uphoff S, Kaufmann R, Dobbie IM. Imaging cellular structures in super-resolution with SIM, STED and localisation microscopy: a practical comparison. Sci Rep. 2016;6:27290. https://doi.org/10.1038/srep27290.
8. Galiani S, Waithe D, Reglinski K, Cruz-Zaragoza LD, Garcia E, Clausen MP, et al. Super-resolution microscopy reveals compartmentalization of peroxisomal membrane proteins. J Biol Chem. 2016;291(33):16948–62. https://doi.org/10.1074/jbc.M116.734038.
9. Liu W, Toussaint KC Jr, Okoro C, Zhu D, Chen Y, Kuang C, Liu X. Breaking the axial diffraction limit: a guide to axial super-resolution fluorescence microscopy. Laser Photonics Rev. 2018;12(8):1700333. https://doi.org/10.1002/lpor.201700333.

10. Yu W, Ji Z, Dong D, Yang X, Xiao Y, Gong Q, et al. Super-resolution deep imaging with hollow Bessel beam STED microscopy. Laser Photonics Rev. 2016;10(1):147–52. https://doi.org/10.1002/lpor.201500151.

11. Willig KI, Nägerl UV. Stimulated emission depletion (STED) imaging of dendritic spines in living hippocampal slices. Cold Spring Harb Protoc. 2012;2012(5):pdb.prot069260. https://doi.org/10.1101/pdb.prot069260.

12. Willig KI, Steffens H, Gregor C, Herholt A, Rossner MJ, Hell SW. Nanoscopy of filamentous actin in cortical dendrites of a living mouse. Biophys J. 2014;106(1):L01–3. https://doi.org/10.1016/j.bpj.2013.11.1119.

13. Urban NT, Willig KI, Hell SW, Nägerl UV. STED nanoscopy of actin dynamics in synapses deep inside living brain slices. Biophys J. 2011;101(5):1277–84.

14. Gould TJ, Burke D, Bewersdorf J, Booth MJ. Adaptive optics enables 3D STED microscopy in aberrating specimens. Opt Express. 2012;20(19):20998–1009. https://doi.org/10.1364/OE.20.020998.

15. Osseforth C, Moffitt JR, Schermelleh L, Michaelis J. Simultaneous dual-color 3D STED microscopy. Opt Exp. 2014;22(6):7028–39. https://doi.org/10.1364/OE.22.007028.

16. Patton BR, Burke D, Owald D, Gould TJ, Bewersdorf J, Booth MJ. Three-dimensional STED microscopy of aberrating tissue using dual adaptive optics. Opt Express. 2016;24(8):8862–76. https://doi.org/10.1364/OE.24.008862.

17. Bethge P, Chereau R, Avignone E, Marsicano G, Nagerl UV. Two-photon excitation STED microscopy in two colors in acute brain slices. Biophys J. 2013;104(4):778–85. https://doi.org/10.1016/j.bpj.2012.12.054.

18. Moneron G, Hell SW. Two-photon excitation STED microscopy. Opt Express. 2009;17(17):14567–73. https://doi.org/10.1364/oe.17.014567.

19. Ter Veer MJT, Pfeiffer T, Nagerl UV. Two-photon STED microscopy for nanoscale imaging of neural morphology in vivo. Methods Mol Biol. 2017;1663:45–64. https://doi.org/10.1007/978-1-4939-7265-4_5.

Two-Photon Imaging

9

Giuseppe Sancataldo, Olga Barrera, and Valeria Vetri

Contents

What You Will Learn in This Chapter
This chapter will provide an overview of two-photon microscopy from elements of the theory underpinning fluorescence phenomena to functioning principles of a two-photon microscope including step-by-step practical advice on how to conduct an experiment using a two-photon microscope. In this context multi-photon excitation is also taken into consideration.

G. Sancataldo · V. Vetri (✉)
Department of Physics and Chemistry – E. Segrè, University of Palermo, Palermo, Italy
e-mail: valeria.vetri@unipa.it

O. Barrera
School of Engineering Computing and Mathematics, Oxford Brookes University, Oxford, UK

Department of Engineering Science, University of Oxford, Oxford, UK

© The Author(s), under exclusive license to Springer Nature Switzerland AG 2022
V. Nechyporuk-Zloy (ed.), *Principles of Light Microscopy: From Basic to Advanced*, https://doi.org/10.1007/978-3-031-04477-9_9

By reading this chapter, you will have a synopsis of the basic principles of two-photon excitation, optical sectioning, and 3D microscopy. Furthermore, fundamentals of promising advanced methods for tissue imaging available for two-photon imaging as second harmonic generation (SHG) and fluorescence lifetime imaging microscopy (FLIM) are briefly described together with classical applications on deep tissue imaging and functional brain imaging.

9.1   Introduction

Vision is one of our primary senses and it is highly informative. This is the reason why the idea of improving the possibilities to see objects and magnifying them to better distinguish shapes, distances, colors, and details has been prioritized in many fields of experimental sciences. Lenses were already described by Seneca about 2000 years ago and first attempts of compound microscopes are reported in XVI century. Optical microscopes comprise a light source, an objective lens, and an eyepiece allowing the observer to see a magnified image with respect to the one accessible to the naked eyes. These three elements are now technologically advanced, but they still constitute the foundations of these magnificent tools. Since the construction of the first microscope, there have been numerous advances in solving questions in life sciences. Scientists became able to see what was previously hidden to their eyes as cells and, with further developments, subcellular organelles. Introducing fluorescence as a contrast method has been an incredible step toward measuring biological and chemical reactions in real time. Fluorescence is an exquisitely a quantistic phenomenon, resulting in the emission of light by a molecule after the absorption of electromagnetic energy. In the last decades, fluorescence microscopy has become a key tool in several biomedical research fields, allowing researchers to localize specific molecules and to characterize cell and tissue morphology at sub-micrometer resolution. The characteristics of the emitted light are critically dependent on the properties of the molecules including their structural, dynamic, and environmental evolution. Therefore, fluorescence signals are the fingerprint of the molecules and of their environment. Fluorescence microscopy images are maps of fluorescent molecules, which can be endogenous in the sample or extrinsic markers and their analysis provides spatial and functional information [1, 2]. Fluorescent molecules can be introduced as a probe into the specimens allowing mapping or tracking them, reporting the properties of the molecules they are interacting with and visualizing physico-chemical processes with high specificity. Due to the availability of measuring multiple observables (intensity, spectral features, lifetime), also as a function of time, the amount of information that can extracted from a single fluorescence experiment dramatically increases. Advances in digital imaging and analysis allow quantitatively monitoring, with high temporal resolution, specific signals from specimens of different nature ranging from proteins, polymeric systems, living cells, living animals [3], or solid state samples like solar panels or synthetic ceramic superconductors [4, 5].

Image resolution using optical microscopy classically remains within submicron to micron scale (hundreds of nanometers), which is still limited with respect to other methods like electron or atomic force microscopy. However, the possibility of investigating living cells/organisms with minimal perturbation, multiple colored markers and of biological structures close to physiological conditions remains a great advantage of fluorescence methods. It is worth mentioning that recently huge effort has been made in order to push fluorescence microscopy to the nanoscopic level by overcoming resolution limits imposed by diffraction. These methods are not discussed in this chapter, and the reader is referred to [6] for a review [6].

In this scenario, two (multi)-photon fluorescence microscopy has been established as a powerful tool for the imaging and the analysis of three-dimensional (3D) samples providing high spatial and temporal resolution ideal for in vivo experiments. The development of this method is a clear sign of the continuous evolution and efforts that are devoted in continuously improving optical and more specifically fluorescence microscopy possibility. Two-photon microscopy is a non-linear method, this means that it uses "higher order" light–matter interactions, involving multiple photons, to generate contrast in images. The features of this non-linear interaction produce several advantages with respect to standard methods that will be described in this chapter.

In conventional fluorescence microscopy, fluorescence is stimulated by the absorption of a single photon by a molecule raising the molecule to an excited energy state. When the molecule returns to its ground state it emits a less energetic photon. In contrast, two (multi)-photon fluorescence microscopy is based upon the simultaneous absorption of two (multiple) low-energy photons by a molecule. Since the excitation depends on the simultaneous absorption of two photons, the probability of triggering a two-photon process is extremely rare to occur at a low photon density. To enhance the probability of simultaneously absorbing two photons, a two-photon microscope concentrates a high dose of light in a small spot within the sample which generates fluorescence only from this confined region. This feature makes two-photon microscopy the most suitable tool for imaging highly scattering samples. It also provides the possibility of three-dimensional (3D) optical microscopy in a non-invasive way of thick samples (mm scale penetration depth). Moreover, compared to single-photon fluorescence microscopy, two-photon microscopy results in an improved image quality with highly reduced background signal, minimized photobleaching and photodamages as lower energy is delivered to the sample [3]. For its properties this method is gaining high popularity, for instance, in analyzing biological tissues that typically strongly scatter light, making high-resolution deep imaging highly challenging.

In the following, we will describe fundamentals of fluorescence microscopy focusing on the rationale of its application in tissue imaging and giving an introduction of functional brain imaging using two-photon microscopy. Furthermore, in this chapter, second harmonic generation (SHG) microscopy and fluorescence lifetime imaging microscopy (FLIM) methods will be described that can be conveniently coupled with two-photon microscopy to increase the level of information accessible in a single experiment.

9.2   Two-Photon Excitation Process

9.2.1   Historical Perspectives

Two-photon excitation process is a relatively old concept in quantum physics. The theoretical basis of two-photon absorption was predicted by Maria Goeppert-Mayer (the second Nobel Prize in physics awarded to a woman) in her doctoral dissertation in 1931 [7]. Since two-photon absorption is an uncommon event at normal light intensities, the experimental evidence of this phenomenon was achieved only in the 1960s mainly due to the advent of laser source capable of delivering a high photon density. The first spectroscopic report on two-photon fluorescence of CaF2:Eu2C was in 1963 by Kaiser and Garret [8]. Only years later, in 1971, two-photon fluorescence of organic dyes was demonstrated. The first applications of two-photon fluorescence in microscopy were presented at the beginning of the 1990s by Denk and colleagues, who demonstrated the potential of imaging two-photon excited fluorescence in a raster scanning microscope coupled to an ultrafast pulsed laser [9]. The development of commercially available mode-locked lasers, with high peak-power, femtosecond pulses, and repetition rates around 100 MHz was then the trigger for a fast uptake of the multi-photon method in biology [10]. Nowadays two-photon microscopy is a widespread technique that, combined with specific labeling technology of biological structures, provides a sensitive means to study a plethora of phenomena in biomedical research [11].

9.2.2   Principles of Two-Photon Excitation

Fluorescence emission commonly takes place when a single photon of the appropriate energy is absorbed by a fluorophore. The single-photon excitation process is schematically illustrated in Fig. 9.1a by means of the Jablonski diagram. This diagram illustrates the electronic states of a fluorescent molecule and the transitions between these states. In Fig. 9.1a, the singlet ground (S_0) and first electronic state (S_1) together with the vibrational energy levels in which they may exist are shown. The electronic states are arranged vertically by the energy level and vertical arrows indicate the transition between levels. The absorption of a photon (occurring in 10^{-15} s time scale) causes the transition of the fluorophore from the ground electronic state to the excited state. A fluorophore is usually excited to high vibrational levels of the excited electronic state so that a rapid relaxation process, named internal conversion (10^{-12} s time scale), occurs to the lowest vibrational level of the excited state. The fluorophore then returns to its ground state by emitting a new photon (10^{-9} s time scale). All of these processes result in a loss of energy; therefore, the emitted photon has less energy and a longer wavelength than the exciting photon. The difference between the excitation and emission wavelengths, known as the Stokes shift, is an important feature to be considered when choosing fluorescent dyes for microscopy experiments. One-photon excitation typically requires photons with energies in the ultraviolet (UV) or visible spectral range. In order to generate

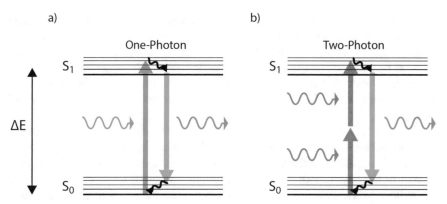

Fig. 9.1 Jablonski scheme for one-photon and two-photon excitation. (**a**) one-photon excitation results from the absorption of single photon (blue arrow).(**b**) Two-photon excitation results from the simultaneous absorption of two low-energy photons by a fluorophore. Green arrows in (**a**, **b**) indicates fluorescence emission

fluorescence by a two-photon process, two different photons must be absorbed simultaneously (within a temporal interval of 10^{-16} s as a consequence of the Heisenberg uncertainty principle) by the same fluorophore. The sum of the energy of the photons required in a two-photon absorption event must be equal to the energy needed for the single-photon absorption. This phenomenon may also occur with multiple photon absorption, e.g. three-photon excitation needs the absorption of three photons and so on. An example of a two-photon excitation process is schematically illustrated in Fig. 9.1b, where the transition to the excited state is induced by two photons of lower energy with respect to the energy gap between the two electronic levels.

For practical reasons, due to laser excitation, the two photons are usually at the same energy about one-half of the energy that is necessary to excite the molecule (two times longer than the wavelength required by a single-photon excitation), more in general the following relation should apply

$$\lambda_{1P} = \left(\frac{1}{\lambda_1} + \frac{1}{\lambda_2} \right)^{-1},$$

where λ_{1P} is the wavelength corresponding to the energy for one-photon excitation and λ_1 and λ_2 are the wavelengths of the two-photon excitation, respectively. For this reason, two-photon absorption process typically requires photons with lower energy and, in the larger part of the experiments, the excitation occurs with high-power laser in the infrared spectral range.

In general, one-photon and two-photon selection rules are different, thus it is possible that electronic transitions (occurring when a molecule absorbs light and electrons are excited to a higher energy level as depicted in Fig. 9.1), forbidden for one of them is allowed for the other one [12–15].

However, a rule of thumb exists for experiments suggesting that, if one-photon excitation occurs at a wavelength of specific excitation, two-photon will occur at about the double of this wavelength. Usually the two-photon absorption spectra are found to be wider. It is important to note that, independently of the way in which the molecule is excited, the emitted light will have the same properties in terms of spectral shape and lifetime since fluorescence occurs from the same excited state [16].

The probability that a two-photon excitation phenomenon occurs is critically lower than the single-photon one. Qualitatively this can be understood by considering that Rhodamine B, which is a common fluorescent dye, absorbs one photon each second in the day-light (roughly) but a two-photon event may occur once every 10 million years. For this reason, two-photon excitation requires a very high concentration of photons in time and, in order to increase the probability of the event, photons should be also highly focused in small spatial regions [17]. This can be obtained with high-power continuous wave lasers or using pulsed laser sources with high-energy short pulses. Two-photon excitation in optical microscopy is usually made possible by laser scanning and using titanium-sapphire lasers; these are tunable lasers which generate ultrashort pulses in the Near Infrared range. By using an ultrashort pulsed laser sources higher peak powers are reached, significantly increasing the probability of two-photon excitation maintaining low incident power [18]. The use of short pulses at high frequency allows image acquisition without irradiating the sample at high-power levels.

Therefore, two-photon excitation becomes a peerless tool for biological imaging as electromagnetic radiation at longer wavelength is delivered to the sample. This facilitates measurements on many endogenous fluorophores in cells such as NAD (P)H, flavins, collagen, and lipopigments [19]. The use of important exogenous markers like DAPI and Hoechst 3342, gold standards for nuclear staining as well as the membrane dye Laurdan, is facilitated as they require UV or blue excitation wavelengths that may induce serious damage to specimen both in terms of photobleaching and phototoxicity [20, 21].

Non-linear light–matter interaction gives rise to other important advantages with respect to conventional light microscopy such as the restriction of the excitation to a tiny volume which allows sample scanning and 3D measurements, higher penetration depth in turbid samples, and improved image quality.

9.2.3   Optical Sectioning and 3D Microscopy

The capability to image biological specimens in three dimensions represents one of the major achievements of optical microscopy. In the past, using conventional instruments, destructive sectioning procedures had to be practiced for thick samples (larger than > circa 30 μm). Indeed, out-of-focus signal can completely obscure the in-focus information and greatly reduces the contrast of acquired images. Sample sectioning in thin slices, besides being a complex and time-consuming procedure, is highly undesirable as it might modify what is observed inducing perturbation or breakage of structural organization of specimens, possibly inducing artifacts.

Two-photon microscopes exploiting non-linear nature of light–matter interaction give the possibility of producing 3D images of thick specimens preserving the structure and functionality. 3D reconstruction is performed collecting and recording a series of two-dimensional images acquired at different planes throughout the specimen. Optical sectioning, namely the ability of the microscope to reject out-of-focus fluorescence background, covers a key role in the process. The elimination of unwanted light provides greater contrast and allows correct three-dimensional (3D) reconstructions.

Optical sectioning is achieved thanks to the high localization of the excitation intrinsically provided by two-photon excitation. Indeed, as reported in the previous section, in order to obtain a significant number of two-photon absorption events, the photon density must be remarkably higher than what is required to generate the same number of one-photon absorptions. Two-photon excitation relies on the simultaneous absorption of two photons so that the number of photons absorbed in the time unit (and thus fluorescence) is proportional to the squared intensity of the excitation light.

In microscopy experiments, the illumination is focused (by the objective) on a point of the sample, this decreases its size and increases the intensity. As a result, for two-photon excitation, the amount of light absorbed across the sample is not constant, it is actually weaker in out-of-focus points. Fluorescence is only excited in a diffraction-limited region centered at the focus point. In two-photon experiments, using high-power laser beams, focused through high numerical aperture (NA) objectives, multi-photon absorption is spatially confined to a tiny femtoliter scale volume. This is significantly different from other microscopy methods which exploit one-photon excitation as confocal fluorescence microscopy which requires the physical screening of unwanted signal using a pinhole. In order to achieve one-photon excitation, laser beam focusing does not change the total amount of light passing through a plane due to linear dependence of absorbed light from excitation light intensity. In Fig. 9.2, the comparison between the two conditions (one-photon and two-photon) is sketched considering a cone shaped (focused beam) illuminated region as usually created by objectives.

In one-photon excitation conditions, fluorescence signal is generated in molecules in the whole cone, whereas in two-photon excitation conditions only molecules localized in a small volume at the focus emit light. To summarize, in two-photon microscopes fluorescence signal comes from molecules localized in a tiny region at the focus. As the focus position moves in the x–y plane, light is delivered to sequential points of the sample and the emission signal reconstructs the distribution of fluorescent objects with high resolution. A 3D reconstruction of the distribution of fluorescent molecules is then obtained through volume-rendering procedures preserving the structure and the functionality of the system. This procedure is sketched in Fig. 9.3.

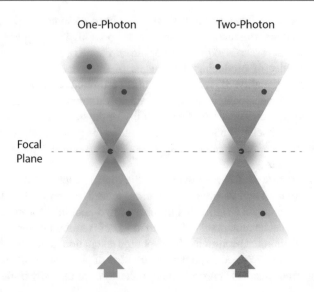

Fig. 9.2 A schematic representation of the localization of one-photon and two-photon excitation. In one-photon excitation, a continuous wave ultraviolet or visible light laser excites fluorophores throughout the volume. Out-of-focus fluorophores are excited and emit. In two-photon microscopy, an infrared laser provides pulsed illumination such that the density of photons sufficient for simultaneous absorption of two photons by fluorophores only occurs at the focal point. Arrows indicate the excitation direction

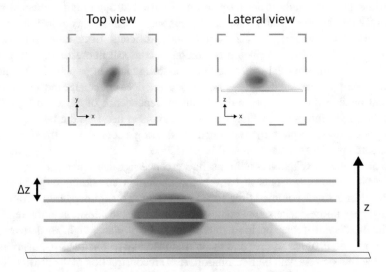

Fig. 9.3 Schematic representation of 3D acquisition by means of optical sectioning. 2D images are acquired in the x, y plane at defined Δz steps changing focus plane position

9.2.4   Scattering and Photobleaching Reduction due to Fluorescence Excitation in the NIR Window

Generally, fluorescent molecules undergo multi-photon excitation when incident light is in the near infrared (NIR) spectral region (700–1000 nm). This long-wavelength illumination light presents advantages also in the achievable penetration depth. Shifting the excitation toward the infrared region of the electromagnetic spectrum allows deeper penetration in high scattering samples like biological tissues as a result of the combination of two different features: (i) minimization of scattering effects and (ii) reduction of losses in the excitation light due to sample absorption. These two properties permit a significant decrease in image degradation, especially when imaging is performed in deep regions of scattering samples. First, the NIR excitation reduces scattering effects which results from the deflection of light from different regions of the specimen due to the inhomogeneity in refraction index. Scattering occurs in most of the samples and may cause artifacts or high reduction of detected signal. As the number of photons which are not scattered reaches the objective, the focus exponentially decreases with depth. In highly turbid sample, the intensity of fluorescence generated at the focal plane can be dramatically reduced by this effect. Scattering intensity is reduced at higher wavelengths which is, in first approximation, inversely proportional to the fourth power of incident light (this is strictly valid in Rayleigh scattering conditions) [22]. Moreover, the analysis of biological tissues takes advantage of the spectral range between 650 and 1350 nm of the excitation, which is often referred as NIR window [23]. In this range, biological tissues generally do not contain endogenous molecules which absorb light (water, hemoglobin, melanin, etc.); consequently, the light penetration is maximized. The reduced excitation volume also implies the reduction of unsought background signal which balances the poorer spatial resolution compared to one-photon measurements (which require shorter excitation wavelengths). Another important advantage of non-linear excitation of fluorescence to consider is the lack of out-of-focus excitation which reduces photodamage during 3D imaging. Photobleaching occurs only at the focal plane, while along the excitation light path absorption is prevented, thus the integrity of the sample at molecular level is also prevented. This makes non-linear excitation of fluorescence extremely suitable for imaging and analyzing biological structures.

9.2.5   The Two-Photon Microscope

Two-photon microscopy is typically realized in laser-scanning microscopes and the main setup shares with confocal microscopes many aspects. The main difference is the light source, two-photon excitation in optical microscopy is usually made possible by titanium-sapphire lasers. These are tunable lasers which generate ultra-short pulses in NIR range. The use of short pulses at high frequency allows image acquisition without irradiating the sample at high-power levels. The pulsed laser is focused to a diffraction-limited spot and scanned in a raster over the sample. When

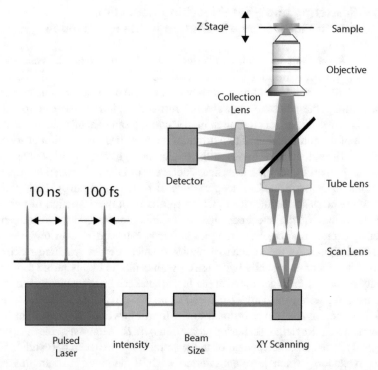

Fig. 9.4 A schematic of typical components in a two-photon scanning microscope. The system contains of a high-peak-power pulsed laser, a scanning microscope, and a high-sensitivity detector

the tiny spot overlaps with fluorophores of the sample, fluorescence signal is generated selectively within the small spot. An image is formed as a well-ordered 2D sequence of points by collecting the fluorescence by a single detector. The temporal signal of the detector is mapped to the corresponding raster-scanned point by the data acquisition computer. Figure 9.4 shows the optical scheme of a typical two-photon microscope.

The light beam from the laser unit is optimized in size and intensity by means of a telescope and by an intensity modulator (intensity and beam size boxes in Fig. 9.4) and directed to the scanning system. This is a fundamental part, composed of two rotating mirrors, that is used to raster the excitation beam in the focal plane of the objective. The objective is generally used both for exciting the sample and for collecting emitted fluorescence. Fluorescence signal is separated from the excitation beam on the path using a wavelength-sensitive dichroic mirror. Finally, a single-element detector such as a photomultiplier tube (PMT) or an avalanche photodiode (APD) collects the signal.

9.3   Advanced Analysis and Applications

9.3.1   Second Harmonic Generation (SHG)

When the energy density at the focal spot of a microscope is sufficiently high, other interesting non-linear optical effects can be observed. These optical phenomena can be used as complementary data in an optical microscope to study biological samples. Among these phenomena, second harmonic generation (SHG) is one of the widespread forms of biological non-linear microscopy. SHG is a second order coherent process in which two lower energy photons are up-converted to exactly twice the incident frequency of an excitation laser. It results from phase matching and sum of light fields induced in ordered non-centrosymmetric structures. This means that a scattered beam is produced at half the wavelength of the illumination one in phase with the input. SHG is essentially an instantaneous process, which makes it distinguishable from other fluorescence phenomena occurring in the nanosecond timescale. SHG imaging was implemented first time to biological imaging in 1986 by Freund [24] and can be performed simultaneously with multi-photon excitation measurements using the same incident light. Since SHG signal does not require the absorption phenomenon, these measurements present reduced photobleaching and phototoxicity effects with respect to fluorescence methods. Given that the signal is generated using same laser sources as the ones of two-photon microscopy, the same advantages discussed for two-photon microscopy hold. In general, molecules that are strongly and directionally affected by electric field generate SHG signal. In biological specimens many types of spatially ordered structures such as muscle myosin lattices or microtubules, polysaccharides as cellulose and starch, are able to generate sufficient amount of SHG and can be identified without performing any sample staining [25–29]. Most studies in biosciences carried out by SHG microscopy is focused on the visualization of collagen fibers in the extracellular matrix and in different kinds of organs and connecting tissues. The signal is highly sensitive to collagen structural organization and it is reported to change in diseased tissues [30]. In the last years, SHG has risen back to prominence as a powerful contrast mechanism for label-free imaging of biological specimens in physiological as well as in disease state enhancing basic research possibilities in biology and medicine and providing quantitative tool for diagnosing a wide range of diseases [31, 32].

9.3.2   Fluorescence Lifetime Imaging Microscopy (FLIM)

The advent of two-photon microscopy has provided an alternative way to obtain improved 3D images due to its optical sectioning capability and its intrinsic enhanced contrast. In most common applications, fluorescence intensity from a chromophore is measured in a specific spectral window. However, the fluorescence is not only characterized by the steady-state emission spectrum, it has also a characteristic lifetime. The fluorescence lifetime is defined as the time the fluorophore remains in the excited state. For a fluorophore, generally the lifetime (τ) is given by

$$\tau = \frac{1}{\Gamma + k_{nr}},$$

where Γ and k_{nr} are the rates of radiative and non-radiative processes from the excited state. In a real measure, as fluorescence is a random process, the fluorescence intensity presents the following exponential time dependence:

$$I(t) = I(0)e^{-t/\tau},$$

where $I(0)$ is the intensity at time zero (upon excitation). The fluorescence lifetime τ can vary from picosecond to nanosecond range and it is a peculiar characteristic of the molecule constituting a fingerprint for every dye and its environment.

Two-photon microscopes, as with other fluorescence microscopes, allow equipment upgrades that make the measure of fluorescence as a function of time (in a sub nanosecond timescale) at each pixel of the image, thus allowing fluorescence lifetime imaging microscopy (FLIM) measurements. FLIM is a fluorescence imaging technique where the contrast is based on the fluorescent lifetime of chromophores in the sample. Two-photon setups take advantage from short pulsed laser sources, to perform fluorescence lifetime measurements recording the time decay of the signal after each short excitation pulse. For example, an 80 MHz pulsed laser, commonly used in two-photon microscopy, provides an excitation pulse every 12.5 ns making accessible measurement of fluorescence decays in this timescale. In order to achieve this aim, time-correlated single-photon counting (TCSPC) is commonly used [33, 34] and fluorescence decays can be measured at discrete locations during raster scan making accessible this valuable observable to imaging. Figure 9.5

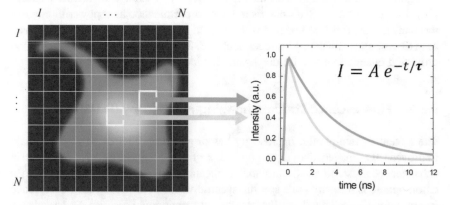

Fig. 9.5 Fluorescence lifetime imaging microscopy (FLIM). Contrast is given by different lifetimes measured at each pixel of the image, this is independent of the fluorescence intensity, number of fluorescence molecule of intensity of excitation light

is a schematic example of using lifetime measurements to discriminate between spectrally similar fluorophores.

FLIM measurements present several advantages when dealing with biological samples which are dynamical and highly heterogeneous. These advantages clearly overcome the apparent difficulty in data analysis and interpretation and the need of sophisticate hardware. A feature that limits the quantitative interpretation of intensity-based fluorescence images is the lack of knowledge of fluorophore concentration at different locations. A fluorescent dye, for instance, can inhomogeneously accumulate in different regions of a cell due to the intrinsic heterogeneous physicochemical features of the environment. As the measured fluorescence intensity linearly depends both on the quantum yield (i.e., the number of emitted photons relative to the number of absorbed photons) and on the number of molecules, it is not always easy to separate the two effects. This means that experiments involving the use of fluorescent dyes which report environmental properties, for example, pH (Fluorescein and carboxyfluorescein) [35] or calcium concentration (Fluo-3, Calcium green) [36] via variation in fluorescence emission intensity are non-trivial in cellular environments. The most important advantage of FLIM over fluorescence intensity imaging is that fluorescence lifetimes are independent of fluorophore concentration and laser excitation intensity. Since the fluorescence lifetime of a fluorophore is sensitive to the local environment (pH, charge, presence of quenchers, refractive index, temperature, and so forth), their measurements under a microscope offer the important advantage of contrast by spatial variations of lifetimes. Fluorescence lifetime measurements can then investigate photophysical events that are difficult or impossible to observe by fluorescence intensity imaging. Furthermore, as fluorescence lifetime provide "absolute" measurements, FLIM is certainly less susceptible to artifacts arising from chromophore inhomogeneity, photobleaching, uneven refraction index, and so on. Importantly, scattering or SHG signals can be easily discriminated from fluorescence as they are instantaneous phenomena. It is importnt to nitice that FRET (Foster Resonance Energy Transfer) is largely employed to evaluate the molecular mechanisms governing diverse cellular processes become simpler as FRET events are marked by the reduction of donor fluorescence lifetime [37].

It should be considered that often, in real experiments, fluorescence decays cannot be described by first-order kinetics, and they may present multiexponential behavior. A variety of reasons can be ascribed to this behavior which may depend, for instance, on molecular structures, inhomogeneous environments, quenching processes, presence of multiple species. The interpretation of fluorescence lifetime measurements results traditionally required specific expertise, complex fitting procedures needing a specific model of the system under studies. However, this has been overcome by recent development of FLIM analysis, introduced by M. Digman and E. Gratton at the beginning of this century [38]. The so-called phasor approach allows fit-free analysis of FLIM data and a rapid extraction of fluorescence lifetime information without the need of prior lifetime knowledge [39–42]. This method provides simple and fast mapping of fluorescence lifetime distribution in the image and, as proved by application in many fields from its

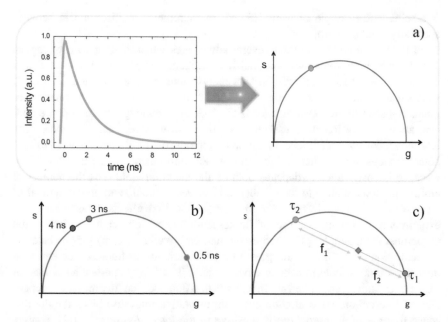

Fig. 9.6 Phasor analysis fundamentals: (**a**) the fluorescence intensity decays are Fourier transformed in a point in the phasor plot which represents the fluorescence lifetime of the fluorophore (called phasor). (**b**) Single exponential decays give rise to phasors lying on the universal circle which becomes a lifetime ruler. The ruler scale can be easily experimentally calibrated from known lifetime of well-known fluorophores (e.g., fluorescein presents a single exponential. (**c**) Complex decays are located inside the universal circle as phasors follow vector algebra. For example, the phasor corresponding to a double exponential decay $I(t) = f_1 e(-t/\tau_1) + f_2 e(-t/\tau_2)$ is located along the line connecting individual species (τ_1) and (τ_2) phasors. The distance between each phasor and the single exponential phasors on the universal circle (f_1 and f_2) represents the fraction of each component

implementation in commercial instrumentation and analysis software, phasor approach makes FLIM data analysis accessible to non-expert audience rapidly becoming mainstream. Despite the apparently complex math involved the concept behind phasor definition as well as the use of phasor analysis is really simple: the basic idea of "phasor approach" is to use phasors, rotating vectors over a polar plane, to describe fluorescence decays.

A schematic representation of phasor plot analysis is reported in Fig. 9.6.

As reported above, it is pretty common that FLIM measurements are acquired using a periodic excitation generated using a pulsed laser (with frequency ω). The fluorescence decay ($I(t)$) measured in each pixel of the image can be Fourier transformed according to the following formula:

$$F(\omega) = \int_{-\infty}^{\infty} I(t) \cdot e^{-i\omega t} dt.$$

The result is a complex number that can be divided into real ($g(\omega)$) and imaginary component ($s(\omega)$) as follows:

$$g(\omega) = \frac{\int_{-0}^{\infty} I(t)\cos(\omega t)dt}{\int_{-0}^{\infty} I(t)dt}$$

$$s(\omega) = \frac{\int_{-0}^{\infty} I(t)\mathrm{sen}(\omega t)dt}{\int_{-0}^{\infty} I(t)dt}$$

$g(\omega)$ and $s(\omega)$ are then used as the x and y components to univocally draw a phasor in a polar plot, this establishing a correlation between the lifetime and the position on of the phasor. So, for each pixel of the image, the phasor analysis transforms fluorescence decay traces to a set of phasor coordinates in a polar plot.

For a single exponential decay with characteristic time τ following simple calculations of the phasor components result:

$$g(\omega) = \frac{1}{1 + \omega\tau^2}.$$

$$s(\omega) = \frac{\omega\tau}{1 + (\omega\tau)^2}.$$

Single exponential lifetimes lie on the so-called universal circle; that is a semicircle with radius ½ going from point (0, 0) to point (1, 0). Long lifetimes are located near the origin (0 on the x axis), while short lifetimes are shifted on the circumference toward the bottom right intersection with the x axis (1 on the x axis) [38, 43] (Fig. 9.6b).

In real measurements, data in the phasor plot appears as clouds of points representing the fluorescence lifetime distributions. There is a one to one correspondence between points in the phasor plot and pixels in the acquired images it is possible to select points in the phasor plot using a colored cursor and map corresponding pixels back with selected color to the images. Importantly, as the phasors follow vector algebra the single exponential components of a complex decay add directly (using weighted sum), so that data analysis is simplified and do not require heavy assumptions. For example, the combination of two single exponential decays components (that lie on the universal circle) will lie on a straight line joining the phasors of the two fixed components. All possible weighting of the two components will result in phasors distributed along the line being quantified by the distance from the single species (Fig. 9.6c). When using phasor analysis, each molecular species is represented in a specific location in the phasor plot becoming essentially the fingerprint of the molecule in a selected environment. This is particularly useful when attempting the identification of specific molecular components and their interactions within biological samples using their intrinsic fluorescence signal. Applications of FLIM coupled with two-photon spectroscopy in biosciences are uncountable, they range from selective visualization of fluorescent molecules with overlapping spectral features, to biosensing applications and tracking of molecular interactions in living tissues [44–46].

9.3.3 Tissue Imaging

Features of two-photon excitation microscopy described in previous sections made this technique a valuable and increasing widespread method for the analysis of biological tissue both in vivo and ex vivo. The most common application is the topological analysis in 3D of tissues architecture [47–49]. Moreover, it is possible to extract additional quantitative information physicochemical features of the tissue component and on functional reaction by coupling the visualization of fluorescence observables with spectroscopic analysis [44, 50, 51].

Biological tissues can be analyzed in 3D and with different levels of accuracy by means of the use of multiple fluorescent markers. Fluorescents dyes are used for a wide range for applications to label biomolecules, mark organelles or specific targets, and to monitor physicochemical properties of the sample allowing dynamic monitoring of different important parameters such as membrane potential, membrane order, reactive oxygen species (ROS) or metal ion presence, mitochondrial membrane potential, pH, and so on. Most of these dyes are accessible to multiphoton microscopy. However, two-photon microscopy has become an incomparable tool for specific applications as it opens the way to the analysis of intrinsic fluorescence of endogenous molecules in biological samples (autofluorescence) that still recently was considered an annoying background when analyzing stained samples. Instead, it is becoming increasingly evident, also owing to FLIM availability, that a large number of information can be extracted from the analysis of these signals. Indeed, as it is possible to infer, changes occurring during physiological processes, variations of the spatial distribution of intrinsic fluorophores, their concentration, chemical modifications, or change in their local environment may occur. These variations result in changes of the detected fluorescence signal. Autofluorescence measurements require minimal or no specimen treatment, thus allowing reduction of artifacts induced by exogenous molecule presence.

Endogenous chromophores are often the main players of molecular reactions of great interest for biological functions/dysfunctions and their fluorescence can be excited by means of IR laser beam and acquired in a broad region in the near-UV-visible range also allowing multicolor experiments. For reference, few examples of endogenous fluorophores in tissues are listed in Table 9.1 together with suitable single-photon excitation wavelengths which fall in the UV side of the electromagnetic spectrum.

The dominant contribution in tissues autofluorescence is given by extracellular matrices collagen and elastin, which are massively present, and they are often organized in micron-scale supramolecular arrangements. The molecular origin of

Table 9.1 Non-exhaustive list of fluorescent endogenous molecules in biological tissues, which can be excited using two-photon excitation using wavelengths between 700 and 850 nm. As a note, aromatic amino acids cannot be detected in fluorescence microscopy without the use of dedicated optics

Fluorophore	λ_{exc}	References
Aromatic amino acids	240–300 nm	[52]
Collagen	330–340 nm	[53]
Elastin	350–420 nm	[54]
Vitamin A	350–380 nm	[55]
Flavins	350–450 nm	[56]
NAD(P)H	330–380 nm	[57]
Folic acid	<400 nm	[58]

their signal is still puzzling possibly due to cross-links [53, 59] their fluorescence has been reported in different conditions and in different spectral range both for excitation and emission. Two-photon microscopy can be used to monitor chemical modification in tissues due to oxidation or AGE effects; for example, lipopigments can be excited in a range between 340 and 400 nm and to have an emission spectrum with two main peaks centered at 450 nm and 600 nm, respectively [60]. Tyrosine oxidation induced by UV can be excited at about 350 nm, these and other oxidation or AGE products contribute to the autofluorescence of biological systems in the visible range. Other interesting fluorescent molecules with relevant application are NAD (P)H and FAD which can be used to determine the redox state of the cell [61–63]. Interestingly, in recent studies, a combination of two-photon imaging, FLIM, and phasor approach was used to image cellular metabolism quantifying variations in NADH lifetime distribution [64].

The use of FLIM methodology is particularly valuable and will certainly soon bring to numerous advances overcoming the main issue of autofluorescence analysis. Indeed, interpretation of data based on autofluorescence is not always straightforward since most of the endogenous molecules are excited in superimposed spectral range in the UV and emit in the far UV or in the green region of the visible spectrum [65]. In this instance and depending on the sample, the molecular fingerprint provided by FLIM and phasor analysis can be a valuable additional information.

In Fig. 9.7, two-photon cross sections (a) and fluorescence spectra of some endogenous fluorophores (b) in biological specimens are reported. These underlay the challenge in the spectral separation, when single species and their variations need to be identified. A valid support in solving this issue is provided by phasor analysis.

The most common use of two-photon fluorescence microscopy is deep tissue imaging performed in combination with SHG that "comes for free" when performing measurements. Two-photon fluorescence microscopy combined with SHG readily provide accurate micro-scale 3D reconstruction of collagen and elastin fibrils architecture and orientation. This approach was successfully applied revealing the structure of bones, tendons, cardiovascular tissue, and cartilage orientation of elastin and collagen bundles [23, 26, 65–75]. Usually elastin is mapped by means of the

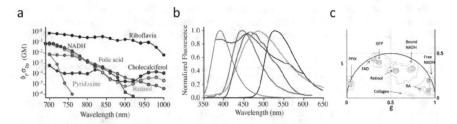

Fig. 9.7 Spectral properties of endogenous fluorophores in biological tissues. (**a**) Two-photon cross sections, (**b**) fluorescence spectra, and (**c**) lifetime distributions of endogenous fluorophores occupy a specific position in the phasor plot. Adapted from Zipfel et al. [65], Stringari et al. [42]

fluorescence signal which is more "red shifted" while SHG is used for efficient collagen detection. Collagen as its triple-helix structure is not centro-symmetric, and it is very effective in generating SHG. The high directionality of SHG provides also information on fibrils.

In Fig. 9.8, as an example, measurements acquired on lateral meniscus tissue. Representative 3D reconstruction is shown. SHG signal (green) is acquired under excitation at 880 nm and acquired at about half of incident wavelength. Fluorescence signal (red) is collected in the range 485–650 nm. Fluorescence signal is not highly specific in this case as it can be attributed to multiple contributions of endogenous molecules; however, elastin fibrils are clearly distinguishable in their beautiful arrangement for their fibrillar morphology and their higher intensity of fluorescence signal. Collagen sheets, characterized by widths between 5 and 10 µm, run along the images. The 3D reconstruction makes possible to appreciate a group of collagen bundles arranged in straight and wavy patterns which are enclosed by elastin fibers oriented along the perpendicular direction.

In tissue studies, two-photon measurements can be acquired at different depths and over different space scale (from micron to nano) to quantify orientation and periodicity of the fibrils and bundles. These measurements are important to relate the architecture of the tissues to its mechanical properties, or to specific pathological conditions.

9.3.4 Functional Brain Imaging

The biological function occurs in such a complex tissue environment that ultimately has to be studied in intact samples [77–79]. Optical recordings are the only means by which the intact brain can be studied with micrometer-spatial resolution [80, 81]. Due to the strong scattering of visible light in tissue, imaging of the deeper layers of cortex had to await the introduction of two-photon excitation fluorescence microscopy into neurobiology [17]. Brain function emerges from the coordinated activity, over time, of large neuronal populations placed in different brain regions. Understanding the relationships of these specific areas and disentangling the contributions of individual neurons to the overall function remain a central goal

Fig. 9.8 Two-photon analysis of the radial section of the central portion of the lateral meniscus. 3D reconstruction of a 60 um depth measurement. The overlap of SHG (green channel) mainly due to collagen fibrils and autofluorescence (red channel) is reported, as can be seen elastin fibers are easily distinguishable for their morphology and stronger intensity. Adapted from Vetri et al. [76]

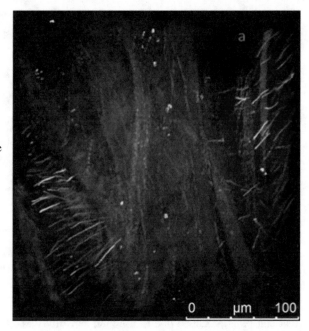

for neuroscience. In this scenario, two-photon microscopy has been proved to be the suitable tool of choice for the in vivo recording of brain activity (Fig. 9.9).

Optical advances combined with genetically encoded indicators allow, nowadays, a large flexibility in terms of spatial-temporal resolution and field of view, while keeping invasiveness in living animals to a minimum [82]. Calcium imaging with fluorescent indicators provides an optical approach to monitor action potentials and is being used systematically to measure in vivo neuronal activity. Recently, much effort has been devoted to the development of new optical architectures for advanced two-photon microscopy. Nowadays, optical architecture that allows imaging of large region (Large Field of View) of model animal brain can be achieved by custom made [83, 84] or commercial [85] microscopes. Emergent optical tools opened the way to fast beam scanning that boosted imaging recording [86–88]. The optical in vivo optical strategies enabled to tackle complementary aspects of neuronal dynamics, encompassing small networks to whole brain. The functionality of deep neurons with subcellular resolution in alive and awake animals can now be achieved which seemed unattainable only a decade ago.

9.4 Step-by-Step Protocol

Microscopes are often thought as instrument easy to manage and quickly suitable for everybody. These beautiful machines open continuously new possibilities pushing forward the number of information that can be obtained and the number of questions

Fig. 9.9 In vivo brain imaging. Schematic of two-photon microscopy through the skull of an awake mouse. Fluorescence from different neurons can be acquired in deep tissue up to hundreds of micrometers

that can be approached due to the high sensitivity and specificity of fluorescence. However, knowing the fundamentals of the methods and details of instrumental setup is of utmost important to avoid error during image acquisition that may lead to wrong results (a short glossary about fluorescence microscopy is reported in table 9.2). In this context, it has also been taken in account the importance of sample preparation, as manipulations and staining procedure may misrepresent the results [89–91]. Thus, the specimen is the most important part of the experiment and its properties should match both the characteristics to answer the scientific question and the instrumentation features.

Designing a two-photon imaging experiment is most often an iterative process. Image acquisition parameters and analysis tools usually need to be set and tested several times until a reproducible standard protocol is validated. A step by step protocol for imaging samples by means of a two-photon microscope is summased below:

- *Setting up the microscope*
  Thanks to the development of commercial turnkey pulsed laser systems, multi-photon microscopy is now available for everyone to use without extreme complexity. Turn on the laser and the microscopy hardware according to the manufactures. Ideally, the laser or any illumination source should be warmed up for a minimum of a couple of hours before imaging to avoid fluctuation of the light intensity.
- *Selecting an objective lens*
  The choice of the suitable objective for two-photon imaging of the sample is determined by three main considerations: numerical aperture (NA), working

distance, and magnification. The NA must be large enough to provide enough spatial resolution and to adequately collect fluorescence from the specimen. The working distance of the objective must be great enough to reach the desired plane of focus within the specimen. Magnification must be chosen in order to image an ample portion of the sample. Choice of immersion medium of the objective is a compromise between accessible resolution and the working distance, depending on the sample, one immersion medium can be more advantageous than others as mismatch between specimen and immersion media refraction index may induce the reduction of collected signal and image aberrations. Usually the ideal objective for typical two-photon imaging is the lowest magnification lens with the highest available numerical aperture.

- *Configuring and optimizing fluorescence channels*
  The photophysical properties of fluorescent probes are critical for two-photon experimental design. It is not so simple to predict two-photon excitation spectra from the one-photon spectra since they follow fundamentally different quantum-mechanical processes and obey very different selection rule. The two-photon spectra of several useful fluorescent molecules have been measured. Compared to one-photon spectra, two-photon spectra tend to be broader and shifted toward the blue.

  The choice of excitation wavelength is generally determined by the fluorophore(s) to be excited. As a rule, select a wavelength that is about two times longer than the wavelength required by a single-photon excitation. Select the appropriate emission range for the fluorophore(s). Commercial microscopes usually allow to select band pass filter or a specific spectral range. In case of multicolor detection, avoid overlaps among bandwidths to reduce crosstalk among different channels.

- *Setting up the specimen*
  Mount the sample on the microscope stage and focus on it using eyepiece in bright-field or epifluorescence mode. A motor in steps of hundreds of nanometers usually controls the microscope focus. Once the sample is in focus it is important to look at two-photon fluorescence. To avoid photobleaching or photodamage keep the laser power to its minimum intensity. Once the region of interest has been found optimize laser intensity till a suitable signal to noise ratio.

- *Optimizing spatiotemporal parameters*
  Commercial two-photon microscopes offer many different acquisition modes. Volumetric imaging and time-lapse modes are most common modalities and allowing the specimen to be imaged over time. Set the frame rate by independently adjusting frame size, scan speed, and number of sections. In case of advanced fluorescence microscopy technique (i.e., spectral detection and FLIM) set the appropriate parameters. Be prepared to start the experiment several times over; the best practice indeed, requires that comparison between samples is made using the same experimental parameters.

- *Data acquisition and visualization*
  It is important to look for compromise between acquisition parameters that allow the faster possible measurements with higher signal to noise ratio. It is advised to

Table 9.2 Glossary

Dichroic mirror	Mirror that selectively reflects one or more parts of the light spectrum while transmitting the rest of the spectrum. In a two-photon microscope it is commonly used to filter the fluorescence from the excitation light. Also called a dichromatic beam splitter.
Field of view (FOV)	The part of the image field that can be imaged by the microscope onto the retina or electronic detector. It depends on the magnification of the optics and on the raster scanning.
Fluorophore	Molecule in which the absorption of a photon triggers the emission of a photon with a longer wavelength, making it visible in a fluorescence microscope. Biological tissues are commonly labeled with fluorophores through immuno-staining or genetic expressed fluorescent proteins
Numerical aperture (NA)	Dimensionless number that characterizes the range of angles over which the system can accept or emit light. It is defined as $\mathbf{NA = n \cdot sin(u)}$, the product of the medium's refractive index (n) and the sine of the half angle (u) of the maximum cone of light that can enter or exit the lens.
Penetration depth	Depth to which a satisfying image quality can be achieved depends on the transparency of the sample and the imaging modality used.
Photobleaching	Process whereby a molecule is rendered non-fluorescent. Almost all fluorophores fade during light exposure.
Raster imaging	Process of rectilinear scanning by which the final image is built up partwise through the sequential illumination (and detection) of small areas (points or lines) required in microscopes that image only a fraction of the image at a time, such as confocal and multi-photon microscopes.
Resolution	Smallest distance between two points in a specimen that can still be distinguished as two separate entities. Theoretically only dependent on the NA and the wavelength of light used to image the object. In practice, the achievable resolution (the resolving power of the microscope system) also depends on additional parameters such as contrast, noise, and sampling. In single-lens microscopy, the axial resolution along the axis of the detection objective is always worse than the lateral resolution.
Scan mirrors	Rapidly moving galvanometer mirrors required in point or line scanning microscopes to sequentially illuminate the whole area of interest.

(if possible) pixel in the image where intensity is out of scale (saturation) in these conditions light arriving to the detector above this limit will not be taken in account certainly affecting image quantification. Image with saturated pixels omit a number of spatial features and they are not pretty! Imaging of samples could span a large range of timescales. In case of long acquisition, check image quality, sample defocusing, and photobleaching.

Once data are collected, use appropriate software for visualizing and analysis.

Take Home Message
- Two (multi)-photon fluorescence microscopy is a powerful tool for imaging and analyzing three-dimensional (3D) samples providing high spatial and temporal resolution ideal for deep tissues analysis of biological tissues also "in vivo."

- Fluorescence images are maps of fluorescent molecules, which can be endogenous in the sample or extrinsic markers and their analysis provide spatial and functional information. Fluorescence signals are the fingerprint of the molecules and of their environment.
- The development of advanced methods for tissue imaging available for two-photon imaging such as second harmonic generation (SHG) and fluorescence lifetime imaging microscopy (FLIM) gives the possibility of coupling spectroscopic information as complementary contrast methods revealing molecular details.

References

1. Diaspro A. Optical fluorescence microscopy: from the spectral to the nano dimension. Berlin: Springer; 2011. https://doi.org/10.1007/978-3-642-15,175-0.
2. Pawley JB. Handbook of biological confocal microscopy. 3rd ed. New York: Springer; 2006. https://doi.org/10.1007/978-0-387-45,524-2.
3. Konig K. Multiphoton microscopy in life sciences. J Microsc. 2000;200:83–104. https://doi.org/10.1046/j.1365-2818.2000.00738.x.
4. Barnard ES, Hoke ET, Connor ST, Groves JR, Kuykendall T, Yan Z, et al. Probing carrier lifetimes in photovoltaic materials using subsurface two-photon microscopy. Sci Rep. 2013;3: 1–9. https://doi.org/10.1038/srep02098.
5. Gaury B, Haney PM. Probing surface recombination velocities in semiconductors using two-photon microscopy. J Appl Phys. 2016;119:125105. https://doi.org/10.1063/1.4944597.
6. Diaspro A. Super-resolution imaging in biomedicine. Boca Raton, FL: CRC Press; 2016. https://doi.org/10.4324/9781315372884.
7. Göppert-Mayer M. Über Elementarakte mit zwei Quantensprüngen. Ann Phys. 1931;401:273–94. https://doi.org/10.1002/andp.19314010303.
8. Kaiser W, Garrett CGB. Two-photon excitation in CaF2: Eu2+. Phys Rev Lett. 1961;7:229–31. https://doi.org/10.1103/PhysRevLett.7.229.
9. Denk W, Strickler JH, Webb WW. Two-photon laser scanning fluorescence microscopy. Science. 1990;248:73–6. https://doi.org/10.1126/science.2321027.
10. Hoover EE, Squier JA. Advances in multiphoton microscopy technology. Nat Photonics. 2013;7:93–101. https://doi.org/10.1038/nphoton.2012.361.
11. Pawley JB. Confocal and two-photon microscopy: foundations, applications and advances. Microsc Res Tech. 2002;59:148–9. https://doi.org/10.1002/jemt.10188.
12. Birge RR. Two-photon spectroscopy of protein-bound chromophores. Acc Chem Res. 1986;19: 138–46. https://doi.org/10.1021/ar00125a003.
13. Birge RR, Pierce BM. A theoretical analysis of the two-photon properties of linear polyenes and the visual chromophores. J Chem Phys. 1979;70:165–78. https://doi.org/10.1063/1.437217.
14. Friedrich DM. Two-photon molecular spectroscopy. J Chem Educ. 1982;59:472–81. https://doi.org/10.1021/ed059p472.
15. Loudon R, von Foerster T. The quantum theory of light. Am J Phys. 1974;42:1041–2. https://doi.org/10.1119/1.1987930.
16. Drobizhev M, Makarov NS, Tillo SE, Hughes TE, Rebane A. Two-photon absorption properties of fluorescent proteins. Nat Methods. 2011;8:393–9. https://doi.org/10.1038/nmeth.1596.
17. Denk W, Svoboda K. Photon upmanship: why multiphoton imaging is more than a gimmick. Neuron. 1997;18:351–7. https://doi.org/10.1016/S0896-6273(00)81237-4.

18. Denk W, Piston DW, Webb WW. Multi-photon molecular excitation in laser-scanning microscopy. In: Handbook of biological confocal microscopy. 3rd ed. Madison: University of Wisconsin; 2006. p. 535–49. https://doi.org/10.1007/978-0-387-45,524-2_28.

19. Aubin JE. Autofluorescence of viable cultured mammalian cells. J Histochem Cytochem. 1979;27:36–43. https://doi.org/10.1177/27.1.220325.

20. Cunningham ML, Johnson JS, Giovanazzi SM, Peak MJ. Photosensitized production of superoxide anion by monochromatic (290–405 nm) ultraviolet irradiation of NADH and NADPH coenzymes. Photochem Photobiol. 1985;42:125–8. https://doi.org/10.1111/j.1751-1097.1985.tb01549.x.

21. Tyrrell RM, Keyse SM. New trends in photobiology the interaction of UVA radiation with cultured cells. J Photochem Photobiol B Biol. 1990;4:349–61. https://doi.org/10.1016/1011-1344(90)85014-N.

22. Pecora R. Dynamic light scattering measurement of nanometer particles in liquids. J Nanoparticle Res. 2000;2:123–31. https://doi.org/10.1023/A:1010067107182.

23. Xu C, Zipfel W, Shear JB, Williams RM, Webb WW. Multiphoton fluorescence excitation: new spectral windows for biological nonlinear microscopy. Proc Natl Acad Sci USA. 1996;93:10763–8. https://doi.org/10.1073/pnas.93.20.10763.

24. Freund I, Deutsch M, Sprecher A. Connective tissue polarity. Optical second-harmonic microscopy, crossed-beam summation, and small-angle scattering in rat-tail tendon. Biophys J. 1986;50:693–712. https://doi.org/10.1016/S0006-3495(86)83510-X.

25. Campagnola PJ, De Wei M, Lewis A, Loew LM. High-resolution nonlinear optical imaging of live cells by second harmonic generation. Biophys J. 1999;77:3341–9. https://doi.org/10.1016/S0006-3495(99)77165-1.

26. Chen X, Nadiarynkh O, Plotnikov S, Campagnola PJ. Second harmonic generation microscopy for quantitative analysis of collagen fibrillar structure. Nat Protoc. 2012;7:654–69. https://doi.org/10.1038/nprot.2012.009.

27. Cox G, Moreno N, Feijó J. Second-harmonic imaging of plant polysaccharides. J Biomed Opt. 2005;10:024013. https://doi.org/10.1117/1.1896005.

28. Gauderon R, Lukins PB, Sheppard CJR. Second-harmonic generation imaging. In: Optics and lasers in biomedicine and culture. Berlin: Springer; 2000. p. 66–9. https://doi.org/10.1007/978-3-642-56,965-4_11.

29. Moreaux L, Sandre O, Mertz J. Membrane imaging by second-harmonic generation microscopy. J Opt Soc Am B. 2000;17:1685. https://doi.org/10.1364/josab.17.001685.

30. Campagnola P. Second harmonic generation imaging microscopy: applications to diseases diagnostics. Anal Chem. 2011;83:3224–31. https://doi.org/10.1021/ac1032325.

31. Campagnola PJ, Loew LM. Second-harmonic imaging microscopy for visualizing biomolecular arrays in cells, tissues and organisms. Nat Biotechnol. 2003;21:1356–60. https://doi.org/10.1038/nbt894.

32. Friedl P, Wolf K, Harms G, von Andrian UH. Biological second and third harmonic generation microscopy. Curr Protoc Cell Biol. 2007;4:4.15. https://doi.org/10.1002/0471143030.cb0415s34.

33. Becker W, Bergmann A, Hink MA, König K, Benndorf K, Biskup C. Fluorescence lifetime imaging by time-correlated single-photon counting. Micros Res Tech. 2004;63:58–66. https://doi.org/10.1002/jemt.10421.

34. Gratton E. Fluorescence lifetime imaging for the two-photon microscope: time-domain and frequency-domain methods. J Biomed Opt. 2003;8:381. https://doi.org/10.1117/1.1586704.

35. Robertson T, Bunel F, Roberts M. Fluorescein derivatives in intravital fluorescence imaging. cells. 2013;2:591–606. https://doi.org/10.3390/cells2030591.

36. Agronskaia AV, Tertoolen L, Gerritsen HC. Fast fluorescence lifetime imaging of calcium in living cells. J Biomed Opt. 2004;9:1230. https://doi.org/10.1117/1.1806472.

37. Wallrabe H, Periasamy A. Imaging protein molecules using FRET and FLIM microscopy. Curr Opin Biotechnol. 2005;16:19–27. https://doi.org/10.1016/j.copbio.2004.12.002.

38. Digman MA, Caiolfa VR, Zamai M, Gratton E. The phasor approach to fluorescence lifetime imaging analysis. Biophys J. 2008;94:L14. https://doi.org/10.1529/biophysj.107.120154.
39. De Luca G, Fennema Galparsoro D, Sancataldo G, Leone M, Foderà V, Vetri V. Probing ensemble polymorphism and single aggregate structural heterogeneity in insulin amyloid self-assembly. J Colloid Interface Sci. 2020;574:229–40. https://doi.org/10.1016/j.jcis.2020.03.107.
40. Fricano A, Librizzi F, Rao E, Alfano C, Vetri V. Blue autofluorescence in protein aggregates "lighted on" by UV induced oxidation. Biochim Biophys Acta Proteins Proteomics. 2019;1867 (140):258. https://doi.org/10.1016/j.bbapap.2019.07.011.
41. Sancataldo G, Anselmo S, Vetri V. Phasor-FLIM analysis of Thioflavin T self-quenching in Concanavalin amyloid fibrils. Microsc Res Tech. 2020;83:811–6. https://doi.org/10.1002/jemt.23472.
42. Stringari C, Cinquin A, Cinquin O, Digman MA, Donovan PJ, Gratton E. Phasor approach to fluorescence lifetime microscopy distinguishes different metabolic states of germ cells in a live tissue. Proc Natl Acad Sci USA. 2011;108:13582–7. https://doi.org/10.1073/pnas.1108161108.
43. Ranjit S, Malacrida L, Jameson DM, Gratton E. Fit-free analysis of fluorescence lifetime imaging data using the phasor approach. Nat Protoc. 2018;13(9):1979–2004. https://doi.org/10.1038/s41596-018-0026-5.
44. Chen Y, Periasamy A. Characterization of two-photon excitation fluorescence lifetime imaging microscopy for protein localization. Microsc Res Tech. 2004;63:72–80. https://doi.org/10.1002/jemt.10430.
45. Stringari C, Abdeladim L, Malkinson G, Mahou P, Solinas X, Lamarre I, et al. Multicolor two-photon imaging of endogenous fluorophores in living tissues by wavelength mixing. Sci Rep. 2017;7:3792. https://doi.org/10.1038/s41598-017-03359-8.
46. Yellen G, Mongeon R. Quantitative two-photon imaging of fluorescent biosensors. Curr Opin Chem Biol. 2015;27:24–30. https://doi.org/10.1016/j.cbpa.2015.05.024.
47. Helmchen F, Denk W. Deep tissue two-photon microscopy. Nat Methods. 2005;2:932–40. https://doi.org/10.1038/nmeth818.
48. Mulligan S, MacVicar B. Two-photon fluorescence microscopy: basic principles, advantages and risks. In: Modern research and educational topics in microscopy. Guadalajara: Formatex; 2007.
49. Wu Z, van Zandvoort M. Two-photon microscopy. In: Comprehensive biomedical physics. Berlin: Springer; 2014. p. 165–74. https://doi.org/10.1016/B978-0-444-53,632-7.00411-1.
50. Meleshina AV, Dudenkova VV, Shirmanova MV, Shcheslavskiy VI, Becker W, Bystrova AS, et al. Probing metabolic states of differentiating stem cells using two-photon FLIM. Sci Rep. 2016;6:21853. https://doi.org/10.1038/srep21853.
51. Rubart M. Two-photon microscopy of cells and tissue. Circ Res. 2004;95:1154–66. https://doi.org/10.1161/01.RES.0000150593.30324.42.
52. Lakowicz JR. Principles of fluorescence spectroscopy. Berlin: Springer; 2006. https://doi.org/10.1007/978-0-387-46,312-4.
53. Fujimoto D, Akiba K, Nakamura N. Isolation and characterization of a fluorescent material in bovine achilles tendon collagen. Biochem Biophys Res Commun. 1977;76:1124–9. https://doi.org/10.1016/0006-291X(77)90972-X.
54. Blomfield J, Farrar JF. The fluorescent properties of maturing arterial elastin. Cardiovasc Res. 1969;3:161–70. https://doi.org/10.1093/cvr/3.2.161.
55. Sobotka H, Kann S, Loewenstein E. The fluorescence of Vitamin A. J Am Chem Soc. 1943;65: 1959–61. https://doi.org/10.1021/ja01250a042.
56. Berg RH. Evaluation of spectral imaging for plant cell analysis. J Microsc. 2004;214:174–81. https://doi.org/10.1111/j.0022-2720.2004.01347.x.
57. Schneckenburger H. Fluorescence decay kinetics and imaging of NAD(P)H and flavins as metabolic indicators. Opt Eng. 1992;31:1447. https://doi.org/10.1117/12.57704.
58. Baibarac M, Smaranda I, Nila A, Serbschi C. Optical properties of folic acid in phosphate buffer solutions: the influence of pH and UV irradiation on the UV-VIS absorption spectra and photoluminescence. Sci Rep. 2019;9:14278. https://doi.org/10.1038/s41598-019-50,721-z.

59. Deyl Z, Macek K, Adam M, Vancikova. Studies on the chemical nature of elastin fluorescence. Biochim Biophys Acta. 1980;625:248–54. https://doi.org/10.1016/0005-2795(80)90288-3.

60. Sohal RS. Assay of lipofuscin/ceroid pigment in vivo during aging. Methods Enzymol. 1984;105:484–7. https://doi.org/10.1016/S0076-6879(84)05067-9.

61. Agati G, Fusi F, Monici M, Pratesi R. Overview on some applications of multispectral imaging autofluorescence microscopy in biology and medicine. In: 19th Congress of the International Commission for Optics: Optics for the Quality of Life. New York: Springer; 2003. p. 992. https://doi.org/10.1117/12.527511.

62. Richards-Kortum R, Sevick-Muraca E. Quantitative optical spectroscopy for tissue diagnosis. Annu Rev Phys Chem. 1996;47:555–606. https://doi.org/10.1146/annurev.physchem.47.1.555.

63. Schomacker KT, Frisoli JK, Compton CC, Flotte TJ, Richter JM, Nishioka NS, et al. Ultraviolet laser-induced fluorescence of colonic tissue: Basic biology and diagnostic potential. Lasers Surg Med. 1992;12:63–78. https://doi.org/10.1002/lsm.1900120111.

64. Gómez CA, Sutin J, Wu W, Fu B, Uhlirova H, Devor A, et al. Phasor analysis of NADH FLIM identifies pharmacological disruptions to mitochondrial metabolic processes in the rodent cerebral cortex. PLoS One. 2018;13:e0194578. https://doi.org/10.1371/journal.pone.0194578.

65. Zipfel WR, Williams RM, Christiet R, Nikitin AY, Hyman BT, Webb WW. Live tissue intrinsic emission microscopy using multiphoton-excited native fluorescence and second harmonic generation. Proc Natl Acad Sci USA. 2003a;100:7075–80. https://doi.org/10.1073/pnas.0832308100.

66. Andrews SHJ, Rattner JB, Abusara Z, Adesida A, Shrive NG, Ronsky JL. Tie-fibre structure and organization in the knee menisci. J Anat. 2014;224:531–7. https://doi.org/10.1111/joa.12170.

67. Berezin MY, Achilefu S. Fluorescence lifetime measurements and biological imaging. Chem Rev. 2010;110:2641–84. https://doi.org/10.1021/cr900343z.

68. Brockbank KGM, MacLellan WR, Xie J, Hamm-Alvarez SF, Chen ZZ, Schenke-Layland K. Quantitative second harmonic generation imaging of cartilage damage. Cell Tissue Bank. 2008;9:299–307. https://doi.org/10.1007/s10561-008-9070-7.

69. Brown CP, Houle M-A, Popov K, Nicklaus M, Couture C-A, Laliberté M, et al. Imaging and modeling collagen architecture from the nano to micro scale. Biomed Opt Express. 2014;5:233. https://doi.org/10.1364/boe.5.000233.

70. Coluccino L, Peres C, Gottardi R, Bianchini P, Diaspro A, Ceseracciu L. Anisotropy in the viscoelastic response of knee meniscus cartilage. J Appl Biomater Funct Mater. 2017;15: e77–83. https://doi.org/10.5301/jabfm.5000319.

71. Georgiadis M, Müller R, Schneider P. Techniques to assess bone ultrastructure organization: orientation and arrangement of mineralized collagen fibrils. J R Soc Interface. 2016;28:405–24. https://doi.org/10.1098/rsif.2016.0088.

72. Pang X, Wu JP, Allison GT, Xu J, Rubenson J, Zheng MH, et al. Three dimensional microstructural network of elastin, collagen, and cells in Achilles tendons. J Orthop Res. 2017;35:1203–14. https://doi.org/10.1002/jor.23240.

73. Schenke-Layland K. Non-invasive multiphoton imaging of extracellular matrix structures. J Biophotonics. 2008;1:451–62. https://doi.org/10.1002/jbio.200810045.

74. Zipfel WR, Williams RM, Webb WW. Nonlinear magic: multiphoton microscopy in the biosciences. Nat Biotechnol. 2003b;21:1369–77. https://doi.org/10.1038/nbt899.

75. Zoumi A, Lu X, Kassab GS, Tromberg BJ. Imaging coronary artery microstructure using second-harmonic and two-photon fluorescence microscopy. Biophys J. 2004;87:2778–86. https://doi.org/10.1529/biophysj.104.042887.

76. Vetri V, Dragnevski K, Tkaczyk M, Zingales M, Marchiori G, Lopomo NF, et al. Advanced microscopy analysis of the micro-nanoscale architecture of human menisci. Sci Rep. 2019;9: 18732. https://doi.org/10.1038/s41598-019-55243-2.

77. Fee MS. Active stabilization of electrodes for intracellular recording in awake behaving animals. Neuron. 2000;27:461–8. https://doi.org/10.1016/S0896-6273(00)00057-X.

78. Svoboda K, Denk W, Kleinfeld D, Tank DW. In vivo dendritic calcium dynamics in neocortical pyramidal neurons. Nature. 1997;385:161–5. https://doi.org/10.1038/385161a0.
79. Venkatachalam S, Fee MS, Kleinfeld D. Ultra-miniature headstage with 6-channel drive and vacuum-assisted micro-wire implantation for chronic recording from the neocortex. J Neurosci Methods. 1999;90:37–46. https://doi.org/10.1016/S0165-0270(99)00065-5.
80. Aharoni D, Hoogland TM. Circuit investigations with open-source miniaturized microscopes: past, present and future. Front Cell Neurosci. 2019;13:141. https://doi.org/10.3389/fncel.2019.00141.
81. Chen S, Wang Z, Zhang D, Wang A, Chen L, Cheng H, et al. Miniature fluorescence microscopy for imaging brain activity in freely-behaving animals. Neurosci Bull. 2020;36: 1182–90. https://doi.org/10.1007/s12264-020-00561-z.
82. Sancataldo G, Silvestri L, Allegra Mascaro AL, Sacconi L, Pavone FS. Advanced fluorescence microscopy for in vivo imaging of neuronal activity. Optica. 2019;6:758. https://doi.org/10.1364/optica.6.000758.
83. Bumstead JR. Designing a large field-of-view two-photon microscope using optical invariant analysis. Neurophotonics. 2018;5:1. https://doi.org/10.1117/1.nph.5.2.025001.
84. Tsai PS, Mateo C, Field JJ, Schaffer CB, Anderson ME, Kleinfeld D. Ultra-large field-of-view two-photon microscopy. Opt Express. 2015;23(13):833. https://doi.org/10.1364/oe.23.013833.
85. Sofroniew NJ, Flickinger D, King J, Svoboda K. A large field of view two-photon mesoscope with subcellular resolution for in vivo imaging. Elife. 2016;5:e14472. https://doi.org/10.7554/eLife.14472.
86. Grewe BF, Voigt FF, van't Hoff M, Helmchen F. Fast two-layer two-photon imaging of neuronal cell populations using an electrically tunable lens. Biomed Opt Express. 2011;2: 2035. https://doi.org/10.1364/boe.2.002035.
87. Reddy GD, Saggau P. Fast three-dimensional laser scanning scheme using acousto-optic deflectors. J Biomed Opt. 2005;10:064038. https://doi.org/10.1117/1.2141504.
88. Yang W, Miller J, Carrillo-Reid L, Pnevmatikakis E, Paninski L, Yuste R, et al. Simultaneous Multi-plane Imaging of Neural Circuits. Neuron. 2016;89:269. https://doi.org/10.1016/j.neuron.2015.12.012.
89. Diel EE, Lichtman JW, Richardson DS. Tutorial: avoiding and correcting sample-induced spherical aberration artifacts in 3D fluorescence microscopy. Nat Protoc. 2020;15(9): 2773–84. https://doi.org/10.1038/s41596-020-0360-2.
90. Jonkman J, Brown CM, Wright GD, Anderson KI, North AJ. Tutorial: guidance for quantitative confocal microscopy. Nat Protoc. 2020;15(5):1585–611. https://doi.org/10.1038/s41596-020-0313-9.
91. North AJ. Seeing is believing? A beginners' guide to practical pitfalls in image acquisition. J Cell Biol. 2006;172:9–18. https://doi.org/10.1083/jcb.200507103.

Modern Microscopy Image Analysis: Quantifying Colocalization on a Mobile Device

10

Vadim Zinchuk and Olga Grossenbacher-Zinchuk

Contents

What You Learn from This Chapter

This chapter aims to introduce modern ways to analyze fluorescence microscopy images, such as extending the image analysis workflow to tablet computers. Improvements in the technical characteristics of mobile devices provide increasingly viable options to supplement and expand the selection of microscopy image analysis tools. As an example, you will learn how to quantify the colocalization of

V. Zinchuk (✉)
Department of Neurobiology and Anatomy, Kochi University Faculty of Medicine, Nankoku, Kochi, Japan
e-mail: zinchuk@kochi-u.ac.jp

O. Grossenbacher-Zinchuk
Clinical and Medical Affairs, Ziemer Ophthalmic Systems AG, Port, Switzerland

fluorescence markers on a tablet computer without sacrificing the reliability of the method and leveraging the benefits of modern mobile computing.

10.1 Introduction

Recent advances in mobile computing and its expansion to medical and research fields are facilitating the adoption of tablet computers to analyze microscopy images. The suitability of mobile devices to visualize and examine intrinsic cellular details is based on the growing technical prowess of mobile devices, particularly on increased screen sizes and enlarged storage options. Combined with wireless capabilities, long battery life, quick boot time, and genuine portability, mobile devices simplify group work and collaboration between researchers.

Although many fields are yet to benefit from mobile computing, the technology has reached the maturity point when it is not only suitable for demanding and computer-intensive image analysis tasks, but can even replace the desktop tools in many cases. Before describing the actual procedure of analyzing fluorescence microscopy images on a tablet computer, let us outline the benefits of doing it.

10.1.1 The Benefits of Using Tablet Computers for Analysis of Light Microscopy Images

10.1.1.1 Ease of Use

When analyzing images, we all wish to have as fuss-free user experience as possible. Conceptually, mobile operating systems are conceived to be easy to use and, compared to desktop operating systems, they are indeed easier to understand and navigate. The ease of use is multiplied by multi-touch interfaces, which rely on natural and efficient hand gestures.

10.1.1.2 Superior Engagement and Improved Work Efficiency

It is agreed that mobile computers provide better user engagement by offering a more fluid user experience, i.e. they work the way people think. This is because mobile devices are designed to combine the freedom of expression with the freedom of movement. The improved work efficiency of tablet computers is due to

1. *Speed*. Tablet computers can be operated significantly faster. There is no need to locate a computer mouse or find the proper keyboard key. Users can just tap what is needed on the screen right away.
2. *User-friendliness*. Pointing at something you want is an instinctive human gesture. That is why touch screens are so intuitive and require very little training to be used. Despite its technological complexity, the touchscreen interface is extremely easy to understand and use.
3. *Simultaneous usage*. Hands can be used to express different gestures by catching, flicking, tilting, and forming any imaginable signs. In addition to the

simultaneous usage of hands, users can also extract various information out of single means of output. Finger touches can vary in pressure sensitivity and angle.

10.1.1.3 More Comfortable Analysis

Since mobile devices are truly portable, they are easy to carry around. The average iPad, for example, is three times lighter than the average laptop, which makes a real difference in practical use.

The portability of tablet computers is enabled by cloud storage. Using the cloud, users can keep microscopy images on remote servers and wirelessly download them when needed. Cloud service ensures that files are kept up-to-date on all devices automatically.

10.1.1.4 Better Affordability

Mobile computers are less costly to acquire and have a longer lifespan. Even tops of the line iPads are usually significantly cheaper than desktop and laptop computers while delivering equal performance. As a rule, professional mobile apps are approximately ten times more affordable than similar software applications for the desktop platform as well. According to statistics, lifespan and buying cycle of iPads are at least several times longer than that of laptops and desktops. This makes purchasing mobile computers a wiser investment.

10.1.1.5 New Possibilities

Currently, technology innovations come to the mobile platform first. Technically, mobile devices were the first to offer high-resolution displays and 64-bit apps. In terms of pure processing power, iPad Pros, for example, surpassed many laptops. Modern machine learning tools are quickly coming to tablet computers, making them more intelligent and capable. Importantly, manufacturers of mobile devices provide advanced development kits to facilitate discoveries in medical and research fields at a scale and pace never seen before. These tools are usually distributed as open-source to encourage researchers and developers to collaborate and share their innovative technologies.

10.2   Performing Microscopy Image Analysis on a Tablet Computer

10.2.1 Analyzing Colocalization of Fluorescence Markers

Let us now look at how we can use a tablet computer, such as an iPad, to analyze the colocalization of fluorescence markers in practice.

Colocalization is a visual phenomenon defined as the presence of two or more color types of fluorescently-labeled molecules at the same location. The information on its appearance can help to understand the properties of interacting proteins [1–3, 4].

To enable image analysis on a tablet computer, we will use software that has versions for both desktop and mobile platforms. On a desktop computer, we will use the app CoLocalizer Pro for Mac, a free download from the site of CoLocalization Research Software: https://colocalizer.com/mac/. On a tablet computer, we will use the app CoLocalizer for iPad, a free download from App Store: https://geo.itunes. apple.com/app/colocalizer/id1116017542?mt=8.

Importantly, at any step of the protocol, the procedure can be switched to the desktop version of the software and then back to the mobile app, if necessary, without any interruption of the workflow.

10.2.2 Setting Up Microscopy Image Analysis on a Mobile Device

At first, we need to enable access to images on a mobile device. We can do it by importing images either directly using USB-C port on iPad Pros or using cloud servers. In the latter case, we perform the following steps (Fig. 10.1):

1. Open an image on a desktop computer.
2. Save the image to the cloud server. Importantly, cloud servers can be accessed using both macOS and Windows PC compatible computers.
3. Re-open an image from the cloud server on a tablet computer.

After saving images to the cloud server, we will be able to access them from both the desktop and mobile platforms (Fig. 10.2).

Fig. 10.1 Image analysis on a tablet computer starts with opening images on a desktop platform and sharing them with the mobile platform via the cloud

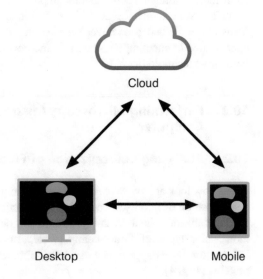

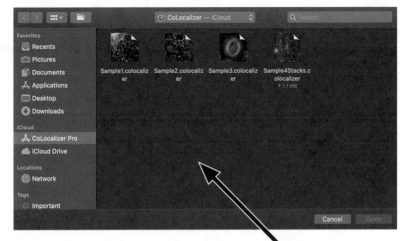

Sharing images between the desktop and mobile platforms

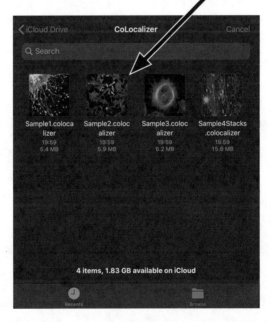

Fig. 10.2 Images, stored on a cloud server, are shared between the versions of the app running on both desktop and mobile computing platforms. Connection to the cloud ensures that changes to the images made on one platform are automatically synchronized with another platform and with all connected devices

10.2.3 Steps of the Workflow

The workflow is identical for the mobile and desktop platforms and consists of the following main steps:

1. Open the image and transform it to super-resolution.
2. Select ROI and intelligently reduce image noise.
3. Calculate coefficients.

We start the analysis by opening an image and accessing the *Tools* option in the app navigation bar (Fig. 10.3).

10.2.4 The Importance of Image Resolution

The vast majority of research studying proteins labeled by fluorescent markers is performed using conventional microscopes. These microscopes provide image resolution which is determined by the physical property of light, such as diffraction, at approximately 250 nm. The limit of diffraction is above the size of most multi-protein complexes, which are in the range of 25–50 nm in diameter. The gap between image resolution and the size of protein complexes means that they remain indistinguishable when visualized. These protein complexes can be visualized using

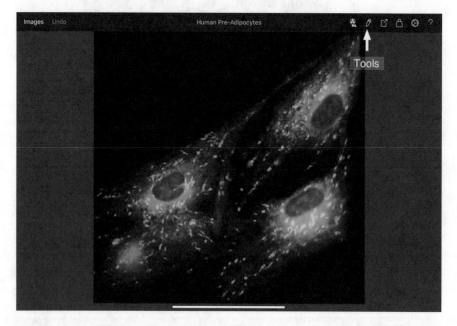

Fig. 10.3 Image analysis starts with opening an image and tapping the *Tools* button (arrow) in the app navigation bar to access specific tools

the so-called super-resolution microscopes, such as structured illumination microscopy (SIM), stimulated emission depletion (STED), photoactivated localization microscopy (PALM), or stochastic optical reconstruction microscopy (STORM). However, these sophisticated microscopes are still rare and very expensive to use and maintain.

10.2.4.1 Using a Machine Learning Model to Transform Conventional Fluorescence Images to Super-Resolution

Modern mobile research apps are capable of closing the gap of image resolution by applying machine learning (ML) models to transform conventional microscopy images to images with resolution comparable to that of super-resolution microscopes. ML is a revolutionary new technology that enables computers to learn and then use that knowledge while improving with experience [5–7]. ML is advantageous for solving complex technical and time demanding research tasks when humans are prone to errors.

We will employ a generative adversarial network (GAN)-based ML model to restore conventional fluorescence microscopy images by increasing their resolution while preserving and enhancing image details [8]. GAN-based models are unsupervised ML models that use less training data and are easier to create. They suit best to image transformation tasks. An increase in image resolution will reveal more structural details in the image and dramatically improve the reliability of colocalization analysis. The protocol also employs another ML model which is less complex and works according to supervised principle. It is applied for the task of image classification to reduce image background noise (see using ML Correct model below).

To transform an image to super-resolution, tap the *ML Super Resolution* icon in the app navigation bar (Fig. 10.4). The application of the ML model will increase the dimension and resolution of the transformed image three times (Fig. 10.5).

The use of the ML Super Resolution model is recommended for the vast majority of images with colocalization to be analyzed (see Current Limitations below).

10.2.5 Selecting a Region of Interest (ROI)

After transforming the image, we can proceed to select a smaller area on the image where we plan to perform the calculation of colocalization coefficients.

The selection of a region of interest (ROI) is a crucial step in performing quantification colocalization on any platform, including mobile. We select ROI to delimit the area of the user's interest to contextualize colocalization analysis. It is necessary to select an area with as little contribution of surrounding pixels as possible. You can perform selection using *Lasso*, *Polygon*, *Oval*, and *Rectangle* types. For achieving the greatest precision in selecting complex biological objects, it is best to use *Lasso* and *Polygon*.

To select ROI, tap *Tools > Select ROI*. On the selection screen, type the selection type matching your research purpose and start selecting (Fig. 10.6). Selection is done

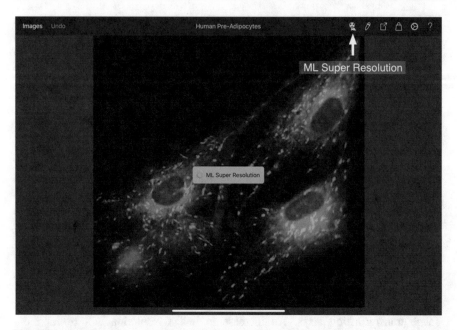

Fig. 10.4 Transformation of an image to super-resolution is done by tapping the *ML Super Resolution* icon in the app navigation bar. Transformation occurs practically in real-time and transformed image replaces a conventional one when the process of transformation is completed

by tapping on the screen and dragging the finger across the areas with colocalization. When the area is defined, tap the ending point of selection to close it.

After you selected an ROI, you can use the resizing points to adjust it closer to the object of your interest. When resizing a selection, you will see the label with the exact pixel size of the selected shape. For moving the selected ROI, tap and hold the selection shape (Fig. 10.7).

For analyzing the whole image, there is no need to use the selection tool for selecting any object. In this case, the entire image will be considered the ROI.

10.2.6 The Importance of Noise Reduction

Before estimating colocalization in an image, we need to reduce image noise. This is crucially important because by its nature fluorescence microscopy produces an inherently weak signal. As a result, raw fluorescence microscopy images are always degraded by noise. This noise appears as random background "crumbliness" throughout the image. Since most colocalization studies focus on tiny objects in the images, background noise can hide crucial structural details and hamper the reliability of image analysis.

Fluorescence images are impacted by two types of noise:

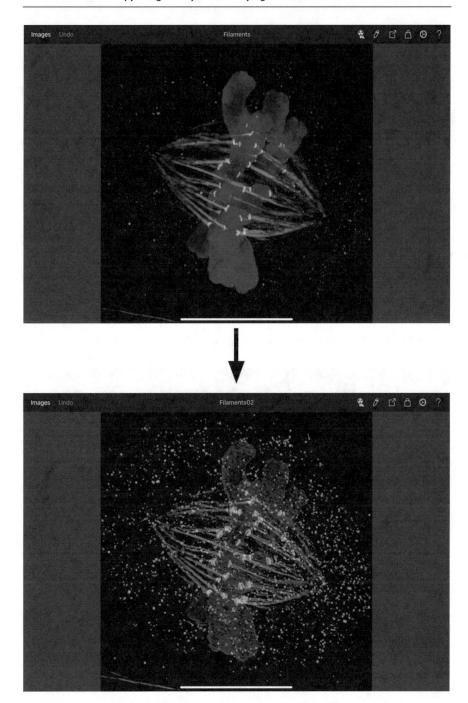

Fig. 10.5 Comparison of a conventional image with an image transformed to super-resolution. Note that the transformed image shows less background haze and reveals fine structural details

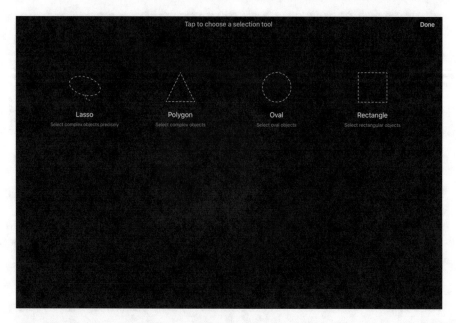

Fig. 10.6 The choice of ROI selection tools. One tool can be selected at a time

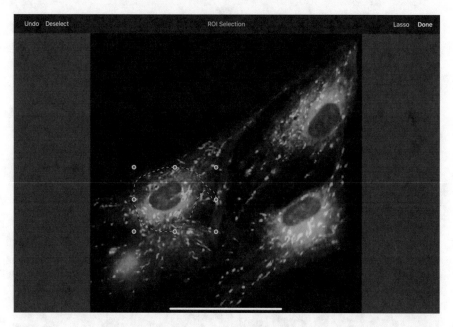

Fig. 10.7 Selection with the *Lasso* tool. Any selected ROI can be adjusted using resizing points

1. *Photon noise.* Signal-dependent. Varies throughout the image. Comes from the emission and detection of the light. Follows a Poisson distribution, in which the standard deviation changes with the local image brightness.
2. *Read noise.* Signal-independent. Depends on the microscope detector. Comes from inaccuracies in quantifying numbers of detected photons. Follows a Gaussian distribution, in which the standard deviation stays the same throughout the image.

The resulting noise, therefore, combines two independent noise types (*photon* and *read*) and its actual pixel value is effectively the sum of the two plus the true (noise-free) rate of photon emission.

1. What types of image noise contain fluorescence microscopy images and why reducing noise is needed prior to image analysis?

10.2.6.1 Using an ML Model to Reduce Background Image Noise

In the CoLocalizer app, you can use a supervised classification ML model to intelligently reduce image noise. The procedure is called background correction. To use ML-powered background correction, tap *Tools > Correct Background* in the app navigation bar (Fig. 10.8) to open the *Background Correction* screen. On this screen, tap the *ML Correct* button to reduce background noise in the image (Fig. 10.9).

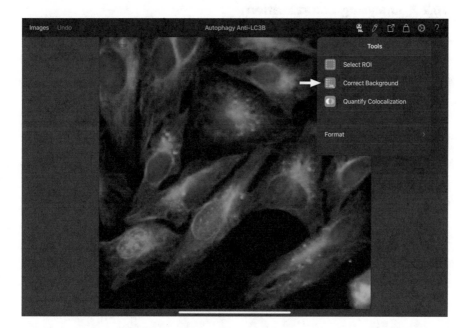

Fig. 10.8 Image noise is reduced using the background correction step. To use it, tap the *Correct Background* icon in the *Tools* popover (arrow)

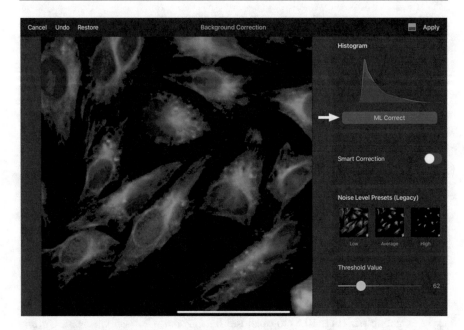

Fig. 10.9 Tapping the *ML Correct* button on the *Background Correction* screen (arrow) reduces noise in the opened image with the help of a specially trained ML model

10.2.7 Analyzing Colocalization

Following noise reduction, we can now proceed to the main task of our procedure, estimating the ratio of colocalized pixels in the images. It is determined by the values called colocalization coefficients.

10.2.7.1 Calculating Colocalization Coefficients
Coefficients are calculated on the selected ROI or the whole image if no ROI was selected (see above). To calculate coefficients, tap *Tools > Quantify Colocalization* in the app navigation bar (Fig. 10.10) to open the *Colocalization* screen. On this screen, you will find an image scattergram (scatterplot) showing the distribution of pixels in the image according to the selected pair of channels, values of coefficients estimating colocalization, and the option to reveal colocalized pixels (Fig. 10.11).

10.2.7.2 Revealing Areas with Colocalization
It is also possible to reveal areas with colocalization on both the scattergram and the image by displaying colocalized pixels. To display colocalized pixels, simply switch the *Reveal Colocalized Pixels* switcher (Fig. 10.11).

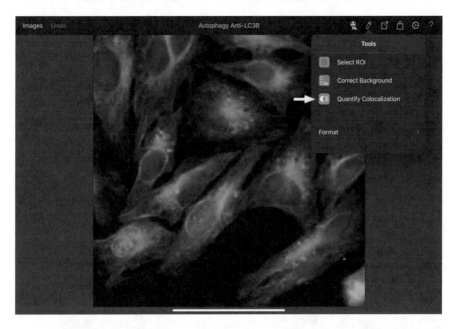

Fig. 10.10 Tapping *the Quantify Colocalization* icon in the *Tools* popover (arrow) will open the *Colocalization* screen

10.2.8 Exporting Results

After we are done with coefficients calculations, we can export the results. Calculation results can be exported in the form of data and in the form of images.

To export the results, tap the *Export* button in the app navigation bar (Fig. 10.12). Choose what to export, either *Data* or *Images* (Fig. 10.13). Exported results can be saved either locally, on the iPad, or to the cloud server and can be accessed from both iPadOS and macOS versions of the CoLocalizer app as well from Android mobile devices and Windows computers.

10.2.9 Documentation

The use of mobile app CoLocalizer for iPad is thoroughly documented and can be accessed within the app by tapping *Settings > CoLocalizer Help* in the app navigation bar without interrupting workflow (Fig. 10.14) as well as by downloading a free eBook from Apple Books store: https://geo.itunes.apple.com/book/colocalizer-for-ipad/id1259842440?mt=11.

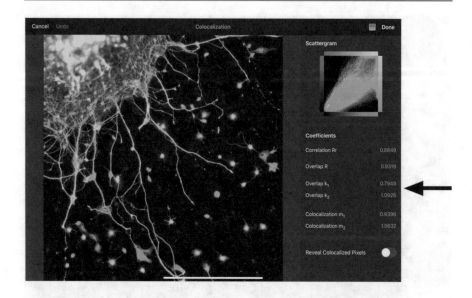

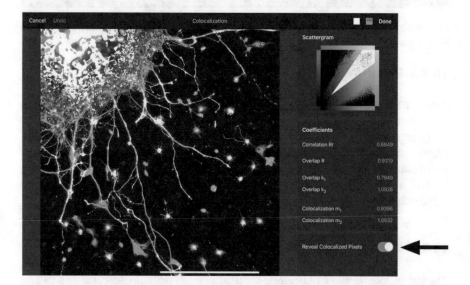

Fig. 10.11 The *Colocalization* screen displays information about images examined with the purpose to quantify the colocalization of fluorescent markers according to the selected pair of channels. It includes a scattergram at the top and the values of coefficients in the middle (top arrow). Switching the *Reveal Colocalized Pixels* switcher at the bottom of the screen will display colocalized pixels on the image and its scattergram (bottom arrow)

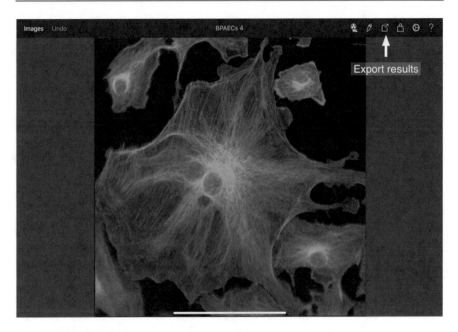

Fig. 10.12 Calculation results can be exported by tapping the *Export* button in the app navigation bar

10.2.10 Current Limitations

Professional scientific image analysis on a mobile device is still a new technique which has several limitations:

1. *Image size.* The biggest limitation of using tablet computers is related to the size of the images to be analyzed. For large images (100 MB and more) the use of mobile computing may not always be practical since downloading and synchronizing them via mobile networks will likely require a long time.
2. *Original image quality.* Another limitation is related to the original quality of the images, particularly when employing the ML Super Resolution option. Out-of-focus, low resolution, and weak fluorescence images will be very difficult to restore reliably, and thus the application of ML model may not be always justified.
3. *Local artifacts.* In some cases, restored images may contain local artifacts. These artifacts are usually detected when restoring complex multi-shaped 3D structures. When artifacts are observed, it is recommended to exclude them from the analysis by using the *Select ROI* tool.

2. What are the limitations of professional scientific image analysis on a mobile device?

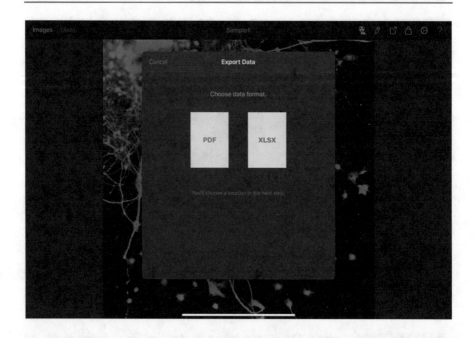

Fig. 10.13 The results of colocalization experiments can be exported as data (PDF and XLSX) (top image) and images (JPEG, PNG, and TIFF) (bottom image) file formats. They can be saved locally, on the device, and to the cloud servers, and then accessed via iPads, Macs, Android, and Windows mobile and desktop devices

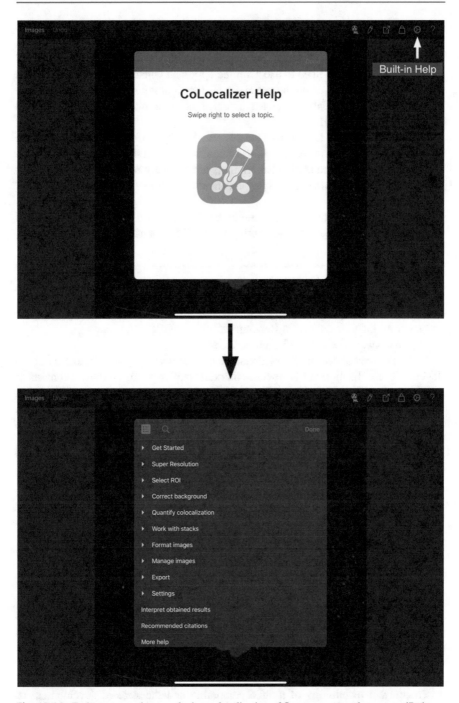

Fig. 10.14 To learn more about analyzing colocalization of fluorescence markers on an iPad, use built-in Help without leaving CoLocalizer app. To access Help, tap the *Settings* icon in the app navigation bar (top arrow). Swipe the Help cover page (bottom arrow) to access specific topics for in-depth information

10.3   Interpretation of Results

To make the results of colocalization experiments understandable to a broad audience of researchers, we need to use a unified approach when interpreting them. This popular approach is based on the use of simple terminology introduced with the help of a set of five linguistic variables obtained using a fuzzy system model and computer simulation [9].

These variables are as follows: *Very Weak*, *Weak*, *Moderate*, *Strong*, and *Very Strong*. They are tied to specific values of coefficients and their use helps to ensure that the results of colocalization experiments are correctly reported and universally understood by all researchers studying localization of fluorescence markers.

3. How the results of colocalization studies can be described in qualitative terms and why a unified approach to interpreting them is important?

10.4   Benchmark Datasets

The results of colocalization experiments can also be compared with reference images from the Colocalization Benchmark Source (CBS):
https://www.colocalization-benchmark.com.

CBS is a free collection of downloadable images for testing and validation of the degree of colocalization of fluorescence markers in microscopy studies. It consists of computer-simulated images with exactly known (pre-defined) values of colocalization ranging from 0% to 90%. They can be downloaded and arranged in the form of image sets as well as separately: https://www.colocalization-benchmark.com/downloads.html.

These benchmark images simulate real fluorescence microscopy images with a sampling rate of 4 pixels per resolution element. If the optical resolution is 200 nm, then the benchmark images correspond to the images with a pixel size of 50 nm. All benchmark images are free of background noise and out-of-focus fluorescence. The size of the images is 1024 × 1024 pixels.

We recommend using these benchmark images as reference points when quantifying colocalization in fluorescence microscopy studies both on mobile and desktop platforms.

Take-Home Message
This chapter describes the benefits and detailed step-by-step procedure of the protocol for using the mobile computing platform to analyze microscopy images, specifically the images with colocalization of fluorescence markers. The main points of the chapter are as follows:

- The current technology of mobile computing and the state of development of mobile apps offer the possibility to perform intensive image analysis tasks previously feasible only on the desktop platform.

- The use of mobile apps greatly improves the efficiency of scientific image analysis by allowing researchers to work in their preferred environment, including from home.
- Meaningful analysis of microscopy images on tablet computers requires the existence of versions of image analysis apps for both mobile and desktop platforms. Mobile and desktop versions of the apps should have the same tools and settings and be capable of providing identical and continuous workflows.
- Mobile apps now offer the latest technology innovations, including state-of-the-art ML-powered image restoration and classification models that dramatically increase the quality of images and the reliability of the analysis.
- By saving exported results of image analysis in universally-compatible data and image file formats on cloud servers, mobile computing helps to bridge together different mobile and desktop computing platforms and extends collaboration between research teams working in different physical locations.

Answers
1. Fluorescence microscopy images contain signal-dependent photon noise and signal-independent read noise. The actual pixel value of image noise combines these two noise types with the addition of the true (noise-free) rate of photon emission. The reduction of image noise facilitates the reliability of colocalization analysis.
2. The biggest limitations of professional scientific image analysis on a mobile device are: (a) image size (too large images are not practical for downloading and synchronizing), (b) original image quality (low-quality images are difficult to restore using ML Super Resolution model), and (c) artifacts (in some cases restored images may contain the local artifacts).
3. The results of colocalization studies can be presented using the following linguistic variables: Very Weak, Weak, Moderate, Strong, and Very Strong. The unified approach to describe them ensures proper reporting and universal understanding by all scientists working in the field.

Acknowledgments   Some of the images used for illustration of the protocol steps were taken from the publicly available Image Gallery of Thermo Fischer Scientific: https://www.thermofisher.com/jp/ja/home/technical-resources/research-tools/image-gallery.html.

References

1. Aaron JS, Taylor AB, Chew T-L. Image co-localization – co-occurrence versus correlation. J Cell Sci. 2018;131(3):jcs211847. https://doi.org/10.1242/jcs.211847.
2. Costes SV, Daelemans D, Cho EH, Dobbin Z, Pavlakis G, Lockett S. Automatic and quantitative measurement of protein-protein colocalization in live cells. Biophys J. 2004;86(6): 3993–4003. https://doi.org/10.1529/biophysj.103.038422.
3. Dunn KW, Kamocka MM, McDonald JH. A practical guide to evaluating colocalization in biological microscopy. Am J Physiol Cell Physiol. 2011;300(4):C723–42. https://doi.org/10.1152/ajpcell.00462.2010.

4. Zinchuk V, Grossenbacher-Zinchuk O. Quantitative colocalization analysis of fluorescence microscopy images. Curr Protoc Cell Biol. 2014;62(1):4.19.11–14.19.14. https://doi.org/10. 1002/0471143030.cb0419s62.
5. Sommer C, Gerlich DW. Machine learning in cell biology – teaching computers to recognize phenotypes. J Cell Sci. 2013;126(24):5529. https://doi.org/10.1242/jcs.123604.
6. Weigert M, Schmidt U, Boothe T, Müller A, Dibrov A, Jain A, Wilhelm B, Schmidt D, Broaddus C, Culley S, Rocha-Martins M, Segovia-Miranda F, Norden C, Henriques R, Zerial M, Solimena M, Rink J, Tomancak P, Royer L, Jug F, Myers EW. Content-aware image restoration: pushing the limits of fluorescence microscopy. Nat Methods. 2018;15(12): 1090–7. https://doi.org/10.1038/s41592-018-0216-7.
7. Zinchuk V, Grossenbacher-Zinchuk O. Machine learning for analysis of microscopy images: a practical guide. Curr Protoc Cell Biol. 2020;86(1):e101. https://doi.org/10.1002/cpcb.101.
8. Wang H, Rivenson Y, Jin Y, Wei Z, Gao R, Günaydın H, Bentolila LA, Kural C, Ozcan A. Deep learning enables cross-modality super-resolution in fluorescence microscopy. Nat Methods. 2019;16(1):103–10. https://doi.org/10.1038/s41592-018-0239-0.
9. Zinchuk V, Wu Y, Grossenbacher-Zinchuk O. Bridging the gap between qualitative and quantitative colocalization results in fluorescence microscopy studies. Sci Rep. 2013;3(1): 1365. https://doi.org/10.1038/srep01365.
10. Li X, Zhang G, Wu J, Xie H, Lin X, Qiao H, Wang H, Dai Q. Unsupervised content-preserving image transformation for optical microscopy. bioRxiv. 2019; https://doi.org/10.1101/848077.

Spinning Disk Microscopy

11

Kim Freeman

Contents

What You Learn from This Chapter

Spinning disk microscopy is a specialized imaging technique utilized with living and light sensitive samples. Arrays of optical pinholes spun at high speeds are used to focus the excitation source and block unfocused emitted fluorescence to derive confocal images.

11.1 Overview

Much of biomedical research is ultimately dependent on using live-cell imaging techniques. This requirement has fueled remarkable advances in microscopy instrumentation as well as the development of state-of-the-art detection systems and fluorescent proteins. The imaging of living samples poses several complications not necessarily seen in fixed tissues or other non-living specimens. Life is neither

K. Freeman (✉)

Olympus Scientific Solutions Americas, Waltham, MA, USA

Indiana University School of Medicine, Indianapolis, IN, USA

two-dimensional nor is it simple. Therefore, to study life while it is alive, we need to do so in a three-dimensional fashion which accommodates for the idiosyncrasies of keeping the environment sustainable for the specimen.

There is an innate thickness and heterogeneity which affects the resultant image: a lipid membrane clouds the surface of the cell, internal organelles of differing shape, size, and optical density all refract incident light, and specific proteins are not visible without contrast. To study that cell while it still living adds another layer of complexity: vesicles and other sub-cellular structures are in constant motion, adding the contrast needed to visualize sub-cellular attributes is commonly toxic, and the necessary high energy light sources are phototoxic to living cells. A single image cannot capture movement; a single image cannot focus in multiple focal plans simultaneously, and a single image cannot provide enough detail to study the aspects of a cell which make it live. Spinning disk microscopy alleviates these issues.

For the highest of imaging specificities, collimated excitations sources, such as lasers, are further focused through the system's optical pathway. Initially, the beam passes through a disk which is made up of a series of microlenses to focus the beam through a dichroic mirror and onto the pinholes in the second (Nipkow) disk. After passing through the Nipkow disk, the process is like standard fluorescent micros-copy as the excitatory beam then passes through the chosen objective and onto the sample being observed. The emitted signal from the specimen is gathered by the objective lens, filtered back through the chosen dichroic mirror, and sent back through the pinholes of the Nipkow disk, filtered through the chosen dichroic mirror and directed toward the camera port for detection.

The entire specimen is scanned by sets of multiple pinholes as the disk spins across the sample leading to images of the focal plane being taken in rapid succes-sion. Using this form of array scanning allows for high spatial and temporal resolu-tion while limiting the phototoxic effects of intense laser stimulation. The technique allows for imaging on a microsecond timescale which is reflective of many biological processes, while doing so in an environment which is much less detri-mental to the processes themselves.

11.2 History

Originally developed for use in televisions, Paul Nipkow's design to spin a disk comprised of identical equidistant pinholes for image reconstruction was patented in 1885 [1]. Future iterations of the disk and scanning device were then incorporated into microscopy, most notably with the help of Mojmir Petráň.

Unlike the original Nipkow disk, the Petráň disk (see Fig. 11.1) was constructed with many Archimedean spirals embedded in the disk, which allows for multiple point illumination and detection, leading to the development of what we now know as spinning disk confocal microscopy. Nesting the Archimedean spirals allows for hundreds of pinholes to be utilized on a disk, hence drastically increasing the light throughput to the specimen.

An Archimedean spiral pinhole pattern is derived from the pinholes being placed in a stepwise fashion along an imaginary line on the disk which rotates around the center to the outermost edge creating an arc of 360 degrees. Imagine if the disk were segmented into concentric circles as well as segmented into equal "pie" portions with radial lines. Starting at the origin, in a clockwise fashion, a pinhole is added where the first radial line intersects the first concentric ring, then the second, and the third. As long as the number of "pie" segments and the number of radial lines are equivalent, the spiral's last radial line will intersect the outermost ring (or edge of the disk area) after a full rotation. See Fig. 11.2.

In the microscope systems, the disk is spun at a constant speed around the origin which provides a constant angular momentum. The distance between the pinholes is also constant by following the spiral pattern. By nesting several spiral pinhole patterns on a single disk, the entirety of the specimen can be scanned very quickly as the disk rotates. Commonly in spinning disk microscopy systems, the disk is segmented into 12 sections each equaling 30° of the 360° disk circumference. These original spin disk systems utilized light sources that were powerful for the time, such as arc lamps. The broad-spectrum incident light would pass through the pinholes and be directed to the specimen. The resultant signal then is filtered for the desired wavelengths through a beam splitter, routed off several mirrors and through the pinholes on the opposite side of the disk where scattered light is limited.

The light path was complicated which made fine-tuning the system to an experimental need difficult. At the time, the added light redirections were necessary to limit glare and reflections from the components of the system itself, such as light bouncing off the metal disk. Further iterations of the spinning disk microscopy systems limited the internal reflections of signal through a variety of ways to allow more efficient light paths to emerge.

Modern spinning disk systems employ technologically advanced disks, light paths, and light sources which are only reminiscent of these early microscopes.

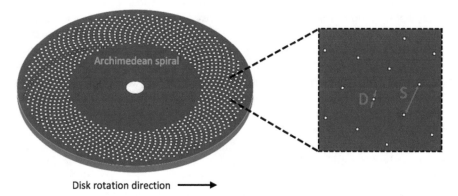

Disk rotation direction ⟶

Fig. 11.1 A standard Nipkow–Petráň disk. The pinholes are arranged in Archimedean spirals and have a set diameter (D) and separation distance (S). Alteration in disk spin speed, (D), and/or (S) the image brightness, contrast, and quality can be optimized for an experiment [2]

Fig. 11.2 The Archimedean spiral

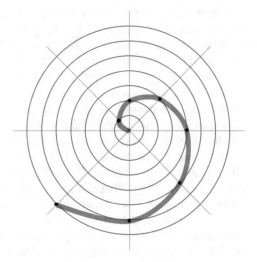

Prevailing terminology often refers to Nipkow or Petráň in reference to any spinning disk microscopy system, regardless of the disk structure actually being used.

11.3 Sample Preparation

As with most scientific endeavors, preparation is of the utmost importance. To the naked eye, an animal cell under standard visible light provides little to no detail. The bulk of a cell's content is water, which is nearly transparent. Some gross cellular structures are discernable, at best. In order to view any detail with specificity, contrast must be added to the sample in the desired locations of study. As covered in the previous chapters of this textbook, standard fluorescent microscopy most commonly utilizes antibodies to target subcellular structures. In living cells, however, that procedure is not useable as it requires the cell to be fixed in place and the membranes damaged, which would kill a living cell. Two of the more common techniques for imparting contrast for in vivo imaging are (1) the use of an expressor which has been genetically introduced to a cell through transfection, such as GFP, or (2) the use of vital dyes, more commonly known as probes, which can be utilized to target specific subcellular compartments (such as organelles), or specific cellular processes (such as reactive oxygen species production, membrane potential, calcium presence, and pH gradients) [3].

Not only must the chosen probes be targeted to what one wants to image, they must also be compatible with each other, having little or no spectral overlap. Spectral overlap occurs when two (or more) fluorophores either excite or emit in the same spectral range. This is especially problematic in a spinning disk system as the cameras used to capture emitted photons are monochromatic and unfiltered to ensure the highest sensitivity. Therefore, if multiple probes are used in the same sample, they are in need of being imaged completely separately to prevent contamination of the signals.

Beyond compatibility with life, and compatibility with each other, the probes must also be imageable with your microscopy system [4]. There are a wide variety of spin disk systems available, most of which can be tailored to specific imaging needs. If a system has a white light excitation source, it can excite multiple fluorophores, but would likely do so simultaneously without sophisticated filtering systems. A system with a single wavelength laser while exciting only the desired fluorophore would limit which fluorophores are useable in an experiment. And if the fluorophore you wish to use only emits in the far-red spectrum, having a camera that is not sensitive at those wavelengths will not permit data accumulation from the experiment as designed. The need for multiple laser lines and high-end cameras, along with proper computational and distinctively elegant software control systems, can lead to overwhelming system costs which often make it impossible for a single investigator to own all of the necessary equipment.

Once a specified cellular process or subcellular target is identified and properly labeled with a probe, and the microscope is set up to equip the desired experiment, then one must also consider keeping the sample alive to do the imaging. Living cells require the maintenance of proper oxygen/carbon dioxide balance, temperatures, nutrients, and humidity to maintain proper functions. Common scientific accessories can balance these provisions readily, but they need to be tailored to work on a microscopy system.

Experimental design is crucial to a favorable experimental outcome. Proper probe choices, proper microscopy configuration, and proper imaging conditions must all be taken into consideration. What seems to be a simple experiment can become a delicate balance with multiple caveats which must simultaneously culminate at the exact time an investigator has access to use the needed microscope.

11.4 Fundamental Microscope Design

Attaining confocality requires tremendous incident energy to attain a limited fluorescent signal from a fairly dark specimen. This requires the unused incident light, and reflections thereof, to be eliminated from the detection pathway along with any light which becomes scattered by passing through the system, disk, or sample itself. This requires a unique light path since there is the inclusion of moving parts, i.e., the spinning disk.

At the time this is written, Yokogawa Electric and their Confocal Scanning Unit (*CSU*) are leading the commercial production of spinning disk microscopy systems. These scanners consist of two aligned disks with a dichromatic mirror positioned between them (Fig. 11.3). This allows the incident light to be filtered between both disks, while the emitted signal also passes through pinholes to further filter out-of-focus light. The upper disk contains microlenses in the pinholes to focus the incident light onto the pinholes of the lower disk (Fig. 11.4). This allows for greater efficiency of incident excitation energy reaching the sample, allowing for lower energies to be

Yokogawa Spinning Disk Unit Optical Configuration

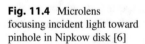

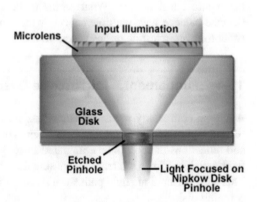

Fig. 11.3 Schematic illustration of the Yokogawa CSU-X1 spinning disk confocal optical pathway [5]

Fig. 11.4 Microlens focusing incident light toward pinhole in Nipkow disk [6]

used and benefits such as lower phototoxicity rates for living specimens and a greater ability to produce a signal from specimens which may only be expressing low levels of a fluorescent reporter.

These spinning disk scanners have rotation speeds of up to 10,000 revolutions per minute, and a pinhole array pattern scanning 12 frames per rotation. This creates a theoretical imaging speed up to 2000 frames per second.

$$10,000\,\frac{\text{revs}}{\text{min}} * 12\,\frac{\text{frames}}{\text{rev}} * \frac{1\,\text{min}}{60\,\text{sec}} = \frac{2,000\text{frames}}{\text{sec}}.$$

Camera sensitivity is paramount in attaining sufficient emitted signal and therefore required exposure time often becomes the limiting factor in speed of image acquisition over disk RPM. With a low-level fluorescence emitting sample, exposure rates can easily slow acquisitions from thousands of frames per second to single digits.

Following the light path of the scanner helps the understanding of how confocality is achieved (Fig. 11.3). Laser light (green) is collimated and projected onto the top disk, which is made up of microlenses. Each microlens (Fig. 11.4) focuses the gathered light through a dichromatic beamsplitter onto a section of the lower/Nipkow type disk.

After being focused through the two disks, the incident light from each pinhole is focused through the back aperture of the objective through to the specimen where it triggers the fluorescent signal emission in the plane of focus. The fluorescent emission then is gathered through the same objective, where it travels back up through the Nipkow type disk. The pinholes in the Nipkow type disk act as confocal apertures, blocking out-of-focus light. The emitted signal passes through the same pinhole as the excitation light which eluded the signal. It then passes to the dichromatic beamsplitter. The beamsplitter allows short wavelengths (excitation) to pass through while reflecting long wavelengths (emission) through a barrier filter which eliminates excess and non-wavelength-specific light. It then continues the path toward the camera for detection.

In order to attain accurate and useful images from a spinning disk microscopy system, the speed of the disk needs to coordinate with the camera acquisition rate, which is dependent on the emitted fluorescence of the sample. With there being 12 imaging frames per revolution and 360° in a revolution, one can easily conclude that each imaging frame occurs during a disk rotation of 30°. Since a complete image of the sample is scanned every time the Nipkow disk rotates 30°, the camera acquisition time has to be based on the time it takes for the disk to rotate that 30° or else distortions will occur. If the image acquisition time is set longer than the time for (or a multiple of the time for) a 30° disk rotation, then the disk and pinhole will be visible in the attained image, appearing as stripes or lines (Fig. 11.5c).

Detecting low light at high speed in a specimen that has moving parts requires extremely sophisticated camera systems. Slow acquisition low quantum efficient photomultipliers are insufficient for this level of imaging. Scientific CMOS cameras capture light so efficiently that low photon producing specimens can be imaged down to the single molecule level. Once the camera's sensor is contacted by a photon emitted by the specimen, the signal is amplified prior to read-out which greatly enhanced the signal-to-noise ratio (essentially the contrast of the image). This is ideal for spinning disk microscopy as the signals produced are most commonly extremely weak due to the need to have fast imaging rates with minimal exposure and low

Unsynchronized Image Capture in Spinning Disk Microscopy

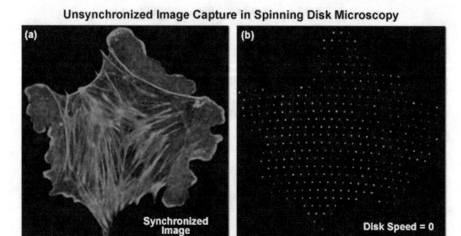

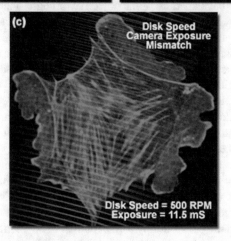

Fig. 11.5 Image comparison from (**a**) camera and disk rotation speed are synchronized, (**b**) camera image when the disk is not in motion, (**c**) visible banding from exposure/disk speed mismatch [5, 6]

excitation energies to avoid phototoxicity. The amplification comes from an additional read-out register. The EM register uses high voltages to add energy to the incident electrons allowing impact ionization; create an additional electro–hole pair and hence a free electron charge which is stored in the subsequent pixel and hence increasing the signal-to-noise ratio. Of course, signal amplification can also mean noise amplification. The electron multiplication will amplify any incident photon signal regardless of origin.

Scientific CMOS cameras are currently gaining sensitivity and have the speed to capture images in real time, aiding biological researchers' understanding of live-cell interactions. CMOS cameras bring the benefit of high frame rate and wide fields of view for imaging faster moving samples.. Choosing the correct camera for the desired experiment is imperative in attaining the best results possible.

11.5 Imaging Resolution

As shown in Fig. 11.1, the pinholes in a Nipkow disk have a set diameter (D) and separation distance (S) between pinholes which is quite a bit larger than the pinhole diameter. While any reflected light from the focal point is gathered through the pinhole, the inter-pinhole disk space physically blocks out-of-focus reflected light, permitting confocality. The smaller the pinhole diameter, the higher the confocality, but a trade-off occurs because less light can be gathered from the sample. In a very bright sample, this may not be an issue; however, live-cell imaging most commonly involves low-level fluorescence along with cellular background autofluorescence leading to a much smaller signal-to-noise ratio than can usually be attained in a fixed and/or stained tissue sample. Optimizing image acquisition occurs not only by synchronizing the camera acquisition speed to the disk rotation speed, but also by using a disk with optimized pinhole diameter and spacing for the highest available brightness and contrast [2].

So how can you tell what the optimized pinhole diameter and spacing would be? Similar to other forms of microscopy, the lateral resolution is directly related to the excitation wavelength and the objective being used. The Abbe equation for spatial resolution depicts this relationship:

$$d = 0.61\left(\frac{\lambda}{NA}\right).$$

Setting up an experiment with a $100\times$ 1.4 NA objective and using a common wavelength green laser light of approximately 488 nm, the diffraction limit of the imaging system would be approximately 212 nm. The same objective being used with a white light source (300–700 nm) would provide an Abbe resolution closer to 305 nm. If the disk being used for the experiment has pinholes smaller than the resolving power based on the Abbe equation, the resultant images will be compromised since too much of the signal will be blocked. Most spinning disk systems will have interchangeable disks with pinholes of varying sizes to allow for this optimization. Disks with 25 µm and 50 µm and 70 µm are commonly available.

Pinhole size is only part of the optical optimization. Confocality can only be achieved by blocking the out-of-focus light, which mainly comes from scattering as the incident or reflected light passes through the sample. The pinholes must be placed far enough apart that scattered light from the emission signal of one pinhole does not reach an adjacent pinhole. Ensuring the proper pinhole size and distribution on the disk along with the previously discussed microscopy factors such as wavelength of excitation light, objective numerical aperture, refractive index of

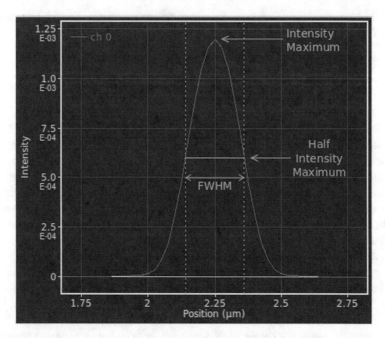

Fig. 11.6 Full width at half maximum (FWHM) graphical representation of a point-spread function intensity plot [7]

immersion medium, and camera functionality are all important factors in attaining quality images with high axial resolution.

Living processes rarely occur solely in a lateral format, however. Three-dimensional organisms have three-dimensional processes which need to be studied in three-dimensions. This is where optical sectioning is employed. A specimen can be scanned through focal planes in the Z-direction. The thickness of each focal plane is determined by the point-spread function (PSFThe PSF is a measure of how a microscope under the experimental parameters images a single point of signal, most commonly a fluorescent bead of a specified size. Laterally, the bead may appear circular and distinct; however when the bead is imaged through a depth there will be a distinctive blurring.

Since biological samples are nonhomogeneous, the emitted signal is refracted in multiple ways while passing through. The refractive index of a cell membrane is different than the refractive index of the intracellular fluids, which is different from the refractive index of neighboring organelles, all of which affect the scattering of the desired signal before it even reached the microscope. The blurring in the PSF in a transparent gel versus the blurring of the PSF in the chosen live sample can be used to mathematically deconvolute the images post-acquisition.

To determine the axial resolution of the microscopy setup, the full width at half maximum (FWHM) of the point-spread function is used. As seen in the PSF intensity plot in Fig. 11.6, this is quite literally the full distance where the intensity of a fluorescent object is half of the most intensely fluorescent point.

The FWHM can be tested experimentally, as with imaging a fluorescent bead. It can also be calculated using the equation:

$$\text{Axial resolution} = \sqrt{\frac{0.88\,\lambda_{\text{exc}}}{\left(n - \sqrt{n^2 - NA^2}\right)} + \left(\frac{D * n * \sqrt{2}}{NA}\right)^2},$$

where λ_{exc} is the excitation wavelength, n is the refractive index of the immersion medium, and NA is the objective numerical aperture and D is the pinhole diameter [8]. A 100× 1.4 *NA* oil immersion objective using a Nipkow disk with 50 μm diameter pinholes and a laser excitation of 488 nm would lead to an axial resolution, or FWHM, of almost 900 nm. Imaging optical sections less than 900 nm in this scenario would not yield any higher quality image but would overexpose the sample to laser intensity and hold a higher risk of photobleaching.

The pinhole in a spinning disk microscope is not adjustable and often is the limiting factor in resolution when considering faint or low signal-to-noise ratio samples, as is common in living specimens. If a system only has a single disk and imaging produces an image smaller than the pinhole, a tube lens can be inserted into the light path to magnify the image as needed. If the produced image is larger than the pinhole, part of the desired signal will be blocked by the pinhole aperture and lost.

Calculating the resolution of a spinning disk microscopy system involves a variety of factors which are both innate to a specimen and innate to the imaging system. Signal mitigating factors such as objective magnification, objective immersion fluid, excitation source, pinhole size, and pinhole separation all have to be taken into account to optimize images.

11.6 Super Resolution via Optical Reassignment Imaging (a.k. a. SoRa)

While standard imaging and even advanced techniques like spinning disk microscopy allow the discovery and exploration in living organisms, the constraints are imparted by the systems' technology creating unsurpassable thresholds; resolution being a drastically limiting factor. Conventional fluorescence microscopy alone is unable to resolve structures less than 200 nm in size. While some organelles in a cell could be visualized, many are below the 200 nm threshold, with proteins being smaller still. In a dormant sample, some super-resolution microscopy can be utilized to break through this imaging barrier. However, living samples are not only sensitive to the phototoxicity resultant from such techniques, but also the timescale of living interactions often occurs at a rate which cannot be captured using standard super-resolution techniques [9]. Theoretically, spinning disk super-resolution microscopy can achieve imaging at 120 nm and do so with a temporal resolution suitable for measuring the rapid dynamics of biological samples [10].

Imaging beyond the diffraction limit is necessary for the further study of living organisms, as many organisms and cellular processes occur at this scale. In order to overcome the barriers of phototoxicity and artifacts from processing, optical reassignment can be used. While extremely complicated to do well, the concept of optical reassignment is quite elegant, and in an oversimplified explanation, essentially equates to adding an additional lens in the emission pathway [11]. Microlenses can also be added to the pinholes in the Nipkow disk [12]. Adding the lens optimizes the light which has passed through the pinhole of the Nipkow disk, to fill the pupil of the objective, maximizing function. The emission signal is then collected as it passes through the lenses in the emission pathway and is optically contracted before reaching the detecting camera system.

Filtering of the acquired image must also occur to reach super-resolution levels. Traditional mathematical algorithms can be applied to images to filter out blur and attempt to resolve objects below the resolution limit of a system. Commonly, filtering algorithms create statistical estimates of an image using a known and measured standard, such as a point-spread function. There is an inherent danger in that these algorithms commonly impart artifact images derived from living samples due to the fact that the algorithms assume the sample and sample noise are stationary [13]. When focusing on living samples, virtually nothing is stationary. To limit artifact production from filtering, algorithms which specifically amplify high spatial frequencies over the low spatial frequencies are used. This creates a focusing of the deconvolution where the highest level of signal is, lessens the artifacts which can develop near the actual signal, and creates a more reliable data set [14] (Fig. 11.7).

11.7 Spinning Disk Confocal Microscopy vs. Laser Scanning Confocal Microscopy

There are other forms of confocal imaging. Laser scanning confocal systems have some similarities to spinning disk confocal systems. Both system types are used to image at a microscopic level in high detail. Both utilize a pinhole system to limit or eliminate out-of-focus light for detailed examination of a specimen. Both utilize or can utilize lasers for specific fluorophore excitation. However, each system is unique in the type of specimens imaged and desired data output.

By utilizing specialized cameras with a high quantum efficiency instead of photomultiplier tubes, faint signals can be accurately detected. This is important for multiple reasons. Firstly, many of the fluorescent dyes used in microscopy are somewhat toxic to living organisms. The less dye which can be introduced to the system, the healthier and less altered the living sample remains. Additionally, the utilization of so many confocal pinholes in a single disk allows for large segments of a specimen to be imaged simultaneously, unlike the raster scan of a laser scanning confocal, which has a single pinhole dragging back and forth across the imaging area. There is nothing static about living processes. Waiting for a pinhole to scan from one corner of an imaging area to the next would provide limited information

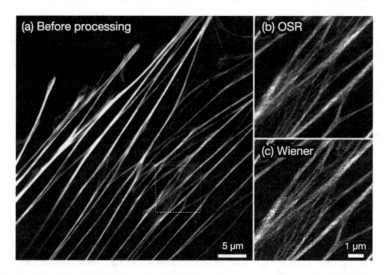

Fig. 11.7 Fluorescence images of stained actin filaments in a fixed cell. (**a**) Confocal fluorescence image acquired by the IXplore SpinSR system. (**b, c**) Magnified image of the highlighted portion (**a**) with the OSR filter (**b**), with a Wiener filter (**c**). Image utilized with direct permission of Olympus America [15]

about a specimen in motion or a process which is occurring in real time. Even if a specimen is scanned fairly quickly and multiple times, the data would still reflect large gaps in time. While spinning disk microscopy could be related to watching a person walk across a room, laser scanning confocal microscopy would be similar to watching that same person walk across a room while a strobe light is flashing. There are gaps in the information gathering due to the use of a single pinhole for image acquisition.

That is not to say that spinning disk microscopy does not have drawbacks. While spinning disk microscopy is very fast compared to laser scanning confocal microscopy, its confocality is limited. This means the optical sections tend to be thicker and less optimizable. Spinning disk microscopy is also limited by depth. Due to imaging speed the excitation signal is extremely brief, while providing for the ability to image with lower toxicity it in turn limits the time frame and signal can be evoked and elicited from a sample. A dim quick pulse of light will dissipate more quickly due to scattering; there is simply less incident light so the light will not travel as deep into the sample. The lessened excitation energy also limits the utilization of some applications. Spinning disk microscopy is excellent for visualizing many cellular processes; however, it also lacks the intensity required for purposeful photobleaching or uncaging of advanced fluorophores. A researcher needs to match the data acquisition needs to the type of imaging system which can best provide that data output.

Spinning disk microscopy is a valuable tool for researchers. When in need of an imaging modality with low toxicity, fast acquisition times, and the ability to pick up confocal signals in a live environment then spinning disk microscopy is the best choice. Geared toward the imaging of cellular processes, protein interactions, vesicular motion, and microorganism interactions, spinning disk microscopy provides for the scientific exploration of how life lives.

11.8 General Questions

1. *What are the advantages of using a spinning disk microscopy system?*
2. *What aspects of the Nipkow disk provide for confocality?*

11.9 Chapter Summary

The need for ultrafast confocal imaging for biomedical research has fueled the optimization of spinning disk microscopy for live-cell applications. Flexibility in light sources and detection systems provides researchers a customizable solution for imaging purposes. The imaging of living samples poses several complications not necessarily seen in fixed tissues or other non-living specimens such as low signal emission, specimens in motion, and the utilized reporters and incident energies being toxic to the living cells which pose the risk of altering the processes being studied. By using spinning disk microscopy, a researcher can modify the system to optimize the experiments and drastically reduce detrimental effects to living samples.

Answers to Questions
1. Spinning disk microscopy offers the advantage of imaging live samples in motion in a biologically friendly manner.
2. The pinholes act as a barrier to out-of-focus light scattered by the sample.

Take-Home Message
- Spinning disk microscopy is used to image living, moving, and/or photo-fragile specimens.
- Spinning disk systems are highly customizable providing a broad spectrum of imaging capabilities.
- Just as the name suggests, all spining disk microscopy systems utilize pinholes in one or more disks which spin through the light path to build an optical image.

Further Reading and Tutorials
- Scientific Volume Imaging (svi.nl/HomePage)
- Molecular Expressions (micro.magnet.fs.edu/micro/about.html)

References

1. Nipkow P. German Patent No. 30105. Berlin: Kaiserliches Patentamt; 1884.
2. Photometrics. https://www.photometrics.com/learn/spinning-disk-confocal-microscopy/what-is-spinning-disk-confocal-microscopy (n.d.)
3. Johnson S, Rabinovitch P. Ex-vivo imaging of excised tissue using vital dyes and confocal microscopy. Curr Protoc Cytom. 2013, 2012:Unit9.39. https://doi.org/10.1002/0471142956.cy0939s61.
4. Ve PI. A guide to choosing fluorescent proteins. Nat Methods. 2005;2:905–9.
5. Yokogawa. CSU-X1 Confocal Scanner Unit. Website. https://www.yokogawa.com/us/solutions/products-platforms/lifescience/spinning-disk-confocal/csu-x1-confocal-scanner-unit/#Details__Live-Cell-Imaging
6. Davidson M, Toomer D, Langhorst M. Introduction to spinning disk microscopy. Image modified with permission from https://zeiss.magnet.fsu.edu/articles/spinningdisk/introduction.html (n.d.).
7. Scientific Volume Imaging. Deconvolution-visualization-analysis. Half-intensity width. Retrieved from https://svi.nl/HalfIntensityWidth (n.d.).
8. Cole RW, Jinadasa T, Brown CM. Measuring and interpreting point spread functions to determine confocal microscope resolution and ensure quality control. Nat Protocols. 2011;6(12):1929–41. https://doi.org/10.1038/nprot.2011.407.
9. Lopez J, Fujii S, Doi A, Utsunomiya H. Microscopy/image processing: a deconvolution revolution for confocal image enhancement. BioOptics World. 2020;2020:1–9.
10. Shinichi Hayashi YO. Ultrafast superresolution fluorescence imaging with spinning disk confocal microscope optics. Mol Biol Cell. 2015;26(9):1743–51. https://doi.org/10.1091/mbc.E14-08-1287.
11. Roth S, Heintzmann R. Optical photon reassignment with increased axial resolution by structured illumination. Methods Appl Fluoresc. 2016;4(4):045005. https://doi.org/10.1088/2050-6120/4/4/045005.
12. Azuma T, Kei T. Super-resolution spinning-disk confocal microscopy using optical photon reassignment. Opt Exp. 2015;23(11):15003. https://doi.org/10.1364/oe.23.015003.
13. Brown NT, Grover R, Hwang PYC. Introduction to random signals and applied Kalman filtering. 3rd ed. Boca Raton, FL: Wiley; 1996.
14. Hayashi S. Resolution doubling using confocal microscopy via analogy with structured illumination microscopy. Jpn J Appl Phys. 2016;55(8):0–7. https://doi.org/10.7567/JJAP.55.082501.
15. Yonemaru, Y. How Olympus Super Resolution and Spinning Disk Technology Achieves Fast, Deep, and Reliable Live Cell Super Resolution Imaging. https://www.olympus-lifescience.com/en/resources/white-papers/super_resolution/

Light-Sheet Fluorescence Microscopy

12

George O. T. Merces and Emmanuel G. Reynaud

Contents

G. O. T. Merces
School of Medicine, University College Dublin, Dublin, Republic of Ireland

UCD Centre for Biomedical Engineering, University College Dublin, Dublin, Republic of Ireland

E. G. Reynaud (✉)
School of Biomolecular and Biomedical Science, University College Dublin, Dublin, Republic of Ireland
e-mail: emmanuel.reynaud@ucd.ie

© The Author(s), under exclusive license to Springer Nature Switzerland AG 2022
V. Nechyporuk-Zloy (ed.), *Principles of Light Microscopy: From Basic to Advanced*, https://doi.org/10.1007/978-3-031-04477-9_12

What You Will Learn

1. The issues with conventional fluorescence microscopy techniques that lead to the development of light-sheet fluorescence microscopy
2. The basic principles of light-sheet fluorescence microscopy illumination, detection, and sample imaging
3. The processes involved in sample preparation for light-sheet fluorescence microscopy
4. The limitations surrounding light-sheet fluorescence microscopy, and the innovations that have been developed to compensate for these limitations

12.1 Introduction

Organisms are three-dimensional objects (3D). Historically, attempts to understand biological processes within organisms have resorted to the imaging of a single slice of tissues, trying to relate these 2D snapshots to the ultimate processes of life. Recently imaging has seen a push toward understanding this microscopic world in its full 3D version. Steps forward have been carried by a mixture of physical sectioning (e.g., histology) and *optical sectioning*. Physical sectioning results in very thin slices of a sample being imaged independently, allowing for transmission of light through an otherwise opaque sample. Optical sectioning relies on optical methods selectively imaging only a single plane of a sample while leaving the physical structure of the whole sample untampered. This innovation has most notably been implemented through the widespread use of confocal microscopy. However, these techniques, and confocal microscopy specifically, come with a range of highly problematic side effects that more modern technologies are now trying to resolve and eliminate completely.

12.1.1 Issues with Conventional Fluorescence Microscopy Techniques

Fluorescence imaging relies on the absorption of light at excitation wavelengths by a fluorophore, followed by emission at a longer wavelength, which is then detected. Conventional approaches to fluorescence microscopy use a basic premise of parallel illumination and detection. A beam of light of a defined wavelength (*excitation*) is shone on a sample. *Emission* light is then collected through the same aperture, with the light diverted along a different path to end up at a detector using dichroic mirrors and specific filters. This generates important issues, mostly overlooked by inexperienced users. Firstly, even though light may be focused onto the specific plane a user is imaging, the beam of light will have to converge through the preceding tissue and diverge into the subsequent tissue. This means that even if only collecting light from a single plane of a sample, the tissues above and below the plane are still exposed to the light. Exposure to light can eventually make these fluorophores unresponsive to excitation, meaning the more light you expose a fluorophore to, the faster it will stop

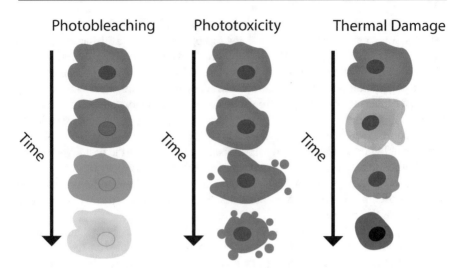

Fig. 12.1 Issues of fluorescence imaging. Fluorophore exposure to excitation light will eventually result in decreased fluorescence emission, meaning that over time the brightness of a fluorescent object will decrease the longer it is exposed. Phototoxicity results from excitation light triggering the generation of free radicals within cells, causing damage to intracellular components ultimately leading to cell death. Light exposure over time imparts thermal energy into a sample, also causing disruption and eventually death to a live sample

emitting fluorescence. This process is known as *photobleaching* (Fig. 12.1). By exposing a whole sample to the excitation light, this process will begin to occur throughout the sample. If you are using microscopy to optically section a sample, as in confocal microscopy, this means that the first layer imaged will be imaged as normal, but the second layer would have already been exposed to one round of excitation light, and the third layer would have been exposed to two rounds before being imaged, and so on for the whole sample. This results in a reduced fluorescent output the more slices you image. This is further compounded if multiple time points are imaged, meaning later slices will appear darker than earlier slices, and later time points will appear darker than earlier time points.

Excessive light exposure can also trigger cellular damage to a sample in two major ways. Firstly, fluorescent excitation always results in the production of free radical oxygen species, which cause intracellular damage to crucial cellular components, including DNA (Fig. 12.1). Free radicals are produced by light interacting with normal oxygen, altering their electron configuration, and turning them into reactive oxygen species. These reactive oxygen species react with intracellular structures and molecules, causing damage. The more fluorescent excitation in a sample, the more free radicals are produced and thus the more damage. Secondly, light transfers energy to a sample in the form of heat, with longer wavelengths of light transferring more heat to a sample. For physiological imaging experiments this poses problems of influencing the environment within a sample, and potential damage to cellular structures sensitive to heat (Fig. 12.1).

A compounding issue, particularly with confocal microscopy, is the time necessary for imaging a sample. Confocal microscopy builds up a planar image of a sample layer pixel by pixel. This is achieved by simply counting the number of photons emitted from each pixel within a plane using a specialized photon detector known as a photomultiplier tube. This process is time-consuming, and also requires a high laser intensity for each pixel to excite enough fluorophores to detect the small levels of light emitted. While the spinning disk confocal microscope was developed to reduce this increased imaging time, it is still relatively time-intensive to acquire even a single image. Cameras have developed significantly in the past few decades, with highly sensitive and photodetector-dense sensors capable of capturing up to hundreds of frames per second under the relatively low light conditions associated with microscopy. However, so far point-scanning confocal microscopy is unable to make use of these cameras.

The quest to understand live organisms in 3D has also been hampered by conventional optical slicing techniques due to their limitations on the size of a sample that can be imaged and the field of view that can be acquired at any one time. As described above, confocal microscopy builds up an image pixel by pixel, and so the more pixels needed per image (larger field of view, or more XY areas scanned) the longer it takes to acquire each image.

To deal with these limitations, a new way of optically slicing samples was required—ideally one that flips the geometry and orientation of illumination and detection.

12.1.2 Introducing Light-Sheet Microscopy

Light-sheet microscopy was first published in 1903, when Henry Siedentopf and Richard Zsigmondy utilized a planar sheet or cone of light to image colloids in solution, for which Richard Zsigmondy was later awarded a Nobel Prize. The technique was mainly used in physical chemistry and as a dark field equivalent for bacteria observation but vanished shortly after 1935. In 1993, this technique was revived and applied for the first time to 3D imaging of the mouse cochlea (Voie et al. 1993). Light-sheet microscopy only illuminates a single Z layer of a sample at a time. This is achieved by uncoupling the illumination and detection axes, illuminating a single plane from a 90° angle relative to the detection system. This means that there is no extra light disrupting the rest of the sample during imaging of an optical section, and thus photobleaching and phototoxicity are significantly reduced compared to confocal and wide-field approaches. This allows for a longer duration of imaging before photobleaching and phototoxicity become destructive, meaning more data and information can be extracted about any biological processes. By using high-quality cameras for image acquisition, and by imaging the whole illuminated plane at once, they allow for much larger fields of view and also for faster imaging, meaning biological processes are easier to visualize within the scale of a sample. Light-sheet microscopy also accommodates for the use of much larger samples than confocal microscopy, meaning small animal organs, tissues samples, and even whole embryos can be imaged many times over the imaging duration to observe real-time processes of biology and physiology. A general visual comparison of wide-field, confocal, and light-sheet microscopy techniques can be found in Fig. 12.2.

The ways in which light-sheet microscopy can counter the problems faced with in confocal microscopy relies on the method of illumination, image acquisition, and motion control systems. These solutions are discussed below.

12.2   Principles of Light-Sheet Microscopy

Light-sheet fluorescent microscopy relies mostly on the same physical principles as other fluorescence microscopy techniques. Excitation light can be shone onto a sample where certain molecules containing fluorophores will then absorb the light and reemit it at a longer wavelength which can be selectively imaged. However, the way the sample is illuminated, imaged, and moved during imaging is distinct from other methods of optical sectioning.

12.2.1 Creating and Using a Light Sheet

Light is composed of particles called photons that will travel in a straight line until they enter a medium of a different refractive index. A few things can happen when light rays are at the interface of two media with different refractive indices. (1) The object reflects the light, sending it away from the object at an angle relative to the angle it hit the object at (e.g., mirror) (Fig. 12.3a). (2) The light (or certain wavelengths) is absorbed by the object (e.g., apple, selective absorbance of light wavelengths is why objects have color perceivable to us) (Fig. 12.3b). (3) The light is transmitted through the object, either at the original angle of movement or at a different angle (e.g., glass) (Fig. 12.3c). The general ways in which light can interact with matter are outlined visually in Fig. 12.4.

Refractive index refers to the velocity of light traveling through the medium relative to the velocity of light traveling in a vacuum. As light travels through one medium (air, for example) and into a new medium with a different refractive index (e.g., glass), the difference in velocity causes its change in direction (refraction). Lenses can use this property in combination with the shape of the lens to bend light in a specific and predictable manner. We can use a lens to focus light onto a specific point, or to convert the non-parallel rays of light into parallel rays.

When light is passed through a cylindrical lens it curves the light in such a way that a single beam is converted into a planar sheet of light (Fig. 12.3d). This sheet behaves like a beam of light as normal; the rays will travel in straight lines, the plane will reach a point of focus followed by divergence away from the focal point, and at certain wavelengths can trigger fluorescence of specific fluorophores. This conversion of a beam into a sheet is what allows for optical sectioning in light-sheet microscopy. We can pass a sheet of light through a sample and it will only illuminate

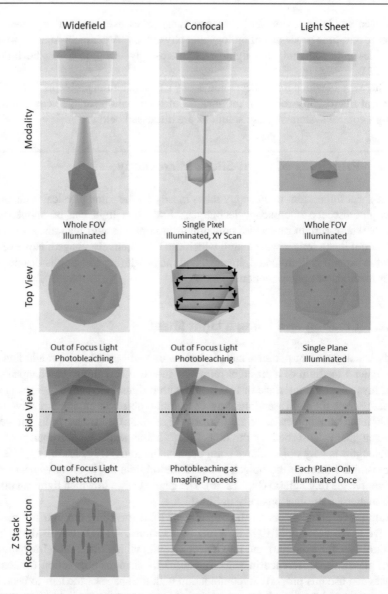

Fig. 12.2 Comparison of wide-field, confocal, and light-sheet microscopy. Wide-field microscopy illuminates the whole sample indiscriminately, confocal microscopy uses a single point of illumination for detection of each pixel independently, and light-sheet microscopy illuminates a single plane of a sample for each acquisition. From the top view, wide-field and light-sheet microscopy illuminate the whole field of view, while confocal microscopy illuminates a single point at a time, building up each Z plane image pixel by pixel. From the side view panels, you can see the excess illumination light (magenta) from each modality, with wide-field and confocal microscopy illuminating all Z planes of a sample while only imaging one plane, in comparison to light-sheet microscopy. When reconstructing Z planes into a 3D structure, wide-field often lacks the Z resolution desired by microscopists, while confocal and light-sheet microscopy show more accurate results

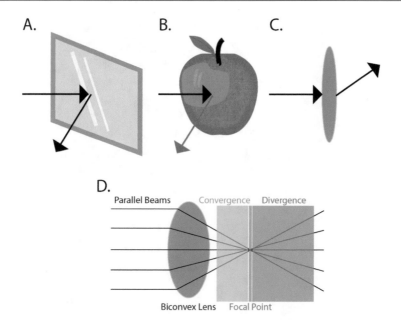

Fig. 12.3 Light behaves differently when hitting different objects. Photons travel in straight lines until they interact with some sort of matter. As a photon hits matter it can be reflected (as with a mirror, **a**), it can be absorbed (as with the non-red wavelengths of light hitting an apple, **b**), or it can be refracted and change direction (as with an optical lens, **c**). It should be noted that absorption is the method by which photons are used for fluorophore excitation. When light rays enter a lens (**d**), they are refracted in a different direction. In the case of a convex lens, the rays are refracted to a common point in space known as the focal point. Prior to this point the rays are converging and after this point they are diverging

a section with the thickness of the light sheet. Fluorophores in this plane will be excited by the light and will fluoresce just as in any other fluorescence microscopy system.

The light sheet will reach a focal point, where the sheet will be optimized to be its thinnest, with thicker regions either side of this point. This means a sample is not necessarily illuminated evenly, and central regions of a sample may be illuminated by a thinner sheet of light than the edges. This is an important consideration when drawing inferences from your imaging data.

If detection of the light were orientated parallel to the illumination, in a similar way to other optical sectioning techniques, this would appear on a camera as a single line of pixels with no detection either side of the light sheet. The acquisition innovation that allows for the detection of the plane of light is the orientation of the camera, using a perpendicular (90°) orientation of the detection to the illumination plane (Fig. 12.5).

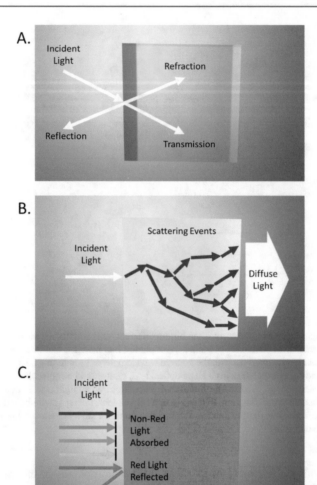

Fig. 12.4 Light behaves differently when interacting with different matter. (**a**) When incident light moves into a substrate with a different refractive index (i.e., from air to glass) the light interacts with the interface in several ways. It can be reflected at an angle relative to its angle of incidence, it can be transmitted and move at in the same direction as it was moving beforehand, or it can be refracted whereby it changes direction at the interface between the two refractive indices. (**b**) As light is traveling through a substrate, individual photons can hit particles within the substrate. These particles can cause the light to alter direction unpredictably, causing scattering of the photons, and diffusing the light. (**c**) When light interacts with an object, the object may be capable of absorbing certain wavelengths of light. In the case of a red object, all light of wavelengths shorter than red light will be absorbed, while red light will be reflected. As red light is the only light not absorbed by the object, our eyes perceive the object as red

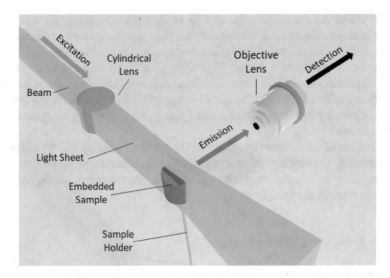

Fig. 12.5 Basic light-sheet microscopy system. A beam of excitation light shone through the curved edge of a cylindrical lens will become a light sheet. This light sheet is orientated to be at the exact focal distance from a known objective lens. The light sheet is passed through a fluorescent sample to allow only a single plane of the sample to be illuminated at any one time, allowing the detection of all fluorescence in one plane of the sample. This can be repeated for all planes within the sample, allowing for later 3D reconstruction of the sample

12.2.2 Capturing an Image

As opposed to confocal microscopy, which makes use of a single photodetector and acquires image pixel by pixel, light-sheet microscopy uses a camera to image a whole plane of a sample. This is advantageous for several reasons. The most prominent advantage is that with only a single acquisition required for each plane sample, image acquisition is faster than for confocal microscopy.

Like our eyes, cameras project the light from an object onto a sensor for detection. Unlike our eyes, cameras do not have an adjustable lens that can selectively focus on either an object close to the camera or far from the camera; they have a fixed point of focus. This means that for a camera detection system, the plane of a sample exposed by a light sheet must be kept at an exact focal distance from the camera. This precise distance must be maintained throughout the imaging procedure, and so the illumination and the detection systems must be physically coupled to ensure focus.

The two main types of cameras used in light-sheet microscopy systems are charge-coupled device (*CCD*) and complementary metal oxide semiconductor (*CMOS*) cameras. Both are composed of a planar array of sensors, and both rely on a conversion of photons into current, and then into voltage. However, they have different methods of extracting data from the sensor photodetectors. In CCD cameras, each individual pixel is responsible for conversion of photons into electrons to form a current. Each row of pixels then outputs its current to another system, where the current is converted to voltage pixel by pixel. By contrast, in CMOS sensors each pixel is responsible for both the conversion of photons into current and the conversion of current into voltage, which can be performed simultaneously by all pixels on a sensor. Therefore, CMOS cameras are much faster at image acquisition than CCD cameras, and can achieve higher framerates under the same illumination conditions. The disadvantage of CMOS cameras is the increased price associated with the newer and more expensive technology compared to the CCD cameras.

Regardless of the camera type used, the framerate of imaging will be higher than for techniques like confocal microscopy. Increased framerate means that fast imaging of the Z layers of a sample can be achieved, but for this to happen the light sheet must pass through the sample.

12.2.3 Imaging a Whole Sample

One of the major advantages of LSFM is the fast imaging of Z planes of a sample. To achieve this, motion control to move the light sheet through a sample is needed. The light sheet needs to always stay at the same distance from the camera to remain in focus, so there are two options for the motion control system. (1) Move the light sheet and the detection components in tandem to image through a sample. (2) Move the sample through the light sheet. This second option is much more frequently used due to its technical ease over the first option.

For automated scanning of a sample, the Z boundaries (e.g., front and back, or dorsal and ventral edges of the sample) need to be defined by the user. The number of slices between these layers can be set, either by maximizing the number of slices (depending on the light sheet thickness), or performing a lower number of slices for a faster imaging process at the expense of sample Z resolution. The light sheet itself will have a certain thickness throughout the sample, and this information is crucial for determining how far to move the sample between each planar acquisition. If you move too little, then overlap will occur between the images, reducing the Z resolution of your final reconstruction. If you move too far, then planes of the sample will not be imaged, again reducing Z resolution, and potentially missing out on crucial data while interpolating.

12.2.4 Image Properties and Analysis

The scanning process outputs a series of 2D images from different Z levels. If you know the physical distance between each Z plane slice, you can project each 2D slice back that exact distance, adding the Z dimension onto each 2D pixel. This creates a voxel with dimensions X, Y, and Z. Lateral resolution (XY) is often better than axial (Z) resolution, due to several factors, one of which being controlled by the light sheet thickness. This means that the Z dimension of each voxel is often larger than the X or

Fig. 12.6 Large samples can be imaged by imaging multiple fields of view. For a sample too large to fit in a single FOV, multiple FOV images can be acquired and subsequently stitched together to generate a reconstruction of the original object

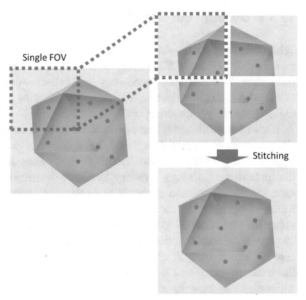

Single FOV

Stitching

Y dimension (termed anisotropic), and thus, each voxel represents a cuboid as opposed to a cube (isotropic).

In cases of very large samples, it may be necessary to image multiple fields of view sequentially for a single sample (Fig. 12.6). This additional process requires tiling, the process of combining multiple FOVs together by stitching and overlaying their common regions.

Some key considerations in the analysis and interpretation of your imaging data result from the design of the system itself. Issues with light penetration may mean that your sample gets progressively darker the further from the light source. Be careful not to conflate this appearance with any physiological importance, for example if you are performing quantitative fluorescence analysis of protein localization. Additionally, if the motion control system is not working with 100% accuracy, you will generate a 3D drift in Z and so the 3D will not be fully accurate. If movement is larger between planes than requested, your voxels will be too small relative to the actual data you acquired, and vice versa, meaning careful calibration and validation of a system should be performed routinely using samples with known geometries and structural landmarks.

12.3 Preparation of a Sample for Light-Sheet Fluorescence Microscopy

So, we have covered how light-sheet microscopy improves upon other techniques for optical sectioning and shown you how to get images from light-sheet microscopy. But how do you prepare a sample for imaging in the first place?

12.3.1 Dyeing Your Sample

When selecting fluorescent dyes for a sample there are three main considerations which are important to all fluorescent microscopy systems. (1) Method of labeling (i.e., antibody conjugation, secondary antibodies, endogenous fluorophore expression, etc.). (2) Illumination/emission wavelengths and system compatibility. (3) Compatibility with living cells/tissues.

If using an antibody system, there needs to be commercial option in the form of either a bulk manufactured antibody or a company willing to generate a specific antibody for your target molecule. This antibody does not need to be conjugated to a fluorophore itself. Using a "primary" antibody (from an animal that you are *not* using as a specimen) specific for your molecule (or antigen) of interest can then be followed by using a "secondary" antibody that is just generally specific for the organism you obtained your primary antibody from. If the secondary antibody is conjugated to an active fluorophore, your molecule of interest will now be fluorescent under the right illumination wavelength. Secondary labeling also increases fluorescence compared to direct labeling (Fig. 12.7), as multiple secondary antibodies can bind to each primary antibody, amplifying the signal. Antibody systems can only be applied in fixed tissues, as they require a lethal amount of membrane permeability to allow antibodies to enter cells and bind to their targets.

If using an endogenously fluorescent system to image a living organism, for example from transient transfection of fluorescent proteins or from stable expression of fluorescent proteins, the main considerations will be the excitation wavelength used and required filters for excitation/emission. This consideration should be reviewed prior to generating/obtaining the fluorescent construct for the sample. Additionally, commercial fluorescent markers exist for routinely stained features, such as DAPI for nuclei, mitotracker for mitochondria, and others, many of which are compatible for use with live tissue.

The excitation and emission wavelengths of your selected fluorophores or live dyes, regardless of the mechanism of expression, must be selected with your specific optics and hardware capabilities in mind, and should precede any experimental steps.

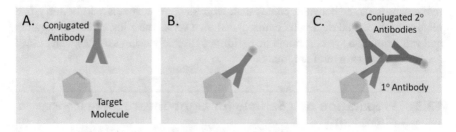

Fig. 12.7 Secondary antibodies can be used to enhance fluorescent output of a sample. (**a**) Adding a fluorescently conjugated antibody specific to a target molecule will result in selective fluorescent labeling of a target molecule (**b**). (**c**) Through using a primary antibody specific for a target molecule, multiple fluorescently conjugated secondary antibodies can bind each primary antibody, increasing the amount of fluorescence output from a single target molecule

12.3.2 Clearing Your Sample

Some samples present difficulties when it comes to imaging. As described earlier, different samples permit transmission of different levels of light, with dense samples attenuating the light earlier than less dense samples. This means that some samples, even if relatively small, are incompatible with light-sheet microscopy. To image these samples, it is necessary to remove the problem of optical density from the sample, which is achieved through the process of clearing. Clearing is the process of removing components such as cell membranes and proteins while leaving the cellular structures intact. This attempts to homogenize the refractive index of the sample and prevent light from being attenuated while passing through. These techniques have shown capabilities in creating transparent tissues. Techniques include the use of organic solvents (BABB and iDISCO), hydrogel embedding, high refractive index solutions, or detergent solutions (Fig. 12.8). Each method has its own advantages and disadvantages, as outlined comprehensively elsewhere [2]. The major issue with all these approaches is their incompatibility with live tissue, meaning imaging of a live sample cannot occur.

12.3.3 Mounting Your Sample

Sample preparation for LSFM depends heavily on the type of sample you are using. For imaging of adherent cell culture monolayers, the height of the sample is often too small to make use of the advantages of light-sheet microscopy. To make use of the benefits of LSFM, the monolayer is placed at a 45° angle relative to the light sheet. This means that the light sheet will pass through the specimen diagonally at a 45° angle while maintaining the 90° orientation of light sheet/detection system. The

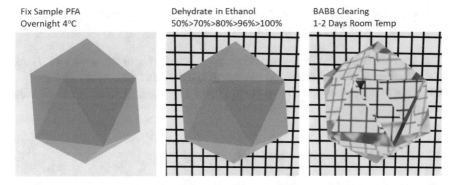

Fig. 12.8 Sample clearing method using BABB. Large biological samples are fixed overnight in paraformaldehyde (PFA) at 4 °C. Following fixation, samples are dehydrated using increasing concentrations of ethanol, ending with 3 rounds of dehydration in 100% ethanol. Samples are then transferred into a solution of BABB (peroxide-free) for 1–2 days until they become transparent. Protocol obtained from Becker et al. [1]

samples themselves are thin enough that light penetration is unlikely to be an issue, but users must take care to adhere their cells to a substrate compatible with light-sheet microscopy imaging.

Larger specimens including organoids, 3D cell cultures, small organs, and embryos need a different approach. One common technique is to embed the object to be images in some kind of material that can be applied as a liquid and then turned solid around the specimen to embed the whole specimen in the middle of a tube of optically transparent solid. Materials like agarose dissolved in water are often used which can be melted to a liquid and solidified by cooling to room temperature. This approach is favored for samples such as whole organs or whole organisms. Thermal considerations must be taken to ensure the embedding material is not excessively hot to the point it will damage a living specimen. A critical consideration when embedding samples is to ensure that the embedding matrix has the same refractive index as whatever fluid you are immersing your sample in. This will prevent detrimental interference with the light sheet when entering the material. Embedding samples is particularly useful in the case of small, live, whole organisms, as you can add nutrients to the embedding matrix to ensure organism survival while also maintaining its physical position and preventing organism movement. This can allow for the long-term imaging of physiological processes, development, and organism functions in stable environmental conditions under relative physiological conditions. This has been applied to the study *C. elegans, D. melanogaster*, and *D. rerio*. If a sample is not alive, and thus does not need to be restrained, you may opt for a hooking approach. This technique physically impales the specimen on a sharp hook, allowing it to be suspended within the FOV in air or in solution. The preparation time needed for this approach is obviously less than for embedding; however, it will not work with a specimen that is large, could possibly move, or lacks sufficient integrity to not be torn through by the hook.

12.4 Trade-Offs and Light-Sheet Microscopy: Development of Novel Systems to Meet Demands

The general principle of light-sheet microscopy is not without its own issues. Here, we will discuss some of the biggest issues with light-sheet microscopy and explore the ways in which people have improved upon the design to solve or compensate for them.

12.4.1 Cost and Commercial Systems

The main commercial system available presently is the Zeiss Lightsheet 7. With commercial microscopy systems becoming more specialized and refined, and often becoming more expensive in the process, many groups are left with no financially accessible method of performing large sample 3D imaging using light-sheet microscopy. Without access to a commercial light sheet, labs are left without the necessary technology. Thus, OpenSPIM set to create an open-source and community-driven group for the development and improvement of open-source LSFM. A full parts list, costings, and construction guide is available with step-by-step instructions on how to build the microscope. It is apparently constructible within 1 h by a non-specialist, according to Pavel Tomancak who helped develop the system. The system constructed in full is estimated at $18,000–$40,000, depending on whether you already have a camera available. This is a significant cost saving compared to the commercial system. Independent of a commercial entity, certain aspects gained with a commercial LSFM will be lost, such as warranty and dedicated customer support. However, the community-driven approach has led to a variety of forum-based problem-solving, aiding users in potential issues with their setups. The quality of this open-source system is limited mainly by the quality of the components used and the accuracy of your construction, which may result in a lower quality microscope than a commercial alternative. However, a major benefit of this approach is the adaptability given to the user to alter and improve the design to cater to their specific research goals, not achievable using commercial systems.

12.4.2 Light Penetration

A beam or sheet of light has a finite number of photons capable of interacting with matter. As light progresses through a sample, some photons will hit structures such as lipids, proteins, organelles, or DNA, and will be absorbed or deflected in a different direction. If this light is disrupted before it reaches a fluorophore, it is unable to cause fluorescence as desired. The deeper the light travels into a sample, the more likely it is that a photon collides with disruptive matter. Meaning, the deeper into a sample the fewer photons remain. Eventually, so few photons will remain traveling through the sample that any fluorescence they produce would be of such low intensity that it is undetectable by camera systems. This means that light can only ever penetrate to a certain extent into a sample and still result in usable fluorescence. Where this limit of penetration is depends on the light interaction with during its passage through a sample. A sample with a generally low optical density will allow more photons to travel through it, but light will always become attenuated the further it travels through a sample. Samples that are too large will not have light illuminate the far end of the sample, meaning the whole plane cannot be imaged. Additionally, samples are rarely homogenous in terms of optical density, with some internal structures absorbing more light than others. Opaque or dense structures absorb more light than low-density structures, and thus less light will reach the structures immediately behind the dense ones. This artifact is called shadowing, due to the characteristic appearance of shadows being cast behind optically dense structures. These two problems can be (for the most part) resolved by the addition of a second light sheet at a 180° angle to the original one (Fig. 12.9). This dual illumination means you can image larger samples than with a single light sheet, while also partly solving the problems of shadowing (Fig. 12.10a, b, d). Shadowing is not completely solved with this method, as there may be other optically dense

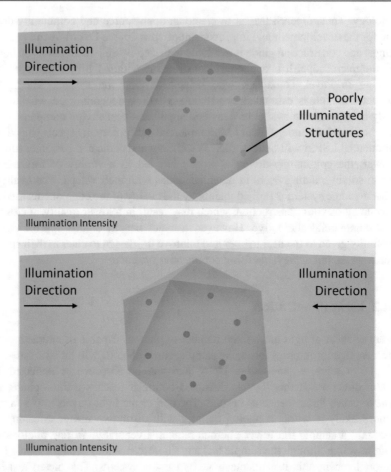

Fig. 12.9 Dual illumination directions can improve illumination uniformity. As light travels through a large sample, it gets attenuated and is thus less able to trigger fluorescence meaning fluorescent objects deeper in the sample show lower fluorescence than those closer to the excitation light source. Addition of a second illumination source from the opposite direction increases uniformity of illumination and reduces this issue. Illumination intensity shown below each image

structures from the opposing angle which will cause shadows that overlap with shadows from the original light sheet; however, it is an improvement. An additional method of addressing shadowing artifacts is the use of a scanning light sheet approach, whereby the light sheet is moved quickly laterally to illuminate the sample from multiple angles as opposed to just one (Fig. 12.10c). The problems caused by a sample's optical density can also be alleviated using clearing agents if possible, but as previously discussed this is not possible when using live samples.

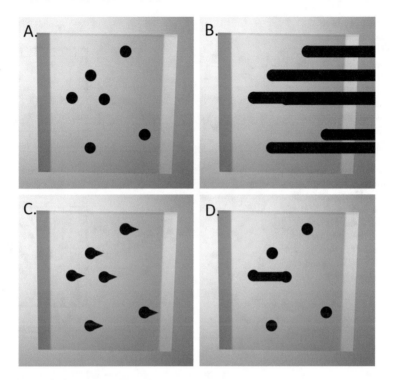

Fig. 12.10 Removing the impact of shadowing on light-sheet microscopy. (**a**) A sample with optically dense structures (black circles) will present issues for light-sheet microscopy. (**b**) Optically dense structures within a sample prevent excitation light from reaching the structures behind them by absorbing the light before it can reach them. (**c**) By vibrating the excitation light laterally, you allow excitation light to reach behind the dense structures, illuminating their features. (**d**) By illuminating from opposite directions, you allow for excitation light to reach the spaces behind the optically dense structures

12.4.3 Resolution Limitations

In microscopy, magnification is the degree to which the size of the acquired image is different from the size of the imaged object. Higher magnifications mean that each pixel on your acquired image represents a smaller area of the object (Fig. 12.11). As a smaller area will be imaged, each pixel on the camera sensor is able to detect a higher number of structures in a certain physical area, and thus resolution will increase. Increasing the magnification infinitely will not infinitely increase the resolution of an acquired image, as resolution of a system is limited by two factors: the optics and the detection system. Light itself has a resolution limit itself, based on the wavelength of the light observed. Cameras have a finite number of pixels on their sensor, meaning for any magnification you will only be able to distinguish two objects if they appear on the sensor with one pixel apart.

Fig. 12.11 Field of view vs. resolution. A single sensor has a set number of pixels capable of detecting light. If an object is projected under low magnification onto the sensor, a large number of features will be hidden as multiple features will take up a single pixel space; however, a large field of view will be imaged at this low resolution. If a sample is projected under high magnification onto the sensor, a very small field of view of the sample will be obtained; however, features will be easier to distinguish as they will now be spread out over multiple pixels

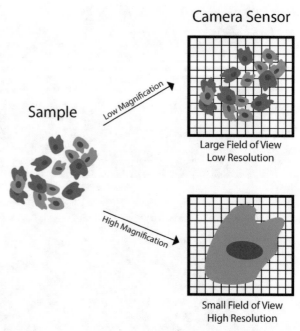

There are several ways to improve the resolving power of a microscope: (1) increase the magnification of the optics, (2) improve camera resolution, and (3) apply super-resolution techniques to the imaging process. Increasing magnification involves using high-magnification objective lenses. However, as these focus onto smaller and smaller regions, the amount of light entering the objective will be lower than for lower magnification objectives. This means that camera sensitivity needs to be higher for smaller areas, and the amount of light an objective can take in needs to be increased. To achieve this, higher intensity light can be applied to a sample, and the objectives used can develop higher numerical apertures, often using immersion liquids like oil. Over time, cameras have improved in maximum pixel density by decreasing the size of photodetectors on the camera sensor. However, this only allows for an increased resolution in the X and Y axes, the Z-axis resolution remains unchanged as this is dependent on the thickness and qualities of the light sheet itself. Applications of alternative imaging modalities to LSFM have also led to improvements in resolution. Combination of stimulated emission depletion (STED) microscopy to create a smaller laser beam for thinner light sheet formation has reportedly resulted in a 60% increase in axial resolution [3]. However, pitfalls to this increased resolution are numerous. STED is notorious for causing high levels of photobleaching and photodamage, and thus negatively impacts two major benefits of LSFM for 3D imaging. The equipment and expertise necessary to establish this system is also intense, putting off lower-expertise users, or facility managers not familiar with self-built optical systems. All these issues account for a 60% increase in axial resolution, which very few users would find essential for their research. In terms of resolution as a trade-off, the resolution needed for an experiment need not be excessively high. If performing macrostructure analysis of tissues, it is not necessary to visualize each organelle within each cell, and so excessively high resolution represents an unnecessary amount of data and information.

12.4.4 Sample Size

A significant trade-off in microscopy is that between resolution and field of view. Typically, the higher the resolution, the smaller the FOV, and the larger the FOV the lower the resolution. Light-sheet microscopy users often desire high-resolution imaging of large samples, meaning this trade-off is incompatible with their research goals. A high-tech innovation in this regard is the MesoSPIM, a light-sheet microscopy system that allows for a much larger FOV with relatively high resolution. One pitfall of this technology is the cost and expertise required to build the open-source system from scratch. Thankfully, the community around this technology is open to assisting in a researcher's goals by offering use of their system, with microscopes available in Switzerland, Germany, and the UK. A significant downside to this equipment is the cost required to build the system, with total cost estimated at between $169,600 and $239,600.

12.4.5 Data Deluge

Light-sheet microscopy has been used to generate full 4D datasets over large time frames of several sample types. This creates a huge amount of data, which is good for getting usable results, but not so good if you need to store or process that data. Using the MesoSPIM, a 3D reconstruction of a full mouse brain has been generated. While the data contained within this reconstruction stands at a modest 12–16 GB for a single time point, time series, multichannel imaging, and numerous equipment users can result in large volumes of data being stored. Datasets of this large size require significant storage space, which can come at a high cost for the user. A dedicated server infrastructure may be necessary if data output is sufficiently high. A further consideration is processing the dataset once imaging has been complete. Even a high-quality computer will struggle to handle the colossal datasets involved in LSFM, so virtual servers may be necessary to even function during normal imaging. Additional problems involve the bandwidth of data able to be transmitted from the device to the storage system. If not high enough, the bottleneck could form in the data transfer, which could abruptly end an imaging run early. All these precautions and additions represent significant financial investment for a group and may not be feasible for a small group with limited funds.

Take-Home Message
Conventional light microscopy systems have major limitations that have led to the development and utilization of LSFM in the life sciences. Light-sheet illumination allows for direct optical sectioning of a sample. Those selected planes can be recorded faster as the technology takes advantage of high-quality cameras with lower illumination power. LSFM can serve as a useful method of optical sectioning of samples for full 3D reconstructive imaging of a wide range of biological sample types, proving its utility in the ever-growing field of LSFM.

References

1. Becker K, Jährling N, Saghafi S, Weiler R, Dodt H-U. Correction: Chemical clearing and dehydration of GFP expressing mouse brains. PLoS One. 2012;7(8):10.1371/annotation/ 17e5ee57-fd17-40d7-a52c-fb6f86980def. https://doi.org/10.1371/annotation/17e5ee57-fd17- 40d7-a52c-fb6f86980def.
2. Ariel P. A beginner's guide to tissue clearing. Int J Biochem Cell Biol. 2017;84:35–9. https://doi. org/10.1016/j.biocel.2016.12.009. Epub 2017 Jan 7. PMID: 28082099; PMCID: PMC5336404.
3. Friedrich M, Gan Q, Ermolayev V, Harms GS. STED-SPIM: stimulated emission depletion improves sheet illumination microscopy resolution. Biophys J. 2011;100(8):L43–5. https://doi. org/10.1016/j.bpj.2010.12.3748. PMID: 21504720; PMCID: PMC3077687.

Further Reading

History

Cahan D. The Zeiss Werke and the ultramicroscope: the creation of a scientific instrument in context. In: Buchwald JZ, editor. Scientific credibility and technical standards in 19th and early 20th century Germany and Britain. Archimedes (New Studies in the History and Philosophy of Science and Technology), vol. 1. Dordrecht: Springer; 1996. https://doi.org/10.1007/978-94- 009-1784-2_3.
Huisken J, Swoger J, Del Bene F, Wittbrodt J, Stelzer EH. Optical sectioning deep inside live embryos by selective plane illumination microscopy. Science. 2004;305:1007–9.
Siedentopf H, Zsigmondy R. Über Sichtbarmachung und Groessenbestimmung ultramikroskopischer Teilchen, mit besonderer Anwendung auf Goldrubinglaesern. Ann Phys. 1903;10:1–39.
Voie AH, Burns DH, Spelman FA. Orthogonal-plane fluorescence optical sectioning: three dimensional imaging of macroscopic biological specimens. J Microsc. 1993;170:229–36.

Reviews

Olarte OE, Andilla J, Gualda EJ, Loza-Alvarez P. Light-sheet microscopy: a tutorial. Adv Opt Photon. 2018;10:111–79.
Power R, Huisken J. A guide to light-sheet fluorescence microscopy for multiscale imaging. Nat Methods. 2017;14:360–73. https://doi.org/10.1038/nmeth.4224.
Reynaud EG, Peychl J, Huisken J, Tomancak P. Guide to light-sheet microscopy for adventurous biologists. Nat Methods. 2015;12(1):30–4. https://doi.org/10.1038/nmeth.3222.

Localization Microscopy: A Review of the Progress in Methods and Applications

Jack W. Shepherd and Mark C. Leake

Contents

What You Will Learn in This Chapter

Here, we report analysis and summary of research in the field of localization microscopy for optical imaging. We introduce the basic elements of super-resolved localization microscopy methods for PALM and STORM, commonly used both in vivo and in vitro, discussing the core essentials of background theory, instrumentation, and computational algorithms. We discuss the resolution limit of light microscopy and the mathematical framework for localizing fluorescent dyes in space beyond this limit, including the precision obtainable as a function of the

J. W. Shepherd · M. C. Leake (✉)
Department of Physics, University of York, York, UK

Department of Biology, University of York, York, UK
e-mail: mark.leake@york.ac.uk

© The Author(s), under exclusive license to Springer Nature Switzerland AG 2022　　299
V. Nechyporuk-Zloy (ed.), *Principles of Light Microscopy: From Basic to Advanced*, https://doi.org/10.1007/978-3-031-04477-9_13

amount of light emitted from a dye, and how it leads to a fundamental compromise between spatial and temporal precision. The properties of a "good dye" are outlined, as are the features of PALM and STORM super-resolution microscopy and adaptations that may need to be made to experimental protocols to perform localization determination. We analyze briefly some of the methods of modern super-resolved optical imaging that work through reshaping point spread functions and how they utilize aspects of localization microscopy, such as stimulated depletion (STED) methods and MINFLUX, and summarize modern methods that push localization into 3D using non-Gaussian point spread functions. We report on current methods for analyzing localization data including determination of 2D and 3D diffusion constants, molecular stoichiometries, and performing cluster analysis with cutting-edge techniques, and finally discuss how these techniques may be used to enable important insight into a range of biological processes.

13.1 The Optical Resolution Limit

Antony van Leeuwenhoek (1632–1723) was the pioneer of the light microscope [1]. Using glass beads with radii of curvature as small as 0.75 mm taken from blown or drawn glass, he managed to construct the seminal optical microscope, with a magnification of 275× and spatial resolution only slightly above one micron. Granted, the microscope had to be in effect jammed into the user's eye and the sample held fractions of an inch from the lens, but it is remarkable that ca. 350 years ago a simple light microscope existed which had a spatial resolution only 3–4 times lower than the so-called diffraction-limited barrier, subsequently known as the optical resolution limit.

Why is it, then, that such a "diffraction limit" exists? The answer emerged ~150 years after van Leeuwenhoek's death from the theoretical deliberations of the German polymath Ernst Abbe: he argued that, when we image an object due to its scattering of light at a large distance from it ("large" being greater than several wavelengths of light), optical diffraction reduces sharpness in direct proportion to the wavelength of light used and in inverse proportion to the largest possible cone of light that can be accepted by the objective lens used (this quantity is characterized by the numerical aperture, or "NA"). Expressed algebraically we find that the minimum resolvable distance using ordinary light microscopy assuming imaging through a rectangular aperture is

$$d = \frac{\lambda}{2\mathrm{NA}}.$$

A variant of this formula as we will see below includes a factor of 1.22 in front of the wavelength λ parameter to account for circular apertures as occur in traditional light microscopy. Visible light has a wavelength approximately in the range 400–700 nm, and the best objective lenses commonly used in single objective lens research microscopes, at least those that avoid toxic organic solvent immersion oil, have an NA of around 1.5, implying that this Abbe limit as denoted above is somewhere between 100 and 250 nm, larger than many viruses but good enough to easily visualize bacteria, mammalian cells, and archaea as well as several subcellular features such as mitochondria, chloroplast, nuclei, endosomes, and vacuoles. However, this spatial resolution is clearly not sufficient to observe the activity of the cell on a single-molecule level whose length scale is 1–2 orders of magnitude smaller.

Fortunately, this apparent hard limit can be softened: if one images an object that has a known shape (or at least that has a shape that has a known functional form), then we may fit an approximate mathematical model to the image obtained from the microscope. A parameter of this fit will be the intensity centroid of the object—and this is the key feature of "localization microscopy." This centroid may be expressed to a sub-pixel resolution albeit with a suitable error related to parameters such as the number of photons sampled, the noise of the detector, and the size of the detector pixels. In brightfield imaging this principle is commonly used to track beads attached to filamentous molecules for tethered particle motion experiments, for example; the first reported use of Gaussian fitting for localization microscopy was actually in 1988 by Jeff Gelles and co-authors, who found the intensity centroid of a plastic bead being rotated by a single kinesin motor to a precision of a few nanometers [2]. With the added binding specificity potential of fluorescence labeling and subsequent imaging, localization microscopy can go much further though.

13.2   Super-Resolved Localization in 2D

Fluorophores imaged onto the Cartesian plane of a 2D camera detector are manifest as a characteristic point spread function (PSF) known as the Airy disk, consisting of an intense central Gaussian-like zeroth order peak surrounded by higher order concentric rings of intensity, as shown in Fig. 13.1a. Physically, the concentric rings arise due to Fraunhofer diffraction as the light propagates through a circular aperture. Mathematically, the intensity distribution due to this effect is given by the modulus of the Fourier transform of the aperture squared. The Rayleigh criterion (though note there are other less used resolution criteria that could be used, such as

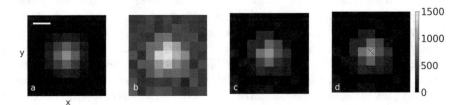

Fig. 13.1 (**a**) A Gaussian PSF; (**b**) A Gaussian PSF as seen with background noise; (**c**) The Gaussian PSF with noise after background correction; (**d**) The results of fitting a 2D Gaussian to (**c**). White cross is the center of the fit Gaussian and the black cross is the true center of the PSF. Bar: 100 nm

the Sparrow limit) specifies that the minimum separation of two resolvable Airy disks is when the intensity peak of one coincides with the first intensity minimum of the other, and for circular lenses we find

$$d = \frac{0.61\lambda}{\text{NA}}.$$

For example, for two green fluorescent protein (GFP) fluorophores [3], a very common dye use in live cell localization microscopy, emitting at a peak wavelength of 507 nm under normal physiological conditions and a typical 1.49 NA objective lens this value of d is 208 nm and therefore to obtain spatial localization information for more than one molecule they must be separated by at least this distance, or alternatively emitting at different times such that each molecule can be analyzed separately.

If we meet these conditions, we will generate an image similar to that in Fig. 13.1b. The diffraction-limited fluorescent "spot" (essentially the zeroth order peak of the Airy disk) is clearly visible spread over multiple pixels, though there is significant background noise, and taking line profiles shows that it is approximately Gaussian in both x and y Cartesian axes. One way to proceed with localization determination using a computational algorithm is the following, exemplified by software that we have developed called ADEMSCode [10], but with a plethora of similar algorithms used by others in this field (see Table 13.1), including probabilistic Bayesian approaches (e.g., 3B [9]) and pre-processing steps to reduce fluorophore density and improve the effectiveness of subsequent analysis [4]. Most of these are capable of analyzing fluorophore localizations, but to the best of our knowledge our own package [10] is the only one which is capable of evaluating localizations, dynamical information such as diffusion coefficients, and utilizing photobleaching dynamics to estimate molecular stoichiometries. ADEMSCode proceeds with localization analysis in the following way: first, find the peak pixel intensity from the camera image and draw a small bounding box around it (typically 17×17 pixels (i.e., one central pixel with a padding of 8 pixels on each side), where for us a pixel is equivalent to 50–60 nm at the sample). Within that square then draw a smaller circle (typically of 5 pixel radius) centered on the maximum of intensity which approximately contains the bulk information of the fluorophore's PSF. The pixels which are then within the bounding box but not within the circle may have their intensities averaged to generate a local background estimate. Each pixel then has the local background value subtracted from it to leave the intensity due only to the fluorophore under examination, and this corrected intensity may now be fitted. An optimization process then occurs involving iterative Gaussian masking to refine the center of the circle and ultimately calculate a sub-pixel precise estimate for the intensity centroid. A similar effect can be achieved by fitting a 2D Gaussian function plus uniform offset to intensity values that have not been background corrected, however, fit parameters often have to be heavily constrained in the low signal-to-noise regimes relevant to imaging single dim fluorescent protein molecules and due to the centroid output, iterative Gaussian masking is often more robust.

Table 13.1 Comparison of some modern super-resolving localization microscopy analysis packages

Algorithm name	Description	Features	Pros	Cons	Notes
Haar wavelet kernel (HAWK) analysis [4]	Data pre-processing method	Decomposes high-density data to create longer, lower density dataset for further analysis	May not collapse structures ~200 nm apart into one structure; experiments easier as high density can be worked around	Depending on imaging, the localization precision may be comparable to lower resolution techniques, e.g. structured illumination microscopy (SIM) [5]	Versatile: can be used with any frame-by-frame localization algorithm
DAOSTORM [6]	STORM-type analysis package	Fits multiple PSF model to pixel cluster to deal with overlapping fluorophores	Increases workable density from ~1 to ~7 molecules/μm^2 dependent on conditions	Requires long STORM-type acquisitions; no temporal information	Adapted from astronomy software
FALCON [7]	Localization for STORM/PALM-type data	Iteratively fits Taylor-series expanded PSFs to data to find best fit	Data treated as a continuum rather than on a grid	Low temporal resolution (~2.5 s/frame)	Localization error ca. 10 to 100 nm depending on imaging conditions
ThunderSTORM [8]	PALM/STORM localization	Toolbox of analysis algorithms user can choose from	Flexible	Many methods need careful selection of parameter values, though some can be set algorithmically	Free ImageJ plugin. Also functions as a data simulator for testing routines
Bayesian analysis of blinking and bleaching (3B) [9]	Factorial hidden Markov model which models whole system as a combination of dark and emitting fluorophores	Produces a probabilistic model of the underlying system; exact positions may have high error bars	Can analyze overlapping fluorophores, eliminates need for traditional user-defined parameters, no dependence on specialized hardware	Results require nuanced interpretation, Bayesian priors must be well known, resolution may vary along and perpendicular to a line of fluorophores	Spatial resolution ca. 50 nm; temporal ~s

(continued)

Table 13.1 (continued)

Algorithm name	Description	Features	Pros	Cons	Notes
ADEMSCode [10]	Matlab toolkit for single-molecule tracking and analysis described in this section	Can analyze stoichiometries, diffusion coefficients, localization	Powerful, flexible	Human-selected parameters require careful treatment	Accesses high temporal resolution (~ms) as well as localization
Single-molecule analysis by unsupervised Gibbs sampling (SMAUG) [11]	Bayesian approach analyzing trajectories of single molecules	Analyzes trajectories to find underlying mobility states and probabilities of moving between states	Non-parametric, reduces bias	Reliant on the quality of input data, needs good priors	Post-processing—Input data generated by other techniques
RainSTORM [12]	MATLAB image reconstruction code	Complete STORM workflow including data simulation	Simple, out-of-the box operation	Weaker with dense data, relatively inflexible, no dynamical information	
QuickPALM [13]	PALM analysis software	3D reconstruction, drift correction	Complete PALM solution	Static reconstruction	Plugin for ImageJ [14]

Although it has an analytic form, for historical reasons relating to past benefits to computational efficiency, the central peak of the Airy disk is commonly approximated as a 2D Gaussian that has equation

$$I(x,y) = I_0 e^{-\left(\frac{(x-x_0)^2}{2\sigma_x^2} + \frac{(y-y_0)^2}{2\sigma_y^2}\right)},$$

where the fittable parameters are I_0, the maximum brightness of the single fluorophore, x_0 and y_0, the co-ordinates of the center of the Gaussian, and σ_x and σ_y which are the Gaussian widths in x and y, respectively (interested readers should read the work of Kim Mortensen and co-workers on the improvements that can be made using a more accurate formulation for the PSF function [15]). Using conventional fluorescence microscopy, assuming any potential polarization effects from the orientation of the dipole axis of the fluorophore are over a time scale that is shorter than the imaging time scale but typically 6–7 orders of magnitude, the Airy disk is radially symmetrical, and so σ_x and σ_y ought to be identical, and they may therefore be used as a sanity check that there is only one molecule under consideration—a chain of individually unresolvable fluorophores will have a far higher spread in x or y. Similarly, the brightness of individual fluorophores in a dataset acquired under exactly the same imaging conditions is a Gaussian distribution about a mean value. After fitting the 2D Gaussian one may usefully plot the fitted I_0 values to check for outliers; indeed, with a well-characterized fluorophore and microscope one may be able to include these checks in the analysis code itself. When iterative masking is used to determine the intensity centroid an initial guess is made for the intensity centroid, and a 2D function is then convolved with the raw pixel intensity data in the vicinity of each fluorescent spot. These convolved intensity data then have a revised intensity centroid, and the process is iterated until convergence. The only limitation on the function used is that it is radially symmetric, and it has a local maximum at the center. In other words, a triangular function would suffice, if the purpose is solely to determine the intensity centroid. However, a Gaussian function has advantages in returning meaningful additional details such as the sigma width and the integrated area.

Having fit the 2D Gaussian, the fitting algorithm will usually report to the user not only the best-fit values but also either estimated errors on these fits or the matrix of covariances which may be trivially used to obtain the error bars. It is tempting to take the error on x_0 and y_0 optimized fitting values as the localization precision, but this reflects the fitting precision that only partially indicates the full error involved. In fact, the error on the centroid needs to be found by considering the full suite of errors involved when taking the experimental measurement. Principally, we must include the so-called dark noise in the camera, the fact that we cannot know which point in a pixel the photon has struck and therefore the PSF is pixelated, and the total number of photons that find their way from the fluorophore to the sensor. Mathematically, the formulation is given in reference [16] as

$$\delta = \sqrt{\left(\frac{s^2 + \frac{(lm)^2}{12}}{N}\right) + \left(\frac{4\sqrt{\pi}s^2b^2}{lmN^2}\right)},$$

where s is the fitted width of the Gaussian PSF and would usually be taken as a mean of σ_x and σ_y, N is the number of photons, b is the camera dark noise, l is the camera pixel edge length, and m is the magnification of the microscope. We can instantly see that a compromise must be reached during experiment. If we image rapidly we will have fewer photons to fit to and the spatial localization precision worsens. The approximate scaling is with the reciprocal of the square root of the number of photons sampled in the case of relatively low dark noise and small camera pixels. If we image for a long time we can see localization precisions as low as 1 nm at the cost of losing dynamical information. In practice we find that imaging on the order of millisecond exposures leads to lateral (i.e., In the 2D plane of the camera detector) localization precisions of 30–50 nm for relatively poor dyes such as fluorescent protein molecules, while imaging for a second or more on far bright organic dyes may give single nanometer precision [17]. This fundamental trade-off has led to two complementary but different forms of super-resolved localization microscopy—long time-course imaging on fixed cells for nm precision localization, and lower spatial resolution imaging that can access temporal information.

If we have obtained a time-series acquisition of a system with mobile fluorophores (either freely diffusing or attached to a translocating molecular machine, for example), we may wish to work out where and how quickly the fluorescent spots are moving, or if their mobility is Brownian or 'anomalous' or confined. The localization information may then be used to infer the underlying types of single-molecule mobility [18]. With localizations in hand this is relatively straightforward and is achieved by comparing successive frames (n and $n + 1$) and accepting that a spot in frame $n + 1$ is the same molecule as a spot in frame n if the two spots are sufficiently close together, have sufficiently comparable intensities, and are of sufficiently comparable shape. The vector between the spots may then be taken and the process repeated for frames $n + 1$ and $n + 2$, iteratively building up a 2D track (Fig. 13.2). This is a threshold-based method and should therefore be used with care, with the threshold determined by converging one physical parameter like diffusion coefficient with respect to the distance cut-off (a useful review of this effect is found in reference [19]). There should also be sufficiently few fluorophores such that a spot in frame $n + 1$ is not (or only rarely could be) accepted as being the same molecule as two or more spots in frame n (or vice versa); should that occur, the track will need to be terminated at that point. Deciding whether spots in two successive frames are the same molecule is clearly fraught with danger; modern methods with Bayesian analysis will be discussed later in our study here.

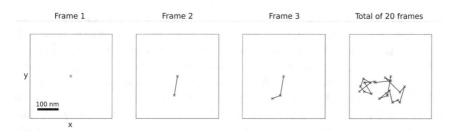

Fig. 13.2   Combining localizations from multiple frames to build a 2D track

13.3   STORM and PALM

Though STORM (STochastic Optical Reconstruction Microscopy) and PALM (PhotoActivated Light Microscopy) have differences in their methods, they work toward the same goal: to spatially and temporally separate the "on" states of the fluorophores used so that the PSFs can be fitted to as described above. By fitting a large population of fluorophores an image of the overall structure or distribution of the system of interest may be generated. Here we will briefly describe each technique and their relative pros and cons.

13.3.1 STORM

STORM (STochastic Optical Reconstruction Microscopy) [20] is a powerful technique which relies on the inherent ability of some fluorophores to switch between "on" (emitting) and "off" (non-emitting) states when irradiated by a high intensity laser. At the beginning of the image acquisition almost all fluorophores will be in the on state. However, as time goes on, the population gradually moves to a combination of fluorophores that have photobleached and are permanently non-emissive, and some that are in the photoblinking state in which they transition between on and off states. At some point, the ratio of these populations will reach the correct state such that individual fluorophores are visible separated from their neighbors, i.e. the mean nearest-neighbor distance between photoactive fluorophores is then greater than the optical resolution limit and so a distinct PSF image associated with each separate fluorophore molecule can be seen. A time series of frames must be acquired of the system in these conditions, and each fluorophore in each frame is localized as described above (Fig. 13.3a–d). The loci found may then be all plotted in one frame, showing the base distribution of the fluorophores and thus the structure of the system. This is "optical reconstruction"—although in each image frame only a few individual foci may be visible, by combining several hundreds or thousands of these image frames enough fluorophores will be captured to give the overall picture in detail far beyond non super-resolution microscopy methods.

For this process to be feasible, several conditions must be met. First, the excitation laser power must be high enough to force the fluorophores into their photoblinking state. Though it is counterintuitive, a low power laser will enable the fluorophores to stay on longer but photobleach permanently afterward, without any blinking. In general, laser excitation intensities at or above ca. 1800 W/cm^2 are effective depending on the dyes used. Secondly, the fluorophores of interest should be capable of photoblinking behavior, and when they do blink their single-molecule brightness must be above the background noise. Fluorophores which are suitable under these constraints will be discussed in Sect. 13.4.

13.3.2 PALM

PALM (PhotoActivated Light Microscopy) [21] takes a second approach to separating fluorophore emissions in space and time. While STORM relies on all fluorophores being excited at the same time but randomly blinking on and off, PALM randomly activates a random subset of the fluorophores in the system with one laser, and then excites them for imaging with a second laser. Activated fluorophores return to the initial state after they are imaged. Then repeat and image a second set of fluorophores (Fig. 13.3f–j). Activation can mean either one of two processes—either the fluorophore is initially dark and switches to a fluorescent state, or under illumination of the activation laser the fluorophore undergoes a color change, commonly from red to green. In either case the activating laser is usually ~long UV at around 400 nm wavelength, while the fluorescence excitation laser is in the ordinary visible range.

The constraints for PALM fluorophores are obvious. Although they do not need to photoblink, they must be capable of switching states in response to UV light exposure, and once again they must be bright enough in their emissive state to be well above the background noise level. As PALM images single molecules, the laser intensity must be relatively high as for STORM.

13.3.3 Pros and Cons

STORM and PALM are powerful techniques that enable reconstruction of tagged systems, for example microtubules in vivo or the architecture of organelles in the cell. In this respect, the information offered by STORM and PALM is unrivaled by other techniques—more detailed information is difficult to find as the crowded cellular environment precludes whole-cell X-ray or neutron diffraction experiments. Imaging tagged substructures using traditional diffraction-limited optical microscopy is possible but gives less detailed information, and even mathematical post-processing techniques such as deconvolution (if they are suitable for the imaging conditions) give a lower resolution than super-resolution imaging itself.

However, these are not "magic bullet" techniques and have their own drawbacks. Principally, both are slow methods. To collect enough information to properly reconstruct the base fluorophore distribution hundreds to thousands of frames at least must be taken, meaning that total imaging times are seconds to minutes. Given that many biological processes occur over millisecond timescales or faster this obviously precludes capturing time-resolved information from these rapid dynamic processes. Further, if there is some biological process restructuring the cellular environment during imaging a false picture may be obtained. For this reason, biological samples are usually "fixed," i.e. rendered static and inert before imaging to ensure that the fluorophore distribution does not change during image acquisition. Photodamage is also of concern. As fluorophores photobleach, they produce free radicals which attack and damage the biological sample. Various imaging buffers exist which minimize this though these can induce lower photoblinking, so a trade-off must be struck.

13.3.4 Techniques Using Modified Point Spread Functions that Use Localization Microscopy at Some Level

STORM and PALM are both powerful techniques in their own right but they are not the only way to generate data that can be processed with a super-resolution algorithm. In 2000, Stefan Hell (who went on to share the 2014 Nobel Prize in Chemistry with William E. Moerner and Eric Betzig for "the development of super-resolved fluorescence microscopy" [22]) published an account of a new super-resolution method based around stimulated emission of fluorophores, and known as STED (Stimulated Emission Depletion) microscopy [23, 24]. In brief, STED involves two lasers that are focused on the same position, one that excites the fluorophores while a donut-shaped beam around this has the effect of suppressing emission from fluorophores in this region, achieved via stimulated emission when an excited-state fluorophore interacts with a photon whose energy is identical to the difference between ground and excited states. This molecule returns to its ground state via stimulated emission before any spontaneous fluorescence emission has time to occur, so in effect the fluorescence is depleted in a ring around the first laser focus. By making the ring volume arbitrarily small the diameter of this un-depleted central region can be made smaller than the standard diffraction-limited PSF width, thus enabling super-resolved precision using standard localization fitting algorithms to pinpoint the centroid of this region, <30 nm being typical at video-rate sampling of a few tens of Hz. Other related STED-like stimulated depletion approaches include Ground State Depletion (GSD) [25], saturated pattern excitation microscopy (SPEM) [26] and saturated structured illumination microscopy (SSIM) [27]. A similar result to STED using reversible photoswitching of fluorescent dyes but not reliant on stimulated emission depletion is known as RESOLFT (reversible saturable/switchable optical linear (fluorescence) transitions) microscopy [28] that also utilizes localization microscopy algorithms.

Similar to but going beyond STED approaches is a recently developed method also from Stefan Hell and colleagues known as MINFLUX (MINimal photon FLUXes) [29] which does not need a depletion laser. Here, the excitation beam is

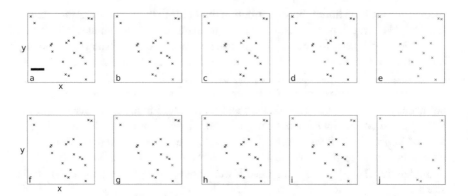

Fig. 13.3 Schematic of STORM and PALM localization. (**a**) The underlying fluorophore distribution. (**b–d**) fluorophores are stochastically excited during STORM. (**e**) reconstructing the original distribution from the emissions observed. (**f–i**) In PALM, fluorophores are first activated with a UV laser (activated fluorophores in blue) and are then excited to fluoresce (red). (**j**) The underlying distribution reconstructed. A given experiment time will produce fewer emission events in PALM and thus sample the underlying distribution slower than STORM. Bar: 200 nm

the donut and so a fluorophore at the center of the beam will not be excited. By spatially scanning the beam and finding the fluorescence emission intensity minimum, the position of a fluorophore can then be found with a nanoscale spatial precision small spatial precision with an exceptionally fast scan rate of up to ca. 10 kHz (Fig. 13.3).

13.4   Choosing a Fluorophore

Selecting the correct fluorophore for the system of interest is clearly a prime concern. Summarized, fluorophores must:

- be bright enough for single molecules to be seen above background noise
- be photoactivatable or photoblink under the correct conditions
- not interfere with ordinary cell processes if imaging in vivo
- not unduly change the structure or function of an in vitro system
- (for multi-color experiments) be sufficiently spectrally separated that they may be imaged individually without cross-excitation

This is a considerable list of necessary attributes, and there are some further desirable ones. For example, some fluorophores are more photodamaging than others, and some laser lines are also more damaging to cells and tissues than others. Fluorophores may be sensitive to pH or ionic strength and thus be inappropriate for the system of interest. Fluorescent proteins that are expressed in vivo are often described as being either definitively "monomeric" or non-monomeric. Non-monomeric fluorophores will have more of a propensity to form homo-

Table 13.2 Table of commonly used classes of fluorophores and their principal applications

Fluorophore class	Example fluorophores	Applications	Notes
First-generation fluorescent proteins	Green fluorescent protein [31], yellow fluorescent protein [32], red fluorescent protein [33]	In vivo protein labeling through genomic integration; FRET, in vitro labeling, STORM	Not all suitable for single-molecule imaging, e.g. cyan fluorescent protein. Derived from sea anemones (RFP) or jellyfish (GFP)
Second-generation red fluorescent proteins (mFruits)	mCherry, mOrange, mStrawberry [34]	As above	Increased brightness over first-generation RFPs
Second-generation GFPs	Enhanced GFP [35], monomeric GFP [30], superfolder GFP [36]	As above	Improved brightness, reduced dimerization, and quickly-maturing, respectively
First-generation cyanine dyes	Cy3 (orange), Cy5 (far-red) [37]	Labeling nucleic acids, proteins, both in vitro and in vivo; FRET	Can be chemically conjugated to proteins, not genomically integrated. Sensitive to local conditions
Alexa Fluor family	Alexa Fluor 488 [38]	As above	Second generation of xanthene, cyanine, and rhodamine dyes with improved brightness and photostability
Hoechst	Hoechst 33342 [39]	Minor groove binding DNA stain	Excited by UV light
Janelia Fluor family	JF525 [40]	Cell permeable dyes, used in vivo with protein labeling such as halo tag/snap tag	Improved quantum yields, ca. 2× brighter than comparable first-generation cyanines
Photoactivatable fluorescent proteins	PAGFP [41], PA-mKate2 [42]	PALM in vitro and in vivo	Enters fluorescent state on application of UV light *PALM in vivo*
Photoconvertible fluorescent proteins	mEos2 [43], Kaede [44], Dendra 2 [45]	As above	Change color on application of UV light

oligomers, and if imaging freely diffusing proteins may seed aggregation. In some cases, therefore, care must be taken to choose the monomeric form of the protein, the nomenclature for which is a lowercase "m" before the fluorescent protein name. For example, the monomeric form of the commonly used green fluorescent protein (GFP) is mGFP [30] that has an A206K mutation that suppresses putative dimerization between GFP molecules. Table 13.2 lists commonly used fluorophores alongside their usual applications.

13.5   Ideal Properties of a Super-Resolution Microscope Relevant to Localization Microscopy

In Sect. 13.4 we discussed the properties of a good fluorophore for super-resolution imaging. Here we will briefly describe the properties of a good super-resolution microscope appropriate for localization microscopy.

Lenses used should be clean and ideally coated with an anti-reflective coating for the wavelengths used, and care must be taken to ensure they are mounted truly perpendicular to the optical axis. Lasers should be able to produce a few milliwatts at a minimum and should produce stable output. Mirrors should be rated for the correct wavelength—what is reflective for infrared may be largely transparent to visible light. The whole system should be mounted on an air table to reduce mechanical vibration. Cameras should be capable of acquiring at the desired speeds, e.g. 10 ms/frame, and should be cooled to reduce shot noise. It is necessary also that the camera have some gain function to amplify the light collected, for example electron multiplying (EM) gain which produces a cascade of electrons to hit the CCD and thus enhance the signal—but also the noise. In general, for best fitting of single-molecule spots the camera should be imaging at a resolution of approximately 40–60 nm/pixel. This is a key consideration and may necessitate additional optics prior to the camera to expand the imaged light.

The objective lens is one of the key components. For best performance this lens should have a high numerical aperture and be ideally oil immersion (that is, oil is placed between the coverslip and the objective lens) to ensure good optical contact and enable high photon capture. That said, for imaging in excess of a few microns depth a water immersion lens may mitigate potential issues of spherical aberration that occur with oil immersion lenses, but with the caveat of a reduced numerical aperture of ~1.2, that reduces the photon capture budget. For dual-color experiments, chromatic aberration can be a problem—red and green light, for example, will come to focus at slightly different distances by a simple non-achromatic lens and therefore at a given height a red fluorophore may be in focus and a red fluorophore slightly defocused. There are four principal ways to get around this. (i) One can measure the chromatic aberration and correct for it in image post-processing. (ii) One can use an automatic stage with multiple settable heights and move between focal distances between acquisitions. (iii) One may acquire all the green fluorescence data, manually refocus, and then take all the red fluorescence data (for example). (iv) One may purchase an objective lens that is apochromatic and has minimal chromatic aberration. The first three techniques have drawbacks—careful calibration is needed for correcting chromatic aberration in post-processing, automatic stages may suffer from drift, and acquiring the data separately is non-trivial on unfixed samples since the acquisition may take some time. Moreover, the first acquisition may damage the system before you get the chance to look at the second fluorophore. Overall, if resources permit, an apochromatic lens is the best method for multi-color experiments. Though each optical microscope is different, the basic principles are similar across all, and a sample schematic is given in Fig. 13.4. For convenience we will refer here only to an epifluorescence microscope, which is one in which the

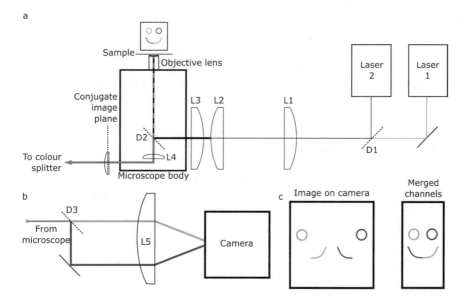

Fig. 13.4 (**a**) Schematic of a super-resolution STORM/PALM microscope. The lasers are combined by the dichroic mirror D1 and the beam is expanded by the lens pair L1 and L2. The lens L3 just before the microscope body forms a telescope with the objective lens and ensures the beam is the correct width and comes out of the objective collimated. Excitation light is directed into the objective using the dichroic mirror D2, which allows the captured fluorescence (pink) though. The imaged light is then focused on the side port of the microscope with the lens L4 within the microscope housing itself, though for convenience we do not image here but recollimated the imaged light with a lens placed at the conjugate image plane (marked). (**b**) Principles of a color splitter. Collected light is passed through the dichroic mirror D3, which separates the two channels (here orange and blue). The distinct channels are focused on the camera chip with the lens L5, so that each color channel hits a separate half of the chip as seen in panel **c**. (**c**) separate channels may be merged to recover the true image

excitation light goes through the back of the objective and enters the sample collimated. The path the light takes is as follows: first, it is emitted from the laser, and if there are multiple lasers present then the fluorescence emissions are combined using dichroic mirrors to form one beam. This beam is then expanded using a telescope and then propagates through a lens that focuses the light on the back focal plane of the objective lens. The beam is directed into the objective lens by way of a dichroic mirror that reflects the excitation light but not the emitted light. The light hitting the objective lens then has originated from one focal distance away from the objective and is therefore collimated after the objective. This process of focusing and collimating with the objective is a second telescope that has the effect of reducing the beam width considerably. To get a beam of the correct width at the sample the expansion of the laser by the first telescope can be varied. After the excited fluorophores have emitted photons, a proportion of these is captured by the objective lens once more and collimated by it. The light path then goes back down the microscope and this time through the dichroic mirror and is focused typically on the back port of the microscope, where it may be recollimated or imaged directly. For a dual-color experiment, the different color light must be split and imaged separately, either on two cameras or on separate parts of the same camera chip, or potentially using time-sharing just as with Alternating Laser Excitation [46] that uses a multi-bandpass filter set. In Fig. 13.4b we show the simplest setup, a static color splitter based on a dichroic mirror imaging each channel on a separate part of the detector.

13.6 3D Localization

Localization in 3D can be approached in one of two ways. Firstly, a sample can be scanned through in the z direction building up a full 3D stack of the entire system of interest. Then the z position can be approximated as being the slice in which a given PSF is most in focus. This has the distinct disadvantage of being extremely slow—the best frame rates are around 2 full stacks per second, but almost all cellular processes happen three orders of magnitude quicker than that. For fixed samples this may be appropriate but for understanding dynamical processes it is simply inappropriate.

Instead, the PSF of the fluorophores can be altered through lenses or spatial light modulators (SLMs) so that they are non-symmetric about the focal point in z. Two principal techniques for this have emerged, namely astigmatism microscopy [47] and double-helix point spread function (DH-PSF) imaging [48].

For astigmatism imaging, the emitted light from the sample propagates through a cylindrical lens between the microscope and camera. This modifies the PSF from being rotationally symmetric—a Gaussian profile—into more of an elliptical profile. The orientation of the ellipse is dependent on whether the fluorophore is above or below the focal plane when imaged, and the ratio of the major to minor axes depends on the specific distance, as shown in Fig. 13.5, which shows a simulated fluorophore's appearance as a function of z position. For this to work in practice, before experiments an in vitro fluorophore sample should ideally be imaged and scanned in z in known increments using an automated nanostage. The ratio of vertical to horizontal axis may then be measured for each slice and plotted against vertical distance. A fluorophore's focal point is where the ratio is 1, so that the relative absolute distance from the focus can be found. When imaging in an experiment, the focal plane is set and kept constant and the z positions of the fluorophores measured relative to that. In practice, this look-up table-based methodology is robust and requires only one additional lens in an existing fluorescence microscope, while the fluorophores themselves have only the same constraints as for 2D imaging. To date, astigmatism imaging with fluorophores in vitro and in vivo has shown an ability to beat the axial resolution limit by approximately a factor of 2–3, with axial spatial precisions of ca. 50 nm being common [49].

DH-PSF is a more complex technique requiring considerable different optics and a reconfiguration of the imaging path of the microscope. In a typical design, emitted light is collected and reflected off an SLM, while the light itself is imaged at an angle of 30° from the emitted light's optical axis. This produces a PSF that forms a double

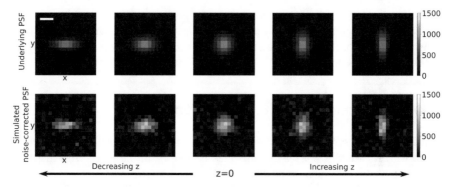

Fig. 13.5 Astigmatism point spread functions with distance from the focal plane. PSF modeled as a 2D Gaussian with x and y sigma values set according to simulated "height." With each height step, σ_x is reduced by 1 and σ_y is increased by 1. For the lower panels, Gaussian noise has been added to each pixel to simulate background, but more complex noise such as camera shot noise are not included. Bar: 200 nm

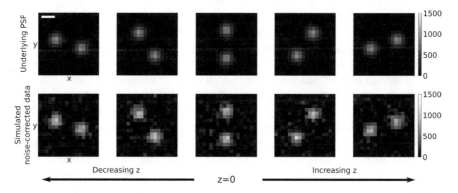

Fig. 13.6 Double-helical point spread functions with distance from the focal plane. PSFs are simulated as two Gaussian distributions at opposite ends of an axis that rotates with each step in z. Again only Gaussian background noise is included in the lower panel simulations. Bar: 200 nm

helix along the optical axis. When imaged in 2D, this appears as two separate dots, whose orientations are dependent on the z position of the fluorophore as seen in Fig. 13.6, which simulates the appearance through z of a double-helical PSF. An angle may be generated by finding the angle between the vector linking the two fluorescent spots and the x axis. Then, as for astigmatism imaging, the angle-to-distance look-up table must be generated ideally from in vitro fluorophores before production data is acquired. The xy position of the fluorophore is taken to be the center of the two spots. This can be calculated by finding the centers of the spots themselves by fitting a 2D Gaussian to each, or by finding the centroid of the two-spot system with a specific centroid algorithm, through this latter technique is computationally more costly. An important drawback of DH-PSF imaging is that to generate the double helix the light is effectively split in two, and each spot has half the brightness of the full fluorophore at a given z position. If one is working in a low signal-to-noise regime, this reduction in brightness may make the spots indistinguishable from background. However, the z axis resolution is excellent—with sufficient photons detected the localization precision can be below 6 nm [50].

13.7   Analyzing 2- and 3D Localization Data

Having found the trajectories of individual spots, a question immediately arises what to do with it. Broadly, we may define three categories of the trajectories that may each give useful information: position, velocity, and brightness.

Analysis of the positions themselves gives access to diffusion coefficients by comparison to Brownian motion, as well as colocalization information between molecules—i.e., tagging different targets with different color fluorophores, measuring the positions of each and identifying if they are in the same place. Simple positional analysis also may tell us if a protein is in the nucleus or cytoplasm, for example. As well as these, the overall spot distribution may be analyzed to determine if there are identifiable distinct regions to which multiple fluorophores·belong. This suite of techniques is known as cluster analysis.

Various methods exist to perform cluster analysis. Most straightforward are distance-based methods such as the Voronoi method [51]. This generates a set of regions around each PSF such that each region is the area closer to the seed PSF than to any other, with small regions then indicating a cluster. Also widely used are density techniques such as density based spatial clustering of applications with noise (DBSCAN) [52] which iterates across all localizations and assesses the local density within a set radius r. If there is a minimum number of spots within the circle defined by r a cluster is accepted. This continues until the boundary of the dense area—and thus the entire cluster—is found. This repeats through all spots to find all clusters. Similarly using density are pure statistics-based methods of measuring clustering, particularly Ripley's H, K, and L functions [53]. These are a group of well-defined statistical transforms which can be applied to the image data and which have minima and maxima correlating to how clustered the data is. The values of the functions however only indicate whether over the spatial extent analyzed clustering is indicated. To classify points into discrete clusters requires analysis beyond the functions. This could be done by using the extended L function as proposed by Getis and Franklin [54, 55].

More recently, Bayesian analysis techniques have been developed which make use of advanced statistical models to evaluate clustering and in general aim to remove the level of human input or parameter selection needed during analysis. Bayesian implementations often make use of the statistical functions described above [56] and the number of clusters is then predicted with reference to the model, usually that the clusters are approximately spherical with molecules inside the cluster distributed according to a Gaussian [57]. Bayesian approaches are also valuable for determining the mode of molecular mobility of tracking data in localization microscopy, for example whether molecules are freely diffusing or their motion is confined [18].

In principle, the outputs of these deterministic methods can also be used to train machine learning models. However, machine learning is often sensitive to the input—if the data to be analyzed is too dissimilar from the training data the output will be unreliable at best. One implementation of neural networks for cluster analysis was published in 2020, which used a neural network trained on a given number of nearest-neighbor distance values, showing efficient computational performance compared to Bayesian methods or DBSCAN on both simulated and experimental data [58]. However, the extensive training needed may offset this gain depending on the size of the dataset to be analyzed.

Velocities may also be characterized, and this is most commonly done in the context of molecular machines, where the step sizes and overall movement speed are difficult to determine by any other means and yet are crucial to biological function. By tracking fluorescently tagged molecular machines or cargoes these parameters can be accurately determined. Similarly, the overall drift of diffusing molecules may be examined to understand whether the Brownian motion they are undergoing is directed (for example facilitated diffusion, or an active process requiring the input of external free energy) or whether it is truly a random walk.

Finally, the intensity (i.e., brightness) of the fluorescent spots contains significant information about the system. Specifically, if we have a population of fluorescently tagged molecules, we may analyze the distribution of intensities to uncover whether they are monomeric or aggregating into clusters. For systems where we know that aggregation happens—for example in liquid–liquid phase separation [59]—we can use the intensity through time to work out the total stoichiometry of molecules within the cluster. Whatever the purpose, the method for this is the same, and is done by taking the initial total intensity and dividing it by the intensity of a single fluorophore. The most important parameter to determine is thus the intensity of a single fluorophore which we denote as the Isingle value. By plotting the total intensity of a cluster through time we will see the decrease in intensity as the fluorophores in the cluster photobleach occurs in a step-wise fashion such that the size of a step, once noise is removed, is an integer multiple of the Isingle value. There may be differences of ca. a few tens of percent with estimates made in vitro for Isingle due to different in local excitation intensity, and buffering conditions inside a cell. Therefore, it is important to determine Isingle in the physiological context. The simplest way to achieve this is to use only the final photobleaching step where there is only one fluorophore that bleaches to leave only the background noise—more accurate than attempting to count all steps since these are limited to a maximum of 6–7 depending on the dye used. Further, steps involving more than one photobleached molecule in a sampling time window will have a higher associated noise due to Poisson sampling of photons at higher intensities. To obtain Isingle, the full intensity through time track can be fitted to a step-wise function usually such as a hidden Markov model [60], or other edge-preserving filters such as Chung-Kennedy [61], and the step sizes extracted and averaged—simulated data of an intensity track is shown in Fig. 13.7a. Alternatively, the intensities of every spot can be plotted. If it

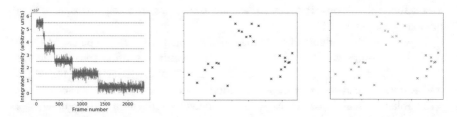

Fig. 13.7 (**a**) Simulation of a step-wise photobleaching of a single intensity track; (**b, c**) schematic of simple cluster analysis

was taken in the truly single-molecule photoblinking regime, the majority of tracked spots should be single molecules, and therefore on a plot of intensity against number of spots a peak would be expected around Isingle. This overall process is known as step-wise photobleaching and is suitable for analyzing either in vivo or in vitro data with the proviso that the Isingle values should ideally be determined separately for each sample. An illustration of simulated intensity-time data is given in Fig. 13.7.

13.8 Applications of Localization Microscopy

Single-molecule localization methods have been extensively applied both in vivo and in vitro to elucidate a wide range of biological processes. These include organization of molecules within the cell, the interplay between various cytoskeleton elements, and measuring diffusion coefficients. These details can tell us about what the key molecular interactions inside cells for specific biological process, as well as insights into mobility of molecular complexes and how these are influenced by the microenvironment of the cell. Here, we briefly present some biological results obtained to date.

DNA and RNA processes are amongst the most important in the cell and they have been studied extensively with single-molecule tools. Yan et al used single-molecule imaging to monitor mRNA translation and measure the switching between translating and non-translating states, finding translation repression due to specific sequences [62]. Also working on replication, Syeda et al. used dual-color imaging of the Rep helicase to demonstrate its dependence on PriC and the helicase's means of negotiating proteins bound to DNA [63]. Wooten et al. recently demonstrated super-resolution imaging could be used for epigenetic studies of chromatin fibers [64] in eukaryotes.

Away from DNA, localization microscopy has been used extensively to image the cellular cytoskeleton such as the organization of actin in 2D [65] and 3D [66] as well as the distribution and degradation of intermediate filaments [67] and intracellular trafficking dynamics where microtubules intersect [68], and a wide range of biomedical questions such as probing cancer biology [69]. Live cell imaging with 3D localization has shown colocalization of proteins and the cell surface [70], as well as the distribution of eukaryote transcription factors which may be used to map the overall genome [49]. 2D localization of synthetic sensors inside living cells has also recently been demonstrated which suggests that localization of proteins may soon be correlated with the physical conditions around them [71], while diffusion coefficient analysis has shown that under osmotic stress eukaryotes experience slower diffusive behavior in the cytosol [72]. Super-resolution microscopy has also been used to observe clustering of key eukaryotic proteins [73], and more generally is being used to understand the currently murky world of liquid–liquid phase separation [59, 74].

Alongside this in vivo work, considerable progress has been made through in vitro experiments also. Protein aggregation can readily be studied in a microscope slide, and amyloid proteins implicated in Alzheimer's disease have been imaged aggregating in human cerebrospinal fluid [75]. Step-wise photobleaching has been used to understand aggregation of amyloid-β [60] in vitro also. DNA origami has been extensively studied for some time, and as well as imaging the structure of the origami tile, it has been used to more robustly characterize protein copy number using immunofluorescence [76].

13.9   Conclusion and Future Perspectives

Overall, it is clear that localization microscopy is an enormously valuable technique, enabling new insight into a range of complex biological process. Both PALM and STORM methods can be used for reconstructing the fine structure of biological structures, while fitting to diffusing molecules exposes diffusion coefficients, and subcellular organization in response to stress or during key biological processes. The detail we are now able to obtain using the methods in this chapter is immense. However, there remain technical challenges that need to be addressed, and there remain drawbacks of STORM/PALM-type experiments, for example in certain instances cells may need to be fixed, losing valuable dynamic information. As we look to the future of biological microscopy, the focus will increasingly be on multi-method and correlative approaches, which promise to give information beyond what is currently possible. Life does not exist separately to physics; rather, cells leverage physical laws to organize and regulate their internal conditions. With cutting-edge physical sensors now available to measure crowding, pH, and ionic strength, we may now begin to correlate our precise local data with the prevailing physical processes and conditions. This integrative understanding across disciplines will be the key battleground in the quest to understand life and develop the medical therapies of the future.

Take Home Message
- A fluorophore may be localized by fitting a PSF to an acquired image.
- The best spatial precision in fixed cells is ~1 nm and in living cells ~30 nm for millisecond to tens of milliseconds time resolution
- Multiple molecules of interest may be labeled and imaged in multiple colors in the same cell that can enable insight into dynamic molecular interactions
- Photoblinking and localization of single molecules requires high laser power and so there is a trade-off with photodamage of samples

- Step-wise photobleaching can be used to determine molecule copy numbers and molecular complex/assembly stoichiometries
- Tracking molecular complexes can yield valuable information about the molecular mobility and the local microenvironment of cells
- Reshaped PSFs can enable 3D spatial information

Acknowledgments This work was supported by the Leverhulme Trust (RPG-2019-156) and EPSRC (EP/T002166/1).

References

1. Wollman AJM, Nudd R, Hedlund EG, Leake MC. From animaculum to single molecules: 300 years of the light microscope. Open Biol. 2015;5(4):150019. https://doi.org/10.1098/rsob. 150019.
2. Gelles J, Schnapp BJ, Sheetz MP. Tracking kinesin-driven movements with nanometre-scale precision. Nature. 1988;331(6155):450–3. https://doi.org/10.1038/331450a0.
3. Ormö M, Cubitt AB, Kallio K, Gross LA, Tsien RY, Remington SJ. Crystal structure of the Aequorea victoria green fluorescent protein. Science. 1996;273(5280):1392–5. https://doi.org/10.1126/science.273.5280.1392.
4. Marsh RJ, et al. Artifact-free high-density localization microscopy analysis. Nat Methods. 2018;15(9):689–92. https://doi.org/10.1038/s41592-018-0072-5.
5. Gustafsson MGL. Surpassing the lateral resolution limit by a factor of two using structured illumination microscopy. J Microsc. 2000;198(2):82–7. https://doi.org/10.1046/j.1365-2818. 2000.00710.x.
6. Holden SJ, Uphoff S, Kapanidis AN. DAOSTORM: an algorithm for high-density super-resolution microscopy. Nat Methods. 2011;8(4):279–80. https://doi.org/10.1038/nmeth0411-279.
7. Min J, et al. FALCON: fast and unbiased reconstruction of high-density super-resolution microscopy data. Sci Rep. 2014;4:4577. https://doi.org/10.1038/srep04577.
8. Ovesný M, Křížek P, Borkovec J, Švindrych Z, Hagen GM. ThunderSTORM: a comprehensive ImageJ plug-in for PALM and STORM data analysis and super-resolution imaging. Bioinformatics. 2014;30(16):2389–90. https://doi.org/10.1093/bioinformatics/btu202.
9. Cox S, et al. Bayesian localization microscopy reveals nanoscale podosome dynamics. Nat Methods. 2012;9(2):195–200. https://doi.org/10.1038/nmeth.1812.
10. Miller H, Zhou Z, Wollman AJM, Leake MC. Superresolution imaging of single DNA molecules using stochastic photoblinking of minor groove and intercalating dyes. Methods. 2015;88:81–8. https://doi.org/10.1016/j.ymeth.2015.01.010.
11. Karslake JD, et al. SMAUG: analyzing single-molecule tracks with nonparametric Bayesian statistics. Methods. 2020;193:16–26. https://doi.org/10.1016/j.ymeth.2020.03.008.
12. Rees EJ, Erdelyi M, Schierle GSK, Knight A, Kaminski CF. Elements of image processing in localization microscopy. J Opt (United Kingdom). 2013;15(9) https://doi.org/10.1088/2040-8978/15/9/094012.
13. Henriques R, Lelek M, Fornasiero EF, Valtorta F, Zimmer C, Mhlanga MM. QuickPALM: 3D real-time photoactivation nanoscopy image processing in ImageJ. Nat Methods. 2010;7(5): 339–40. https://doi.org/10.1038/nmeth0510-339.
14. Schneider CA, Rasband WS, Eliceiri KW. NIH image to ImageJ: 25 years of image analysis. Nat Methods. 2012;9(7):671–5. https://doi.org/10.1038/nmeth.2089.
15. Mortensen KI, Churchman LS, Spudich JA, Flyvbjerg H. Optimized localization analysis for single-molecule tracking and super-resolution microscopy. Nat Methods. 2010;7(5):377–81. https://doi.org/10.1038/nmeth.1447.

16. Thompson RE, Larson DR, Webb WW. Precise nanometer localization analysis for individual fluorescent probes. Biophys J. 2002;82(5):2775–83. https://doi.org/10.1016/S0006-3495(02)75618-X.

17. Wang Y, Cai E, Sheung J, Lee SH, Teng KW, Selvin PR. Fluorescence imaging with one-nanometer accuracy (fiona). J Vis Exp. 2014;91:51774. https://doi.org/10.3791/51774.

18. Robson A, Burrage K, Leake MC. Inferring diffusion in single live cells at the single-molecule level. Philos Trans R Soc B Biol Sci. 2013;368(1611):20120029. https://doi.org/10.1098/rstb.2012.0029.

19. Leake MC. Analytical tools for single-molecule fluorescence imaging in cellulo. Phys Chem Chem Phys. 2014;16(25):12635–47. https://doi.org/10.1039/c4cp00219a.

20. Rust MJ, Bates M, Zhuang X. Sub-diffraction-limit imaging by stochastic optical reconstruction microscopy (STORM). Nat Methods. 2006;3(10):793–5. https://doi.org/10.1038/nmeth929.

21. Shroff H, Galbraith CG, Galbraith JA, Betzig E. Live-cell photoactivated localization microscopy of nanoscale adhesion dynamics. Nat Methods. 2008;5(5):417–23. https://doi.org/10.1038/nmeth.1202.

22. Royal Swedish Academy of Sciences. Nobel Prize in Physics press release 2018, Nobel prize press releases, 2018. [Online]. https://www.nobelprize.org/nobel_prizes/physics/laureates/1952/

23. Hell SW. Improvement of lateral resolution in far-field fluorescence light microscopy by using two-photon excitation with offset beams. Opt Commun. 1994;106(1–3):19–24. https://doi.org/10.1016/0030-4018(94)90050-7.

24. Klar TA, Jakobs S, Dyba M, Egner A, Hell SW. Fluorescence microscopy with diffraction resolution barrier broken by stimulated emission. Proc Natl Acad Sci U S A. 2000;97(15):8206–10. https://doi.org/10.1073/pnas.97.15.8206.

25. Fölling J, et al. Fluorescence nanoscopy by ground-state depletion and single-molecule return. Nat Methods. 2008;5(11):943–5. https://doi.org/10.1038/nmeth.1257.

26. Heintzmann R, Jovin TM, Cremer C. Saturated patterned excitation microscopy—a concept for optical resolution improvement. J Opt Soc Am A. 2002;19(8):1599. https://doi.org/10.1364/josaa.19.001599.

27. Gustafsson MGL. Nonlinear structured-illumination microscopy: wide-field fluorescence imaging with theoretically unlimited resolution. Proc Natl Acad Sci U S A. 2005;102(37):13081–6. https://doi.org/10.1073/pnas.0406877102.

28. Hofmann M, Eggeling C, Jakobs S, Hell SW. Breaking the diffraction barrier in fluorescence microscopy at low light intensities by using reversibly photoswitchable proteins. Proc Natl Acad Sci U S A. 2005;102(49):17565–9. https://doi.org/10.1073/pnas.0506010102.

29. Balzarotti F, et al. Nanometer resolution imaging and tracking of fluorescent molecules with minimal photon fluxes. Science. 2017;355(6325):606–12. https://doi.org/10.1126/science.aak9913.

30. Zacharias DA, Violin JD, Newton AC, Tsien RY. Partitioning of lipid-modified monomeric GFPs into membrane microdomains of live cells. Science. 2002;296(5569):913–6. https://doi.org/10.1126/science.1068539.

31. Chalfie M, Tu Y, Euskirchen G, Ward WW, Prasher DC. Green fluorescent protein as a marker for gene expression. Science. 1994;263(5148):802–5. https://doi.org/10.1126/science.8303295.

32. Nagai T, Ibata K, Park ES, Kubota M, Mikoshiba K, Miyawaki A. A variant of yellow fluorescent protein with fast and efficient maturation for cell-biological applications. Nat Biotechnol. 2002;20(1):87–90. https://doi.org/10.1038/nbt0102-87.

33. Campbell RE, et al. A monomeric red fluorescent protein. Proc Natl Acad Sci U S A. 2002;99(12):7877–82. https://doi.org/10.1073/pnas.082243699.

34. Shaner NC, Campbell RE, Steinbach PA, Giepmans BNG, Palmer AE, Tsien RY. Improved monomeric red, orange and yellow fluorescent proteins derived from Discosoma sp. red fluorescent protein. Nat Biotechnol. 2004;22(12):1567–72. https://doi.org/10.1038/nbt1037.

35. Zhang G, Gurtu V, Kain SR. An enhanced green fluorescent protein allows sensitive detection of gene transfer in mammalian cells. Biochem Biophys Res Commun. 1996;227(3):707–11. https://doi.org/10.1006/bbrc.1996.1573.

36. Pédelacq JD, Cabantous S, Tran T, Terwilliger TC, Waldo GS. Engineering and characterization of a superfolder green fluorescent protein. Nat Biotechnol. 2006;24(1):79–88. https://doi.org/10.1038/nbt1172.

37. Ernst LA, Gupta RK, Mujumdar RB, Waggoner AS. Cyanine dye labeling reagents for sulfhydryl groups. Cytometry. 1989;10(1):3–10. https://doi.org/10.1002/cyto.990100103.

38. Invitrogen. Alexa Fluor 488 Phalloidin - Product Page. [Online]. https://www.thermofisher.com/order/catalog/product/A12379. Accessed 03 Oct 2020.

39. Lalande ME, Ling V, Miller RG. Hoechst 33342 dye uptake as a probe of membrane permeability changes in mammalian cells. Proc Natl Acad Sci U S A. 1981;78:363–7. https://doi.org/10.1073/pnas.78.1.363.

40. Grimm JB, et al. A general method to improve fluorophores for live-cell and single-molecule microscopy. Nat Methods. 2015;12(3):244–50. https://doi.org/10.1038/nmeth.3256.

41. Patterson GH, Lippincott-Schwartz J. A photoactivatable GFP for selective photolabeling of proteins and cells. Science. 2002;297(5588):1873–7. https://doi.org/10.1126/science.1074952.

42. Gunewardene MS, et al. Superresolution imaging of multiple fluorescent proteins with highly overlapping emission spectra in living cells. Biophys J. 2011;101(6):1522–8. https://doi.org/10.1016/j.bpj.2011.07.049.

43. Wang S, Moffitt JR, Dempsey GT, Xie XS, Zhuang X. Characterization and development of photoactivatable fluorescent proteins for single-molecule-based superresolution imaging. Proc Natl Acad Sci U S A. 2014;111(23):8452–7. https://doi.org/10.1073/pnas.1406593111.

44. Ando R, Hama H, Yamamoto-Hino M, Mizuno H, Miyawaki A. An optical marker based on the UV-induced green-to-red photoconversion of a fluorescent protein. Proc Natl Acad Sci U S A. 2002;99(20):12651–6. https://doi.org/10.1073/pnas.202320599.

45. Adam V, Nienhaus K, Bourgeois D, Nienhaus GU. Structural basis of enhanced photoconversion yield in green fluorescent protein-like protein Dendra2. Biochemistry. 2009;48(22):4905–15. https://doi.org/10.1021/bi900383a.

46. Kapanidis AN, Lee NK, Laurence TA, Doose S, Margeat E, Weiss S. Fluorescence-aided molecule sorting: analysis of structure and interactions by alternating-laser excitation of single molecules. Proc Natl Acad Sci U S A. 2004;101(24):8936–41. https://doi.org/10.1073/pnas.0401690101.

47. Huang B, Wang W, Bates M, Zhuang X. Three-dimensional super-resolution imaging by stochastic optical reconstruction microscopy. Science. 2008;319(5864):810–3. https://doi.org/10.1126/science.1153529.

48. Pavani SRP, et al. Three-dimensional, single-molecule fluorescence imaging beyond the diffraction limit by using a double-helix point spread function. Proc Natl Acad Sci U S A. 2009;106(9):2995–9. https://doi.org/10.1073/pnas.0900245106.

49. Wollman AJM, Hedlund EG, Shashkova S, Leake MC. Towards mapping the 3D genome through high speed single-molecule tracking of functional transcription factors in single living cells. Methods. 2020;170:82–9. https://doi.org/10.1016/j.ymeth.2019.06.021.

50. Lew MD, Lee SF, Badieirostami M, Moerner WE. Corkscrew point spread function for far-field three-dimensional nanoscale localization of pointlike objects. Opt Lett. 2011;36(2):202. https://doi.org/10.1364/ol.36.000202.

51. Andronov L, Orlov I, Lutz Y, Vonesch JL, Klaholz BP. ClusterViSu, a method for clustering of protein complexes by Voronoi tessellation in super-resolution microscopy. Sci Rep. 2016;6:24084. https://doi.org/10.1038/srep24084.

52. Ester M, Kriegel H-P, Sander J, Xu X. A density-based algorithm for discovering clusters in large spatial databases with noise. In: Proceedings of the 2nd international conference on knowledge discovery and data mining, vol. 96. New York: ACM Digital Library; 1996. p. 226–31.

53. Kiskowski MA, Hancock JF, Kenworthy AK. On the use of Ripley's K-function and its derivatives to analyze domain size. Biophys J. 2009;97(4):1095–103. https://doi.org/10.1016/j.bpj.2009.05.039.
54. Getis A, Franklin J. Second-order neighborhood analysis of mapped point patterns. Ecology. 1987;68(3):473–7. https://doi.org/10.2307/1938452.
55. Lopes FB, et al. Membrane nanoclusters of FcγRI segregate from inhibitory SIRPα upon activation of human macrophages. J Cell Biol. 2017;216(4):1123–41. https://doi.org/10.1083/jcb.201608094.
56. Khater IM, Nabi IR, Hamarneh G. A review of super-resolution single-molecule localization microscopy cluster analysis and quantification methods. Patterns. 2020;1(3):100038. https://doi.org/10.1016/j.patter.2020.100038.
57. Griffié J, et al. 3D Bayesian cluster analysis of super-resolution data reveals LAT recruitment to the T cell synapse. Sci Rep. 2017;7(1):4077. https://doi.org/10.1038/s41598-017-04450-w.
58. Williamson DJ, et al. Machine learning for cluster analysis of localization microscopy data. Nat Commun. 2020;11(1):1493. https://doi.org/10.1038/s41467-020-15293-x.
59. Hyman AA, Weber CA, Jülicher F. Liquid-liquid phase separation in biology. Annu Rev Cell Dev Biol. 2014;30(1):39–58. https://doi.org/10.1146/annurev-cellbio-100913-013325.
60. Dresser L, et al. Amyloid-β oligomerization monitored by single-molecule stepwise photobleaching. Methods. 2020;193:80–95. https://doi.org/10.1016/j.ymeth.2020.06.007.
61. Leake MC, Wilson D, Bullard B, Simmons RM, Bubb MR. The elasticity of single kettin molecules using a two-bead laser-tweezers assay. FEBS Lett. 2003;535(1–3):55–60. https://doi.org/10.1016/S0014-5793(02)03857-7.
62. Yan X, Hoek TA, Vale RD, Tanenbaum ME. Dynamics of translation of single mRNA molecules in vivo. Cell. 2016;165(4):976–89. https://doi.org/10.1016/j.cell.2016.04.034.
63. Syeda AH, et al. Single-molecule live cell imaging of Rep reveals the dynamic interplay between an accessory replicative helicase and the replisome. Nucleic Acids Res. 2019;47(12):6287–98. https://doi.org/10.1093/nar/gkz298.
64. Wooten M, Li Y, Snedeker J, Nizami ZF, Gall JG, Chen X. Superresolution imaging of chromatin fibers to visualize epigenetic information on replicative DNA. Nat Protoc. 2020;15(3):1188–208. https://doi.org/10.1038/s41596-019-0283-y.
65. Xu K, Zhong G, Zhuang X. Actin, spectrin, and associated proteins form a periodic cytoskeletal structure in axons. Science. 2013;339(6118):452–6. https://doi.org/10.1126/science.1232251.
66. Xu K, Babcock HP, Zhuang X. Dual-objective STORM reveals three-dimensional filament organization in the actin cytoskeleton. Nat Methods. 2012;9(2):185–8. https://doi.org/10.1038/nmeth.1841.
67. Pan L, et al. Hypotonic stress induces fast, reversible degradation of the vimentin cytoskeleton via intracellular calcium release. Adv Sci. 2019;6(18):1900865. https://doi.org/10.1002/advs.201900865.
68. Bálint Š, Vilanova IV, Álvarez ÁS, Lakadamyali M. Correlative live-cell and superresolution microscopy reveals cargo transport dynamics at microtubule intersections. Proc Natl Acad Sci U S A. 2013;110(9):3375–80. https://doi.org/10.1073/pnas.1219206110.
69. Wollman AJM, et al. Critical roles for EGFR and EGFR-HER2 clusters in EGF binding of SW620 human carcinoma cells. J R Soc Interface. 2022;19(190):20220088. https://doi.org/10.1098/rsif.2022.0088.
70. Lew MD, et al. Three-dimensional superresolution colocalization of intracellular protein superstructures and the cell surface in live Caulobacter crescentus. Proc Natl Acad Sci USA. 2011;108(46):E1102–10. https://doi.org/10.1073/pnas.1114444108.
71. Shepherd JW, Lecinski S, Wragg J, Shashkova S, MacDonald C, Leake MC. Molecular crowding in single eukaryotic cells: using cell environment biosensing and single-molecule optical microscopy to probe dependence on extracellular ionic strength, local glucose conditions, and sensor copy number. bioRxiv. 2021;193:54–61. https://doi.org/10.1101/2020.08.14.251363.

72. Babazadeh R, Adiels CB, Smedh M, Petelenz-Kurdziel E, Goksör M, Hohmann S. Osmostress-induced cell volume loss delays yeast Hog1 signaling by limiting diffusion processes and by Hog1-specific effects. PLoS One. 2013;8(11):e80901. https://doi.org/10.1371/journal.pone.0080901.

73. Wollman AJM, Shashkova S, Hedlund EG, Friemann R, Hohmann S, Leake MC. Transcription factor clusters regulate genes in eukaryotic cells. Elife. 2017;6:1–36. https://doi.org/10.7554/eLife.27451.

74. Jin X, et al. Membraneless organelles formed by liquid-liquid phase separation increase bacterial fitness. Sci Adv. 2021;7(43):eabh2929. https://doi.org/10.1126/sciadv.abh2929.

75. Horrocks MH, et al. Single-molecule imaging of individual amyloid protein aggregates in human biofluids. ACS Chem Neurosci. 2016;7(3):399–406. https://doi.org/10.1021/acschemneuro.5b00324.

76. Cella Zanacchi F, Manzo C, Alvarez AS, Derr ND, Garcia-Parajo MF, Lakadamyali M. A DNA origami platform for quantifying protein copy number in super-resolution. Nat Methods. 2017;14(8):789–92. https://doi.org/10.1038/nmeth.4342.

Correction to: Live-Cell Imaging: A Balancing Act Between Speed, Sensitivity, and Resolution

Jeroen Kole, Haysam Ahmed, Nabanita Chatterjee,
Gražvydas Lukinavičius, and René Musters

Correction to
Chapter 6 in: V. Nechyporuk-Zloy (ed.),
Principles of Light Microscopy: From Basic to Advanced,
https://doi.org/10.1007/978-3-031-04477-9_6

The original version of this chapter was inadvertently published with errors. The following corrections have been made after publication:

1. Fig. 6.10 & 6.11: source detail has been updated as Source: https://phiab.com
2. Page 161, Box. 6.3: line 10: Tomocube has been replaced with Phase Holographic Imaging PHI AB, Lund, Sweden
3. Page 169, Appendix:
   a. 'Phase Holographic Imaging (PHI)—https:/phiab.com/' has been included
   b. Tomocube—www.tomocube.com has been removed

The updated original version for this chapter can be found at
https://doi.org/10.1007/978-3-031-04477-9_6

Printed in the United States
by Baker & Taylor Publisher Services